Changing Land Use, Changing Livelihoods

Special Issue Editors

Claudia A. Radel
Jacqueline M. Vadjunec

MDPI

Special Issue Editors
Claudia A. Radel
Utah State University
USA

Jacqueline M. Vadjunec
Oklahoma State University
USA

Editorial Office
MDPI AG
St. Alban-Anlage 66
Basel, Switzerland

This edition is a reprint of the Special Issue published online in the open access journal *Land* (ISSN 2073-445X) from 2015–2016 (available at: http://www.mdpi.com/journal/land/special_issues/landuse_livelihoods).

For citation purposes, cite each article independently as indicated on the article page online and as indicated below:

Author 1; Author 2; Author 3 etc. Article title. *Journal Name*. **Year**. Article number/page range.

ISBN 978-3-03842-342-3 (Pbk)
ISBN 978-3-03842-343-0 (PDF)

Table of Contents

Section 1: Smallholder Practices and Environmental Sustainability

Section 2: Dynamic Smallholder Livelihoods

Section 3: Smallholder Vulnerability and Adaptation

Section 4: Beyond Smallholders

About the Guest Editors

Claudia Radel is an Associate Professor of Human Geography in the Department of Environment and Society at Utah State University. Her research explores the changing nature of natural resource-based livelihood strategies for individuals, households, and communities in the rural Global South, especially in Mexico and Central America. She is interested particularly in how gender ideologies and practices relate to livelihood strategies and land-use/environmental change. Her research also examines the gendered nature of resource access, control, and decision making in smallholder farming systems. Prior to receiving her PhD in Geography from Clark University, Dr. Radel worked as a community development consultant in Colombia. She holds a Master's in Public Affairs from Princeton University's Woodrow Wilson School of Public and International Affairs, where she studied international development. She received her Bachelor's degree from Brown University, with dual concentration in Environmental Studies and Comparative Development Studies.

Jacqueline M. Vadjunec is an Associate Professor of Geography at Oklahoma State University in Stillwater, Oklahoma, USA. Her research explores issues surrounding agriculture, land use, and livelihoods in the Americas. She is especially interested in using mixed methods and participatory approaches for studying the impact of governance on both livelihoods and land-use land-cover change (LULCC). Jacqueline received her PhD and MA in Geography from Clark University, and her BA in both Geography and English from Bucknell University.

Preface to "Changing Land Use, Changing Livelihoods"

The papers in this volume were inspired by lively discussions regarding relationships between smallholder livelihoods and land-use land-cover change (LULCC) at the Land System Science Symposium (LSSS) at the Annual Meeting of the Association of American Geographers (AAG) in Chicago (2015). Those engaging discussions at the LSSS, along with our own research interests at the time, led naturally to the development of a Special Issue on the place of smallholders today, their changing land use and changing livelihoods. This edited volume has benefited from our own related research (and vice versa) funded by National Science Foundation (NSF) research grants (1266381 and 1056811). The editors thank Andrew Millington, the Academic Editor-in-Chief of *Land*, for his patience and support during this long process. We also thank Elvis Wang, the Managing Editor for *Land*, as well as the other countless very helpful support staff at MDPI. We are grateful for the collegiality of the contributors, as well as the many helpful reviewers, whose comments greatly improved the quality of the Special Issue. B.L. Turner II was especially helpful at various stages along the way. Lastly, we thank smallholders and farmers everywhere for sharing their time and knowledge with us. Without their generosity and support there would be no Special Issue. We hope in these pages you will see the continued importance of smallholders today to human–environment studies and to the wellbeing of humans and their land systems.

Claudia A. Radel and Jacqueline M. Vadjunec
Guest Editors

land

MDPI

Editorial

Introduction: The Continued Importance of Smallholders Today

Jacqueline M. Vadjunec [1,*], Claudia Radel [2] and B. L. Turner II [3]

1 Department of Geography, Oklahoma State University, 324 Murray Hall, Stillwater, OK 74078, USA
2 Department of Environment and Society, S.J. & Jessie E. Quinney College of Natural Resources, Utah State University, 5215 Old Main Hill, Logan, UT 84322, USA; claudia.radel@usu.edu
3 School of Geographical Sciences and Urban Sciences and School of Sustainability, Arizona State University, PO Box 875302, Tempe, AZ 85287, USA; Billie.L.Turner@asu.edu
* Correspondence: jacqueline.vadjunec@okstate.edu; Tel.: +1-405-744-3293; Fax: +1-405-744-5620

Academic Editor: Sheona Shackleton
Received: 5 October 2016; Accepted: 20 October 2016; Published: 25 October 2016

Abstract: Smallholders remain an important part of human-environment research, particularly in cultural and political ecology, peasant and development studies, and increasingly in land system and sustainability science. This introduction to the edited volume explores land use and livelihood issues among smallholders, in several disciplinary and subfield traditions. Specifically, we provide a short history of smallholder livelihood research in the human-environment tradition. We reflect on why, in an age of rapid globalization, smallholder land use and livelihoods still matter, both for land system science and as a reflection of concerns with inequality and poverty. Key themes that emerge from the papers in this volume include the importance of smallholder farming and land-use practices to questions of environmental sustainability, the dynamic reality of smallholder livelihoods, the challenges of vulnerability and adaptation in contemporary human-environment systems, and the structural and relative nature of the term "smallholder." Overall these contributions show that smallholder studies are more pertinent than ever, especially in the face of global environmental change. Additionally, we argue that questions of smallholder identity, social difference, and teleconnections provide fertile areas of future research. We conclude that we need to re-envision who the smallholder is today and how this translates into modern human-environment smallholder studies.

Keywords: cultural and political ecology; global environmental change; identity; inequality; land system science; smallholders; sustainability

1. Introduction

Smallholders remain relevant today, if for no other reason than their continued sheer numbers and the total area devoted to their production [1]. Approximately, 570 million smallholder households exist worldwide, and those farming two hectares or less take up about 75% of agricultural land globally [2]. Overall, "the world's present day population of 2.0–2.5 billion smallholders and their access to food and resources, combined with non- and off-farm linkages" make them key actors influencing land-use/land-cover change (LULCC) and biodiversity ([3]; p. 4, this volume). As a result, smallholders remain an important part of human-environment research, particularly in cultural and political ecology, peasant and development studies, and increasingly in land system and sustainability science. This edited volume explores land use and livelihood issues among smallholders, in a diversity of disciplinary and subfield traditions and mixed methods approaches.

Despite their importance and research attention, neither a standard definition nor an explicit, systematic set of criteria to define smallholders exists [1,2,4,5]. Rather, different researchers, research communities, and data compiling units (i.e., countries to international agencies) utilize different criteria or parameters for criteria to define smallholders. These differences, in turn, can lead to variable estimates of the smallholder population and area farmed. Commonly, however, smallholders are considered to be those agriculturalists cultivating from less than one up to 10 hectares of land ([6] and [7], this volume), the size often reflecting differing land pressures, agro-ecological zones, or alternative economic opportunities [1,2]. In addition, smallholders are typically considered to engage in some measure of subsistence production, with ephemeral to incomplete ties to commercial production owing to constraints on land and social and economic capital, to poor market access, to gender roles, or to cultural norms [8–15]. This last factor, cultural norms, is often linked to the term peasant or peasant production, in which societal values engender community obligations that inhibit individual behavior that might otherwise be captured by the state or the market [16,17].

Past work identified smallholder agriculturalists as engaged in risk-averse production [18,19], although innovative [20,21], in which market ties are mainly limited to sales of normal surplus (i.e., production beyond subsistence needs) [11,12]. Currently, however, smallholders engage the market in some way, if at different levels of intensity, while maintaining some element of subsistence [5,17]. Indeed, it is difficult today to identify many smallholder households that are purely subsistence in kind and not engaged in a cash economy in some way. Even if cultivation is largely for subsistence, household members are often engaged in wage labor off the farm, in many cases at distance, sending remittances back to the household [22,23].

This reality notwithstanding, understanding changes in smallholder livelihoods is essential to comprehending some types of LULCC. Contemporary changes in smallholder livelihood strategies, however, have not been at the forefront in more recent LULCC studies, especially as articulated in land system science (LSS) [24,25]. Nor have understandings of household livelihood change been regularly linked to regional or larger patterns of LULCC. The papers in this edited volume seek to bridge this gap by highlighting the role of changing livelihoods to changing land systems. By way of introducing this bridging effort, we provide a short history of smallholder livelihood research in the human-environment tradition. Second, we reflect on why, in an age of rapid globalization, smallholder land use and livelihoods still matter, as a reflection of concerns over inequality, poverty, and poverty traps (e.g., [26]). Next, we discuss the primary themes that emerge from the edited volume, followed by other fertile areas of future research. We conclude that not only is smallholder research as important as ever (or perhaps even more so), but also the current realities of smallholders require that we rethink how we conceptualize "the smallholder."

2. Smallholder Livelihood Research in the Human-Environment Tradition—Past to Present

Explicit attention to smallholder production systems, from the lens of human-environment interactions, can be traced back centuries in anthropology and geography, though the rationale for this attention has differed through time and among disciplines and transdisciplinary subfields. Anthropology has long examined these interactions to gain understanding of the role of culture in shaping group behavior, foremost focused on non-western societies, which includes subsistence-based smallholders [27]. This attention generated questions about "Homo economicus" and the search for behavioral rationales differing from those undergirding classical western economic thought [28]. In contrast, geography, at least as influenced by Humboldt [29], sought to understand the landscapes generated by smallholders, a theme passed down and applied to non-western smallholders in various guises, from Sauer's [30] landscape morphology to Brookfield's [9] examinations of the production rationales leading to those landscapes. Throughout, the influence of the two disciplines on one another was large.

Fast forwarding to the middle of the 20th century, variants of these traditions merged in the transdisciplinary subfield of cultural ecology (CE), which strongly focused on smallholders, foremost those residing outside western, industrialized countries. Maintaining the different core objectives of their disciplines, the anthropological and geographical cohorts shared a desire to reshape human-environment research to be consistent with (eco)system- and theory-based science [31,32].

Much attention was given to system assessments of various types of smallholder production (e.g., [33,34]) and to why and where smallholder agriculture intensified, which required attention to subsistence production rationales. Cultural ecologists borrowed from the agricultural economists Chayanov [19] and Boserup [20], providing the empirical support for Boserup's thesis about land pressures and intensification, which involved, at least partly, a non-market rationale, and paid explicit attention to how site-situation, especially environmental conditions, modified the intensification process [14,35].

As this and ancillary research progressed, vibrations from critical geography and anthropology gave rise to political ecology [36], invigorated in part by challenges to the risk/hazard subfield addressing smallholders in the developing world [37]. Political ecology (PE) emphasized the role of power relationships and politics on smallholder human-environment relationships that challenged the applicability of western-designed development strategies [38]. Initially focused on structural, explicitly critical, perspectives to problem framing (e.g., [39,40]), PE subsequently expanded to include constructivists' perspectives focused on various social theory interests [41,42]. The questions posed of smallholders shifted to those of equity, marginalization, and gender inequality, among others (e.g., [43,44]), and the previous distinctions in the core objectives of practitioners from geography and anthropology faded.

As political ecology expanded in topics and approaches [45], practitioners aligning more with the science-oriented dimensions of these disciplines loosely merged with their risk/hazard counterparts to enter the global environmental change arena, with an eye on questions of sustainability [46]. The smallholder research cohort of this alignment increasingly positioned itself with the emerging LSS subfield, one in which remote sensing, GIScience, and modeling methods were integrated to tackle the causes and human-environment consequences of land change [24,25]. This subfield, in turn, interacted strongly, and in some case integrated with, environmental and resource economists focused on land and smallholders, political scientists and sociologists interested in smallholder governance and the environment, foremost common property regimes (e.g., [47]), and ecologists interested in social–environmental (aka human-environment) systems, many studies of which involved smallholders (e.g., [48]).

Attention to the human-environment relationships of smallholders no longer resides primarily with practitioners of anthropology and geography linked to the subfield traditions noted above. The currents of global environmental change and sustainability have marshalled in a new era of attention to the social–environmental (or ecological) system. Championed by ecology, this new label for human-environment systems has attracted a wide array of practitioners, largely focused on system resilience and vulnerability [49–51], poverty traps (e.g., [26]), and ecosystem services, including payments for services otherwise lost in land change [3,52]. Increasingly, this problem orientation has been applied to smallholders worldwide (e.g., [48,53]), with linkages to political ecology (e.g., [54]).

In many ways, sustainability and resilience-vulnerability research have become the new transdisciplinary labels for a science-based formulation of human-environment or nature–society science. This is not say that alternative explanatory perspectives, such as structural (critical) and constructivist formulations of smallholder human-environment dynamics, will not remain vibrant and, as noted above, interact with, around, and in spite of this emergent problem framing. For instance, the more historical-, enthnographic-, and cultural-laden version of CE remains intact, interacting with all of the human-environment subfield orientations that have emerged since the middle of the last century. Likewise, PE and LSS increasingly share methods, such as remote sensing, potentially engendering more cooperation among them while maintaining their subfield interests.

This edited volume brings together a diversity of perspectives, including sustainability and resilience-vulnerability framed research along with alternative critical and constructivist framings. Collectively, the contributions in this volume demonstrate how the difference among the approaches is not always that great after all. What remains different is the base set of issues and questions that garners the interest of each cohort. There would appear to be significant synergies in understanding that could be gained by an improved appreciation and melding of interests among the cohorts.

3. Importance of Smallholders Today: Considering Inequality and Smallholder Poverty

Considerations of LULCC are not the only reason why scholars engaging in human-environment smallholder studies continue to do so. As the above history demonstrates, much of the research on smallholders has been concerned with livelihoods or environments in their own right, and some of this has been motivated by concerns with smallholder well-being and environmental services. Perhaps no rationale for this engagement has been stronger than that dealing with development.

Many early smallholder studies seeking to understand smallholder behavior in the environmental and natural resource arenas were motivated by the desire to inform and improve development efforts. Initially, this work focused on smallholders as rational actors, even if that rationality differed from profit maximization or optimization, rather than smallholders as irrational actors whose production confined them to poverty. This stage of smallholder research, however, did not challenge the notion of development itself. Over the last 50 years, scholarly perspectives on the role of academics and practitioners in development and poverty-alleviation efforts have changed. Frequently informed by more critical theoretical development perspectives, many academics cast a skeptical and even negative eye on efforts to promote agricultural development and/or resource management in the smallholder sector [55,56].

Yet "we cannot not desire development" ([56]; p. 10). The realities of smallholders' poverty and their occupation of a particular position within political–economic systems of often intense inequality that traps them in that position (e.g., [26]) requires that we continue to keep a focus on smallholder studies, even apart from their key role as actors in land systems. The sociopolitical and economic constraints on smallholders are increasingly amplified by climate and socioeconomic and political change, which will continue to push smallholders into precarious positions, raising pressing issues regarding adaptation, vulnerability, and resilience. Not only do these issues necessitate an understanding of change in smallholder systems, as we consider vulnerability and adaptation within these systems, they also ask that we look to smallholder systems as sources of potential dynamism and adaptive possibility, for application to a variety of other production systems [57]. In addition, focusing on smallholders can help us extend our conceptualization of land use in LSS beyond the coupling of land-use change with land-cover change. It can encourage simultaneous attention to both a more intimate scale of land and resource practice, and a multi-site scale of larger production systems and chains, of interest to sustainability science, development studies, and other fields.

4. Smallholder Studies Revisited: Rethinking Land Use and Livelihoods

The papers in this volume illustrate a wide diversity and increasing complexity of smallholder livelihood studies. The research detailed is largely empirical and draws on a wide spectrum of mixed qualitative and quantitative methodologies. The case studies that follow cover a range of geographic locations, including Brazil, Burkina Faso, South Africa, Botswana, Malawi, Madagascar, Vietnam, and the USA, with the greatest emphasis on sub-Saharan Africa. Key themes that emerge include the importance of smallholder farming and land-use practices to questions of environmental sustainability, the dynamic reality of smallholder livelihoods, the challenges of vulnerability and adaptation in contemporary human-environment systems, and the structural and relative nature of "smallholder" as a category. Overall these case studies show that smallholder studies are more pertinent than ever, especially in the face of finite resources and global environmental change.

4.1. *Smallholder Practices and Environmental Sustainability*

Linkages between smallholder practices and sustainability are explored in depth in the volume's first set of papers with implications for sustainability science. First, sociologist Thomas Rudel and colleagues [7] draw on Netting [5] and others to explore if and how diverse smallholder livelihood trajectories encourage sustainable land use practices through mixed crop–livestock practices. They create an index of sustainable agricultural practices and perform a meta-analysis of previous studies. Results show that, in general, smallholders more frequently practice sustainable agricultural techniques than do large-holders, although this is not always the case. Ultimately, they press for agricultural policies that support the mixed crop–livestock systems frequently characteristic of smallholder livelihoods and land use.

Geographers Karl Zimmerer and Steven J. Vanek [3] also use a meta-analysis to draw complex linkages and feedbacks between smallholder agrobiodiversity, livelihood diversification, social-ecological systems, and sustainability. Their work expands upon and reframes previous land use and livelihood studies to also include the impacts of livelihood diversification on both above- and below-ground biodiversity, and soil and water resources, with implications for sustainability science. This re-framing contributes to better understandings of processes related to Sustainable Intensification (SI) and Ecological Intensification (EI).

Agriculture and forestry expert Daniel Etongo and colleagues [58] test the poverty and degradation vicious circle hypothesis through a case study approach in southern Burkina Faso. Drawing on participatory poverty assessment methods, the team created indicators to categorize wealth status, environmental degradation, and land management practices. Results suggest ambiguity in the relationship between poverty and degradation. For instance, farmers who are better off tend to be involved in the most environmentally degrading activities, such as deforestation and overgrazing. However, results also reveal that the poorest farmers are the most constrained, and are less likely to adopt land management practices to improve their land. Their findings suggest that, in specifying theories of how poverty and class shape land-use and management practices, there is a need to differentiate among these practices.

4.2. Dynamic Smallholder Livelihoods

The next three papers focus on shifting smallholder livelihoods. Geographer Ritaumaria Pereira and colleagues [59] explore how smallholder farming in the Brazilian Amazon has changed over time. Drawing on mixed-methods field research in southeastern Pará, Pereira and colleagues show how smallholders are influenced and exploited by transnational cattle corporations and global production networks. As a result, smallholders have reconfigured their livelihood systems away from diversified agriculture to increasing engagement with the cattle economy through processes of contract ranching "land grabs." As Pereira and colleagues show, such shifting livelihood processes leave Amazonian smallholders increasingly vulnerable to various kinds of shocks, with implications for long-term socioeconomic and environmental sustainability.

Geographers Sarah Turner and Thi-Thanh-Hiền Pham [60] used remote sensing and ethnographic fieldwork to examine rapid land-use/land-cover change (LULCC) in three borderland provinces in northern Vietnam. Results show that government interventions, state policies, market opportunities, and changing agricultural and ecological conditions led to highly heterogeneous LULCC in the region between 2000 and 2009. Turner and Pham's results reveal the complexity of LULCC mapping, especially in areas undergoing rapid change. This empirical case study demonstrates the need to look at land use and livelihood dynamics using mixed methods across multiple scales, including the importance of on-the-ground ethnographic fieldwork to better inform policymakers.

Geographer William J. McConnell and colleagues [61] use LSS informed by political ecological contexts to critically re-think dominant narratives of smallholders as deforesters and land degraders in Madagascar's highlands. In particular, remote sensing analysis reveals increased patchy patterns of tree growth on mainly smallholder farms and village lands, illustrating that increasing land pressures elsewhere have led to tree planting of mostly exotic tree species as a means of adopting more resilient livelihoods. Such trees (including eucalyptus, pine, acacia, and a variety of fruit trees) have become central to rural livelihoods, helping to secure rural land tenure, while also helping to fulfill increasing urban needs. At the same time, the long-term environmental consequences of exotic tree planting need to be further studied. What McConnell and colleagues suggest is a more nuanced understanding of the costs and benefits from shifts in rural livelihood strategies on social-environmental systems.

4.3. Smallholder Vulnerability and Adaptation

The next three papers focus on smallholder adaptation, resilience, and vulnerability. Geographer William Moseley's [62] research in Botswana explores smallholder vulnerability in the face of climate change. More specifically, Moseley's research reveals that a middle-income, relatively prosperous country also has high rates of poverty, food insecurity, and income inequality. Using semi-structured key-informant interviews and smallholder household surveys, Moseley's case study shows the importance of looking beyond climate change to include the local, regional and international political economy when considering issues of smallholder marginalization, vulnerability and household food security.

Geographer Brian King and colleagues [63] make a strong argument regarding the need for more studies related to environmental perception and smallholder adaptation in the face of global climate change. Using qualitative methods and a regional case study approach, King and associates studied adaptation in three communities in the Okavango Delta of Botswana. Results reveal that while flooding and precipitation patterns are heterogeneous and fluid, some livelihood responses are rigid due to local regulations, suggesting the need for evolving environmental governance.

Environmental scientist Sheona Shackleton and forest economist Marty Luckert [64] synthesize their research and experiences, providing lessons learned from a large mixed-methods research project dealing with smallholders in the Eastern Cape, South Africa. They argue that as agricultural production decreases, rural Eastern Cape smallholders are increasingly vulnerable to adverse outcomes following livelihood shocks. They identify multi-scalar and complex drivers ranging from historical and national to local processes that influence a smallholder's capacity to adapt. Specifically, Shacketon and Luckert show that being female-headed or HIV-impaired, and having low education levels, along with polarized historical–political legacies, may drive some households into a "trap," thus making it difficult to respond to current and future stressors on the social-ecological system. Overall, they make a strong case for using multiple methods and time trajectories to better assess the complex drivers influencing smallholder vulnerability and adaptation, with policy implications.

4.4. Beyond Smallholders

Finally, the last two contributions in this volume push us to think beyond traditional smallholder categories, to grapple with smallholder definitions that are more fluid and relational in nature. These studies stretch us to consider that what makes a smallholder a smallholder is often more a state of mind or the result of complex power dynamics, than an easy definition related to the number of hectares in their individual land holding. First, geographer Christian Brannstrom and colleagues [65] use GIS and secondary data sources to study the spatial distribution of estimated wind power royalties in West Texas. Results reveal a "property advantage" and a "royalty paradox" when it comes to the unequal distribution of wind power royalties, where large ranchers seem to have a much greater advantage over small-farm cotton growers in the region. This study demonstrates how family farm producers in this context, whom we may not have otherwise thought of as "smallholders," are positioned more weakly within the wind economy. Here, being a smallholder is clearly relative to the larger ranchers.

Last, development economist Ward Anseeuw and colleagues [66] explore the discourse surrounding the seeming transformation of smallholder farmers into medium-scale farmers in Malawi. Surveys, however, reveal that just less than half of the medium-scale farmers are actually urban-based professionals. Anseeuw and colleagues, therefore, challenge the idea that smallholder agriculturalists are emerging through economic growth and development as producers of a larger scale, but instead argue that much of the growth of medium farmers is due to outside urban investment. Instead of transformation, such changes lead to significant land consolidation, thus endangering some smallholders. Ultimately, this research makes a strong case regarding the need to explore the "Other" with regards to smallholder impacts for sounder policy development.

5. The Road Forward: Fertile Areas for Future Research on Smallholders

While the papers in this volume emphasize the importance of smallholder farming and land-use practices to questions of environmental sustainability, the dynamic reality of smallholder livelihoods, the challenges of vulnerability and adaptation in contemporary human-environment systems, and the structural and relative nature of "smallholder" as a category, here we emphasize a few related but pressing areas for smallholder studies. These include smallholder definitions and identities; gender, race, ethnicity, and difference; and the place of smallholders in teleconnections and global environmental change research.

5.1. Rethinking Smallholder Definitions and Identities

The contributions in this volume encourage us to rethink standard definitions of smallholders and smallholder identities. Considering smallholders in terms of land holding size alone restricts what is observed and how research problems are formulated, limiting the implications for land change and related issues. The assessment of landholding size—once a common-sense matter of knowing a smallholder when you see one—is changing as, for example, the number of medium-scale farms begins to grow in many developing countries, especially if they retain subsistence elements of production. In addition, the breakdown of a conceptual distinction between agricultural systems and situations in developed versus developing countries brings into question traditional definitions of smallholders by holding size and subsistence-orientation.

What these differences suggest is that it may be better to think of smallholder definitions as more fluid and/or relational in nature. As such, important questions remain to be explored: How does a more fluid concept of smallholder relate to that of the "peasant" and the "moral economy" today? How does it relate to a distance from the state and considerations of citizenship? To contemporary incorporation into markets?

Another way to understand relationships between smallholders, land use, and livelihoods is to focus on understanding smallholder identity. Collectively, the contributions in this volume, in some ways, challenge us to reimagine smallholders in a more contemporary way. With the varying smallholder definitions used throughout this volume, the term can be seen as many things: a moving target, as something to be defined in relation to the Other, or, perhaps, even as a state of mind.

For instance, work in the western Brazilian Amazon shows the importance of identity and 'cattle culture' as influencing land use not only among cattle ranchers, but also among more traditional swidden agriculturalists and rubber tappers [67]. Diving into these issues further, it becomes apparent that changing smallholder identities have serious implications for sustainable resource management.

Furthermore, Brannstrom and colleagues ([65], this volume) push us to make space in the discussion for "family farmers and small ranchers" in the USA and elsewhere that may have increasingly more in common with traditional smallholders than with corporate farms and transnational agricultural conglomerates. Traditional family farms in the industrialized west are neither purely subsistence-oriented nor even remotely similar to smallholder land holdings in size alone. Yet as these family farms continue to decrease in number, and are replaced with substantially larger corporate industrial farms, they appear to be moving in some ways culturally closer to smallholders with regards to both their production attributes and their relative position of power. For instance, family farmers tend to be risk-averse, and share strong social and community obligations, similar to smallholders. While not wanting to understate the vulnerability of those smallholders who eke out a living on less than two hectares of land, we do note that family farmers feel increasingly marginalized and vulnerable to the larger structural and economic processes going on around them [68]. For instance, family farmers and ranchers in the High Plains of the American West share a strong sense of identity as agriculturalists and Dust Bowl survivors, with a growing concern about the death of a way of life [69]. As a result, both land use and livelihoods are deeply embedded in identity.

5.2. Gender, Race, Ethnicity, and Difference in Smallholder Studies

Noticeable in its absence from the contributions to this volume is the explicit treatment of socially-constructed difference, including, for example, gender and race. Several of the contributions do consider the question of ethnicity, through an examination of how different ethnic groups are positioned differently within local land systems ([58], this volume) or socioeconomic systems ([60], this volume). In many smallholder studies, ethnicity functions as a more naturalized and therefore often unexamined category of constructed difference. The research in this volume by Shackleton and Luckert [64] highlights smallholders in a highly racialized context; however, race is not a key aspect of the authors' analysis. Nonetheless, structures of race and ethnicity are arguably central to the systems in which smallholders must operate, conditioning, for example, their access to land, the state, and markets. In many cases, understanding the linked processes of livelihood and land-use change requires that we better understand the construction of racial and ethnic difference, at various scales, and its outcomes for smallholders.

Gender difference is likewise mostly invisible in the majority of human-environment smallholder studies, and this volume is little different (although Zimmerer and Vanek [3] include in their meta-analysis an assessment of the examined studies' inclusion of gender analysis). Women often become visible in the context of female-headed households (see [64], this volume) or as specific or dominant categories of smallholders (see [62], this volume). These glimpses of gender difference invite us to consider further how we might better incorporate into smallholder studies the progress made in gender scholarship in other fields of study. A particularly promising avenue may be to further explore how identities of gender intersect with smallholder-related identities to shape land use and livelihood practices. However, considerations of gender-based differences in access to agricultural inputs, networks, and information remain key to understanding women and men as differentially situated in relation to markets, institutions, and states.

Sociologists have long debated whether the family should function as the unit of socioeconomic class stratification, with the conventional view of the family or household as the appropriate unit challenged by feminist scholars who argue that individual work situations lead to individual (and potentially different) class status for men and women in the same family [70]. Likewise, we suggest that men and women agriculturalists in the same households should be understood individually, and potentially differently, vis-à-vis smallholder status. Thus we must consider the likelihood of a currently invisible group of smallholders, whose own structural position relative to the state and to markets is hidden by their situation within a dual-agriculturalist household or within a household in which another member holds enough land to no longer be considered a smallholder.

5.3. The Place of Smallholders in Teleconnections Research

Urban impacts on smallholders within the urban hinterland are well known and documented. Increasingly, however, the forces external to the smallholder or smallholder communities emanate from well beyond the urban hinterland, constituting teleconnections. Research that explicitly considers the impact of teleconnections on smallholder land use and livelihoods is key to better understanding broader LULCC processes and patterns, as suggested in this special feature (see [59,60]). With increasing calls for studies related to teleconnections in land systems and sustainability science [71,72], it is worth stressing the importance of smallholders in teleconnections research. Changes in smallholder livelihoods and land use often reflect or respond to processes emanating from distant urban and rural areas.

For instance, smallholder households play a key role in understanding complex dynamics around the food, energy, and water nexus [73]. In another example, a teleconnections framework is used to explore vulnerabilities to climate change, and global transmissions of severe acute respiratory syndrome, while focusing on the livelihoods and adaptations of coffee farmers [74]. Lastly, "land grabbing"—corporate or country entities taking large areas of distant land—is a form of teleconnections that directly affects smallholders not only through the loss of land, but also the access to water on that land [75]. While some smallholder communities have successfully negotiated this access to their benefit, perhaps a larger segment of them have lost land and water access owing to their inadequate power to prevent the loss. Regardless of which outcome is involved, the character of land use is changed, as are the livelihoods of smallholders.

Overall, more explicit focus on smallholders in teleconnections research is needed, to understand questions including what agency smallholders have within different kinds of production chains; how smallholders are differentially impacted by various kinds of teleconnections; how the nature of various teleconnections shape LULCC, as well as smallholder social difference and identity; and how smallholder practices can, in turn, shape processes and outcomes elsewhere.

6. Conclusions and Implications

Smallholder studies are more pertinent than ever, given finite resources, climate change, and political and economic globalization. Many land change science models of smallholders treat these groups as static or fixed, primarily as regards their material base. We suggest that solely material conceptualizations of smallholders may be problematic for a variety of reasons. The contributions in this volume challenge us to reimagine smallholders in a different way, inviting us to rethink land use in a way that considers an intimate scale of land and resource practices. This rethinking holds important implications for understanding smallholder farming and land-use practices relative to questions of environmental sustainability, the dynamic reality of smallholder livelihoods, the challenges of vulnerability and adaptation in contemporary human-environment systems, and the structural and relative nature of "smallholder" as a category. Questions of smallholder identity, social difference, and teleconnections provide fertile areas of future research. Together, these re-imaginings and roads forward call us to re-envision who the smallholder is today and how this translates into modern human-environment smallholder studies.

Acknowledgments: The papers in this volume were, in part, inspired by rich discussions regarding smallholders at the Land System Science Symposium (LSSS) at the Annual Meeting of the Association of American Geographers (AAG) in Chicago (2015). We are grateful to the many anonymous reviewers whose comments greatly improved the quality of the special issue. The editors thank Andrew Millington, the editor of *Land*, for his patience and support in this endeavor. We also thank Richard Aspinall, Andrew Millington, and Burak Guneralp, co-organizers, along with Jacqueline Vadjunec, of the LSSS Chicago meeting. This edited volume has benefitted from generous funding from National Science Foundation (NSF) research grants (1266381 and 1056811). Lastly, we are grateful to smallholders and farmers everywhere for sharing their time and knowledge with us. Without their generosity and support there would be no special issue.

Author Contributions: Jacqueline M. Vadjunec, Claudia Radel, and B.L. Turner II conceptualized and wrote this manuscript.

Conflicts of Interest: The authors declare no conflict of interest.

References

1. Morton, J.F. The impact of climate change on smallholder and subsistence agriculture. *Proc. Natl. Acad. Sci. USA* **2007**, *104*, 19680–19685. [CrossRef] [PubMed]
2. Lowder, S.K.; Skoet, J.; Raney, T. The number, size, and distribution of farms, smallholder farms, and family farms worldwide. *World Dev.* **2016**, *87*, 16–29. [CrossRef]
3. Zimmerer, K.S.; Vanek, S.J. Toward the integrated framework analysis of linkages among agrobiodiversity, livelihood diversification, ecological systems, and sustainability amid global change. *Land* **2016**, *5*, 10. [CrossRef]
4. Dixon, J.; Taniguchi, K.; Wattenbach, H.; Tanyeri-Abur, A. (Eds.) *Smallholders, Globalization and Policy Analysis*; Food and Agriculture Organization of the United Nations: Rome, Italy, 2004; Available online: http://www.fao.org/docrep/007/y5784e/y5784e00.htm#Contents (accessed on 1 October 2016).

5. Netting, R.M. *Smallholders, Householders: Farm Families and the Ecology of Intensive, Sustainable Agriculture*; Stanford University Press: Stanford, CA, USA, 1993.
6. Food and Agriculture Organization (FAO). *Smallholders and Family Farmers Factsheet*; Food and Agriculture Organization of the United Nations: Rome, Italy, 2012; Available online: http://www.fao.org/fileadmin/templates/nr/sustainability_pathways/docs/Factsheet_SMALLHOLDERS.pdf (accessed on 1 October 2016).
7. Rudel, T.K.; Kwon, O.; Paul, B.K.; Boval, M.; Rao, I.M.; Burbano, D.; McGroddy, M.; Lerner, A.M.; White, D.; Cuchillo, M.; et al. Do smallholder, mixed crop-livestock livelihoods encourage sustainable agricultural practices? A meta-analysis. *Land* **2016**, *5*, 6. [CrossRef]
8. Boserup, E. *Woman's Role in Economic Development*; St. Martin's Press: New York, NY, USA, 1970.
9. Brookfield, H.C. Questions on the human frontiers of geography. *Econ. Geogr.* **1964**, *40*, 283–303. [CrossRef]
10. Foster, G.M. Peasant society and the image of the limited good. *Am. Anthropol.* **1965**, *67*, 293–315. [CrossRef]
11. Kroeber, A.L. *Anthropology*; Harcourt, Brace and Co.: New York, NY, USA, 1948.
12. Netting, R. *Cultural Ecology*; Waveland Press: Prospect Heights, IL, USA, 1986.
13. Chowdhury, R.R. Differentiation and concordance in smallholder land use strategies in southern Mexico's conservation frontier. *Proc. Natl. Acad. Sci. USA* **2010**, *107*, 5780–5785. [CrossRef] [PubMed]
14. Turner, B.L., II; Shajaat Ali, A.M. Induced intensification: Agricultural change in Bangladesh with implications for Malthus and Boserup. *Proc. Natl. Acad. Sci. USA* **1996**, *93*, 14984–14991. [CrossRef] [PubMed]
15. Wolf, E.R. Closed corporate peasant communities in Mesoamerican and Central Java. *S. J. Anthropol.* **1957**, *13*, 1–18.
16. Hydén, G. *Beyond Ujamaa in Tanzania: Underdevelopment and an Uncaptured Peasantry*; University of California Press: Berkeley, CA, USA, 1980.
17. Laney, R.; Turner, B.L., II. The persistence of self-provisioning among smallholder farmers in northeast Madagascar. *Human Ecol.* **2015**, *43*, 811–826. [CrossRef] [PubMed]
18. Sutton, M.Q.; Anderson, E.N. *Introduction to Cultural Ecology*; Altimira Press: Lanham, MD, USA, 2010.
19. Chayanov, A.V.; Thorner, D.; Kerblay, B.H.; Smith, R.E. *AV Chayanov on the Theory of Peasant Economy*; Manchester University Press: Manchester, UK, 1986.
20. Boserup, E. *The Conditions of Agricultural Growth: The Economics of Agrarian Change under Population Pressure*; University of Chicago Press: Chicago, IL, USA, 1965.
21. Richards, P. *Indigenous Agricultural Revolution: Ecology and Food Crops in West Africa*; Westview Press: Boulder, CO, USA, 1985.
22. Jokisch, B.D. Migration and agricultural change: The case of smallholder agriculture in highland Ecuador. *Hum. Ecol.* **2002**, *30*, 523–550. [CrossRef]
23. Radel, C.; Schmook, B.; McEvoy, J.; Méndez, C.; Petrzelka, P. Labour migration and gendered agricultural relations: The feminization of agriculture in the ejidal sector of Calakmul, Mexico. *J. Agrar. Chang.* **2012**, *12*, 98–119. [CrossRef]
24. Turner, B.L.; Lambin, E.F.; Reenberg, A. The emergence of land change science for global environmental change and sustainability. *Proc. Natl. Acad. Sci. USA* **2007**, *104*, 20666–20671. [CrossRef] [PubMed]
25. Verburg, P.H.; Erb, K.H.; Mertz, O.; Espindola, G. Land System Science: Between global challenges and local realities. *Curr. Opin. Environ. Sustain.* **2013**, *5*, 433–437. [CrossRef] [PubMed]
26. Coomes, O.T.; Takasaki, Y.; Rhemtulla, J.M. Land-use poverty traps identified in shifting cultivation systems shape long-term tropical forest cover. *Proc. Natl. Acad. Sci. USA* **2011**, *108*, 13925–13930. [CrossRef] [PubMed]
27. Steward, J.H. *Theory of Cultural Change: The Methodology of Multilinear Evolution*; University of Illinois Press: Urbana, IL, USA, 1955.
28. Schneider, H.K. *Economic Man: The Anthropology of Economics*; The Free Press: New York, NY, USA, 1974.
29. Walls, L.D. *The Passage to Cosmos: Alexander von Humboldt and the Shaping of America*; University of Chicago Press: Chicago, IL, USA, 2009.
30. Sauer, C.O. The morphology of the landscape. *Univ. Calif. Publ. Geogr.* **1938**, *2*, 19–54.
31. Moran, E.F. *The Ecosystem Approach in Anthropology: From Concept to Practice*; University of Michigan Press: Ann Harbor, MI, USA, 1990.
32. Turner, B.L., II. The specialist–synthesis approach to the revival of geography: The case of cultural ecology. *Ann. Assoc. Am. Geogr.* **1989**, *79*, 88–100. [CrossRef]

33. Nietschmann, B. *Between Land and Water: The Subsistence Ecology of the Miskito Indians, Eastern Nicaragua*; Seminar Press: New York, NY, USA, 1973.
34. Rappaport, R.A. *Pigs for the Ancestors: Ritual in the Ecology of a New Guinea People, a New Enlarged Edition*; Yale University Press: New Haven, CT, USA, 1984.
35. Netting, R.M.; Stone, M.P.; Stone, G.D. Kofyar cash-cropping: Choice and change in indigenous agricultural development. *Human Ecol.* **1989**, *17*, 299–319. [CrossRef]
36. Blaikie, P.; Brookfield, H.C. *Land Degradation and Society*; Methuen: London, UK, 1987.
37. Hewitt, K. (Ed.) *Interpretations of Calamity: From the Viewpoint of Human Ecology*; Allen & Unwin: London, UK, 1983.
38. Peet, R.; Watts, M.J. (Eds.) *Liberating Ecologies: Environment, Development, Social Movements*; Routledge: New York, NY, USA, 1996.
39. Forsyth, T. *Critical Political Ecology: The Politics of Environmental Science*; Routledge: New York, NY, USA, 2003.
40. Zimmerer, K.S.; Bassett, T.J. (Eds.) *Political Ecology: An Integrative Approach to Geography and Environment-Development Studies*; Guilford Press: New York, NY, USA, 2003.
41. Brannstrom, C.; Vadjunec, J.M. (Eds.) *Land Change Science, Political Ecology, and Sustainability: Synergies and Divergences*; Routledge: New York, NY, USA, 2014.
42. Robbins, P. *Political Ecology: A Critical Introduction*; John Wiley & Sons: Malden, MA, USA, 2011.
43. Radel, C.; Schmook, B.; Haenn, N.; Green, L. The gender dynamics of conditional cash transfers and smallholder farming in Calakmul, Mexico. *Women's Studies International Forum* **2016**. [CrossRef]
44. Rocheleau, D.; Thomas-Slayter, B.; Wangari, E. *Feminist Political Ecology: Global Issues and Local Experience*; Routledge: New York, NY, USA, 1996.
45. Turner, B.L.; Robbins, P. Land-change science and political ecology: Similarities, differences, and implications for sustainability science. *Annu. Rev. Environ. Resour.* **2008**, *33*, 295–316. [CrossRef]
46. Kates, R.W.; Clark, W.C.; Corell, R.; Hall, J.M.; Jaeger, C.C.; Lowe, I.; McCarthy, J.J.; Schellnhuber, H.J.; Bolin, B.; Dickson, N.M.; et al. Sustainability science. *Science* **2001**, *292*, 641–642. [CrossRef] [PubMed]
47. Ostrom, E.; Nagendra, H. Insights on linking forests, trees, and people from the air, on the ground, and in the laboratory. *Proc. Natl. Acad. Sci. USA* **2006**, *103*, 19224–19231. [CrossRef] [PubMed]
48. Forbes, B.C.; Stammler, F.; Kumpula, T.; Meschtyb, N.; Pajunen, A.; Kaarlejärvi, E. High resilience in the Yamal-Nenets social–ecological system, West Siberian Arctic, Russia. *Proc. Natl. Acad. Sci. USA* **2009**, *106*, 22041–22048. [CrossRef] [PubMed]
49. Berkes, F.; Colding, J.; Folke, C. (Eds.) *Navigating Social-Ecological Systems: Building Resilience for Complexity and Change*; Cambridge University Press: Cambridge, UK, 2008.
50. Ostrom, E. A general framework for analyzing sustainability of social-ecological systems. *Science* **2009**, *325*, 419–422. [CrossRef] [PubMed]
51. Walker, B.; Holling, C.S.; Carpenter, S.R.; Kinzig, A. Resilience, adaptability and transformability in social-ecological systems. *Ecol. Soc.* **2004**, *9*, 5.
52. Bulte, E.H.; Lipper, L.; Stringer, R.; Zilberman, D. Payments for ecosystem services and poverty reduction: Concepts, issues, and empirical perspectives. *Environ. Dev. Econ.* **2008**, *13*, 245–254. [CrossRef]
53. Speranza, C.I. Buffer capacity: Capturing a dimension of resilience to climate change in African smallholder agriculture. *Reg. Environ. Change* **2013**, *13*, 521–535. [CrossRef]
54. Peterson, G. Political ecology and ecological resilience: An integration of human and ecological dynamics. *Ecol. Econ.* **2000**, *35*, 323–336. [CrossRef]
55. Escobar, A. *Encountering Development: The Making and Unmaking of the Third World*; Princeton University Press: Princeton, NJ, USA, 1995.
56. Wainwright, J. *Decolonizing Development: Colonial Power and the Maya*; Blackwell Publishing: Malden, MA, USA, 2008.
57. Zimmerer, K.S. The compatibility of agricultural intensification in a global hotspot of smallholder agrobiodiversity (Bolivia). *Proc. Natl. Acad. Sci. USA* **2013**, *110*, 2769–2774. [CrossRef] [PubMed]
58. Etongo, D.; Djenontin, I.N.S.; Kanninen, M. Poverty and environmental degradation in southern Burkina Faso: An assessment based on participatory methods. *Land* **2016**, *5*, 20. [CrossRef]
59. Pereira, R.; Simmons, C.; Walker, R. Smallholders, agrarian reform, and globalization in the Brazilian Amazon: Cattle versus the environment. *Land* **2016**, *5*, 24. [CrossRef]

60. Turner, S.; Pham, T. "Nothing is like it was before": The dynamics between land use and land cover, and livelihood strategies in the northern Vietnam borderlands. *Land* **2015**, *4*, 1030–1059. [CrossRef]
61. McConnell, W.J.; Viña, A.; Kull, C.; Batko, C. Forest transition in Madagascar's highlands: Initial evidence and implications. *Land* **2015**, *4*, 1155–1181. [CrossRef]
62. Moseley, W. Agriculture on the brink: Climate change, labor and smallholder farming in Botswana. *Land* **2016**, *5*, 21. [CrossRef]
63. King, B.; Shinn, J.E.; Crews, K.A.; Young, K.R. Fluid waters and rigid livelihoods in the Okavango Delta of Botswana. *Land* **2016**, *5*, 16. [CrossRef]
64. Shackleton, S.; Luckert, M. Changing livelihoods and landscapes in the rural Eastern Cape, South Africa: Past influences and future trajectories. *Land* **2015**, *4*, 1060–1089. [CrossRef]
65. Brannstrom, C.; Tilton, M.; Klein, A.; Jepson, W. Spatial distribution of estimated wind-power royalties in West Texas. *Land* **2015**, *4*, 1182–1199. [CrossRef]
66. Anseeuw, W.; Jayne, T.; Kachule, R.; Kotsopoulos, J. The quiet rise of medium-scale farms in Malawi. *Land* **2016**, *5*, 19. [CrossRef]
67. Hoelle, J. *Rainforest Cowboys: The Rise of Ranching and Cattle Culture in Western Amazonia*; University of Texas Press: Austin, TX, USA, 2015.
68. Vadjunec, J.M.; Sheehan, R. Ranching and state school lands in Cimarron County, Oklahoma. *Great Plains Research* **2010**, *20*. Available online: http://digitalcommons.unl.edu/greatplainsresearch/1111/ (accessed on 1 October 2016).
69. Sheehan, R.; Vadjunec, J.M. Placing community through Actor-Network Theory in Oklahoma's 'No Man's Land'. *Soc. Cult. Geogr.* **2012**, *13*, 915–936. [CrossRef]
70. Sorensen, A. Women, family and class. *Annu. Rev. Sociol.* **1994**, *20*, 27–47. [CrossRef]
71. Muller, D.; Munroe, D.K. Current and future challenges in land-use science. *J. Land Use Sci.* **2014**, *9*, 133–142. [CrossRef]
72. Seto, K.C.; Reenberg, A.; Boone, C.G.; Fragkias, M.; Haase, D.; Langanke, T.; Marcotullio, P.; Munroe, D.K.; Olah, B.; Simon, D. Urban land teleconnections and sustainability. *Proc. Natl. Acad. Sci. USA* **2012**, *109*, 7687–7692. [CrossRef] [PubMed]
73. Giampietro, M.; Aspinall, R.J.; Ramos-Martin, J.; Bukkens, S.G.F. (Eds.) *Resource Accounting for Sustainability Assessment: The Nexus between Energy, Food, Water and Land Use*; Routledge: New York, NY, USA, 2014.
74. Adgar, W.N.; Eakin, H.; Winkels, A. Nested and teleconnected vulnerabilities to environmental change. *Fron. Ecol. Environ.* **2009**, *7*, 150–157. [CrossRef]
75. Rulli, M.C.; Saviori, A.; D'Odorico, P. Global land and water grabbing. *Proc. Natl. Acad. Sci. USA* **2013**, *110*, 892–897. [CrossRef] [PubMed]

Section 1:

Smallholder Practices and Environmental Sustainability

land

MDPI

Article

Do Smallholder, Mixed Crop-Livestock Livelihoods Encourage Sustainable Agricultural Practices? A Meta-Analysis

Thomas K. Rudel [1,*], Oh-Jung Kwon [2], Birthe K. Paul [3,†], Maryline Boval [4,†], Idupulapati M. Rao [3,†], Diana Burbano [5,†], Megan McGroddy [6,†], Amy M. Lerner [7,†], Douglas White [8,†], Mario Cuchillo [9,†], Manuel Luna [10,†] and Michael Peters [3,†]

1 Departments of Human Ecology and Sociology, Rutgers University, 55 Dudley Rd., New Brunswick, NJ 08901, USA
2 The Commission for SSK Multicultural Research, Hanyang University, Seoul 04763, Korea; ziankwon@gmail.com
3 Tropical Forages Program, International Center for Tropical Agriculture (CIAT), Cali, AA 6713, Colombia; b.paul@cgiar.org (B.P.); i.rao@cgiar.org (I.M.R.); m.peters@cgiar.org (M.P.)
4 Livestock Research Program, INRA (L'Institut National de la Recherche Agronomique), Guadeloupe 97170, France; Maryline.Boval@antilles.inra.fr
5 Department of Geography, McGill University, Montreal, QC H3A 0G4, Canada; dianaburbanon@live.com
6 Department of Environmental Sciences, University of Virginia, Charlottesville, VA 22904, USA; mmcgroddy@gmail.com
7 Woodrow Wilson School, Princeton University, Princeton, NJ 08540, USA; amy.m.lerner@gmail.com
8 Research for Development and Conservation, Burlington, VT 05401, USA; dwhite.r4dc@gmail.com
9 Instituto Nacional de Ciencias Medicas y Nutricion, Tlalpan, DF 14080, Mexico; mario.cuchilloh@incmnsz.mx
10 Instituto Nacional de Investigaciones Forestales Agrícolas y Pecuarias, Tecoman 28930, Mexico; silva.manuel@inifap.gob.mx
* Correspondence: rudel@aesop.rutgers.edu; Tel.: +1-848-932-9238; Fax: +1-732-932-6667
† These authors contributed equally to this work.

Academic Editors: Claudia A. Radel and Jacqueline M. Vadjunec
Received: 15 July 2015; Accepted: 1 February 2016; Published: 6 February 2016

Abstract: As calls for bolstering ecosystem services from croplands have grown more insistent during the past two decades, the search for ways to foster these agriculture-sustaining services has become more urgent. In this context we examine by means of a meta-analysis the argument, proposed by Robert McC. Netting, that small-scale, mixed crop-livestock farming, a common livelihood among poor rural peoples, leads to environmentally sustainable agricultural practices. As predicted, mixed crop-livestock farms exhibit more sustainable practices, but, contrary to predictions, a small scale of operation does not predict sustainability. Many smallholders on mixed crop-livestock farms use sustainable practices, but other smallholders practice a degrading, input-scarce agriculture. Some large farm operators use soil-conserving, minimum-tillage techniques while other large operators ignore soil-conserving techniques and practice an industrialized, high chemical input agriculture. The strength and pervasiveness of the link in the data between mixed crop-livestock farming and sustainable agricultural practices argues for agricultural policies that promote mixed crop-livestock livelihoods.

Keywords: mixed crop-livestock farming; sustainable agriculture; conservation agriculture; smallholders; agro-ecology

1. Introduction

Over the past decade as the intertwined threats of food security, biodiversity losses, and climate change have become more apparent, scientists and policymakers have shown a renewed interest in sustainable agriculture. They have called for the sustainable intensification of agriculture through the adoption of agricultural practices that both increase yields and provide ecosystem services for plants and animals [1]. Practices like mulching, manuring, minimum tillage, and crop rotations foster more biodiverse soils, reduce soil erosion, increase nutrient cycling in soils, and make the food system more resilient in the face of climate shocks [2–4]. The circumstances under which farmers adopt these sustainable practices remain open to question. We address this question through a meta-analysis of data from farms where farmers have or have not adopted sustainable agricultural practices.

More specifically, the meta-analysis assesses the hypothesis that a crop-livestock product mix on small landholdings leads to environmentally sustainable agricultural practices. For this task, we use cases of sustainable and unsustainable agricultural practices which come from published studies of farming in many different geographical locales. We begin by discussing what we mean by "sustainable agriculture." Then we describe a theory about the social and ecological conditions that give rise to sustainable agricultural livelihoods. Under this theory, articulated by Robert McC. Netting, sustainable agriculture emerges most frequently among small-scale farmers who grow crops and raise livestock on the same landholding [5]. An assessment of the potential of mixed crop-livestock farming for fostering sustainable agriculture would test the accuracy of Netting's argument, and also raise questions about the likelihood that mixed crop-livestock farms could become a "seed bed" for the spread of sustainable practices to other farms. This prospect currently seems unlikely because the growing bifurcation in global agriculture between large mechanized operations and very small farms seems to have reduced the numbers of mixed crop-livestock farms. Their decline in numbers would diminish the likelihood that mixed crop-livestock farms could provide the bases for a more sustainable agriculture.

In this uncertain context, we address questions about the contemporary bases for sustainable agriculture. Following a discussion of definitions, we describe Netting's theory about the social and ecological conditions that foster sustainable agricultural practices and present an argument about the ways in which trends in contemporary agriculture may be making sustainable agriculture more difficult to practice. Then we describe our meta-analytic methods and present the findings from the data analyses. The final sections of the paper summarize the socio-ecological conditions that enable sustainable agriculture and identify policy interventions that would expedite its spread.

2. What Is "Sustainable Agriculture"?

Any effort to analyze sustainable agriculture as a unitary phenomenon has to contend with the variable definitions of the term [6]. It can mean an agriculture that produces enough food for consumers and enough income for farmers. It can also refer to a collection of agricultural practices that maintain agro-ecosystem services like pollination, soil fertility, and pest control on tracts of land at levels sufficient to sustain agricultural activities year after year without heavy doses of chemical inputs. This paper focuses on the determinants of this last, environmental dimension of sustainable agriculture.

Farmers who mulch, minimize tillage, and rotate crops practice "conservation agriculture" (CA) [7]. Sustainable agriculture refers to a wider set of practices. It includes conservation agriculture, but it also includes manuring, because manuring bolsters agro-ecologies (the ecological processes associated with agricultural production systems) by promoting nutrient cycling and building up organic matter in soils [8–10]. A comprehensive index of sustainable agricultural practices would, therefore, have to include manuring as well as rotating crops, leaving crop residues in the fields (mulching), and tilling soils less. The analyses reported below use an index of sustainable agriculture, measured by the presence or absence on farms of these practices, plus an additional presence/absence score for the self-conscious practice of conservation agriculture (CA). This last, rhetorical dimension of the index is meant to capture the discursive element in the practice of sustainable agriculture, which in turn may contribute to other farm-specific, and less frequently cited sustainable practices. Box 1 describes the structure of this index more completely.

Box 1. An Index of Sustainable Agricultural Practices.

This index focuses on the most frequently mentioned agricultural practices in discussions of sustainable agriculture. In this respect the index represents a central tendency in sustainable agricultural practices, rather than a comprehensive index which incorporates all possible types of sustainable practices. The index focuses on practices associated with soils and rainfed agriculture. It does not include the extensive conservation measures sometimes found in irrigation systems.

The component practices in the index are as follows:

1. Minimum Tillage (includes no-till, ridge till, and strip till variants). It forestalls soil erosion [7].
2. Mulching with crop residues (includes cover crops and other green manure options). It increases nitrogen in soils and reduces soil erosion [6].
3. Crop rotations (includes strip cropping techniques and other techniques designed to manage pest infestations). It improves control of pests and weeds [6].
4. Manures (deposited by grazing livestock on croplands or by spreading manures from confined livestock on croplands). It increases soil nitrogen availability and therefore soil fertility [10].
5. Professed practitioner of conservation agriculture or organic agriculture (this orientation leads to farm-specific measures, not included above, that would contribute to the environmental sustainability of farm operations).

3. A Theory: The Ecology of Small-Scale, Sustainable Agriculture

A broad explanation for the variable adoption of sustainable agricultural practices comes from an agricultural anthropologist. Robert McC. Netting outlined the sources for sustainable agricultural practices among the world's small-scale farmers in *Smallholders, Householders: The Ecology of Small Scale, Sustainable Agriculture* [5]. Netting drew upon an enormous body of previous work on the economy and ecology of peasant smallholders. He outlined a socio-ecological dynamic in which farmers and other members of their households engaged in laborious routines in which they cultivated an agro-biodiverse set of crops, recycled household, livestock, and agricultural wastes back into production processes, and made a multitude of improvements to their land, like terracing it, irrigating it, and planting fertilizer trees. From Netting's point of view, mixed crop-livestock livelihoods might not be necessary for the practice of sustainable agriculture, but there was a clear affinity between crop-livestock product mixes and sustainable agricultural practices because these practices often entailed the tight integration of animal wastes into crop production processes. For example, almost all of Netting's smallholders used manure from their animals to maintain the fertility of their croplands.

Netting advanced his argument through analyses of similar agro-ecologies in different places, rice cultivators in central China, potato growers in the Swiss Alps, and hill farmers in central Nigeria. A singular pattern runs through agriculture in all of these places. It involves labor intensive activities that, through biological controls and recycling of wastes, refurbish agro-ecologies and promote the long-term sustainability of agriculture. Although Netting does not say so explicitly, his work provides implicit support for the contention that sustainable agriculture is scale dependent, that it occurs most frequently on small farms. In Netting's theory of smallholder agro-ecologies, the different factors are hypothesized to interact in key ways (see Figure 1). The favorable labor to land ratios that frequently prevail on small farms make it possible for households to carry out the labor intensive agricultural tasks, like spreading manure or mulch, that make agriculture sustainable.

Netting's argument had global reach. He saw the theorized dynamic occurring among small farmers in both the Global North and the Global South, among state-supported small farmers in

Europe as well impoverished rice cultivators in Southeast Asia. In the more than two decades since Netting published his book, a series of scientists, some of them his students [11], have spelled out its implications for environmental change in places where smallholders predominate, further expanding the theory in the process. One field study of smallholders in northern Mexico found that the greater agro-biodiversity associated with smallholder agro-ecologies worked to the benefit of smallholders during times of drought [2]. Other researchers have investigated the association between social capital and smallholder agro-ecology with the idea that social capital might facilitate the creation and maintenance of common property institutions that in turn would maintain a common pool resource like an irrigation system [12]. The emergence of political ecology as a sub-field in Geography and Anthropology has led to a greater appreciation for the political struggles and conflicts that disturb and periodically transform smallholder agro-ecologies [13]. Agricultural scientists have explored similar crop-livestock synergies on somewhat larger farms in the United States [14,15]. Despite all of this work there has never been a quantitative, global-scale assessment of Netting's argument, so the utility of his thesis for understanding global trends in sustainable agriculture remains uncertain.

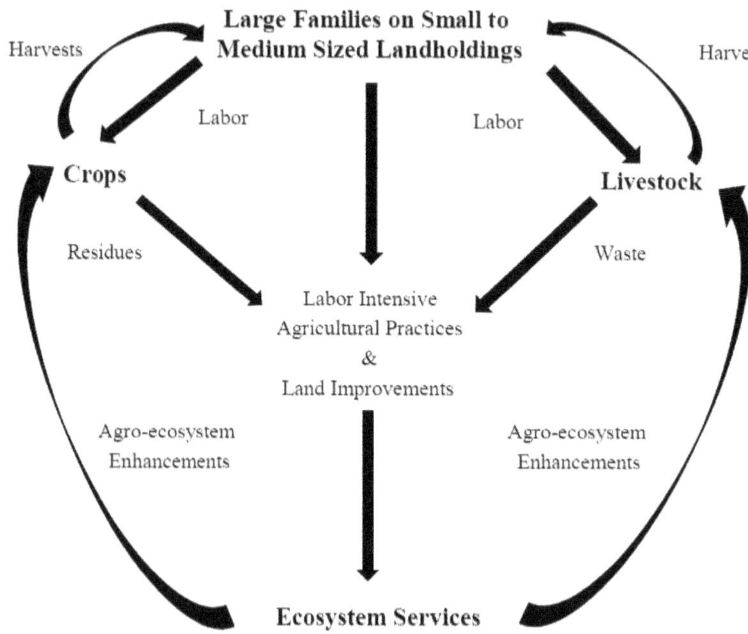

Figure 1. The ecology of small-scale, sustainable agriculture (after Netting, 1993).

4. A Changing Context: Sustainable Agriculture in a Diverging Agrarian Order

Two recent trends in agriculture call into question the link between mixed crop-livestock farming and sustainable agricultural practices. First, conservation agriculture (CA), in which farmers pursue environmental sustainability without contributions from livestock, has become more common. Second, small farmers in Sub-Saharan Africa, pressed for additional sources of income, sell crop residues and manure rather than returning these materials to the land. These specific trends derive from a larger, long-established trend towards the bifurcation of the global agricultural sector into large-scale, mechanized farms and small, family-run agricultural plots. This trend may have reduced the proportion of landholdings where people might be able to practice the mixed crop-livestock agro-ecology described by Netting. This diverging dynamic is described below, beginning with the continuing growth in the numbers of large-scale farms.

While large-scale plantations in coastal locations in the tropics and sub-tropics date from the 17th century [16], the trend in the United States' Middle West and Plains states towards larger farms dates from the early 20th century when homesteading ended. Farmers have specialized, as their farms have gotten larger, with the larger farms focused increasingly on only cultivating crops [17]. The segregation of crops from animals eliminated the synergies between crop cultivation and animal waste that had characterized so much sustainable agriculture. Farmers benefited from the segregation of crops from animals on large farms because they could obtain economies of scale in the purchase of larger quantities of the smaller range of inputs needed to grow crops. The per-unit costs of inputs like fertilizer declined for these farmers, even as the total amount of their expenditures on inputs climbed with the increase in the scale of their farms. The decline in per unit input costs proved very important to farmers during the 1920 to 2005 period when the price of agricultural products rose more slowly than manufactured goods. For many family farmers it was get big or get out [17]. All of these operations became more specialized and more mechanized as they got larger, so the large operators spent more money on a few specialized inputs, like machinery. In this context some innovations central to CA, like reduced tillage, grew more attractive to large operators because they promised to reduce further the costs of inputs of machinery, fuel, and labor [7].

This dynamic changed during the 2008 economic crisis when some rice-producing countries suddenly banned the export of rice [18]. Driven by concerns about food security or by desires to profit from food security concerns, groups of private investors began to engage in "land grabs," in which they acquired large tracts of land which they then dedicated to the production of foodstuffs for markets overseas [19]. The expansion in the scale of agricultural operations continued, but new groups of operators became the owners of agricultural lands.

CA typically has three components, minimum tillage, mulching of agricultural residues, and the planting of nitrogen-fixing cover crops when croplands are not being cultivated [20]. These agricultural practices were first widely adopted during the Dust Bowl era [21], but they only became recognized as a package of sustainable practices when farmers began to talk about them and adopt them on fairly large, erosion prone, crop-only farms in Kentucky during the 1960s [7]. Governments have generally not provided incentives for CA, but it has appealed to farmers because, in the places where it can be practiced, CA limits the numbers of times that farmers have to turn the earth, and this reduction in ploughing reduces the costs of labor and gasoline to run agricultural machinery [6,7]. Because CA adopters have typically had large areas of cropland, they have had the economic capital to invest in minimum tillage machinery and in herbicides to prevent weed infestations. The CA regimen also boosts nutrient recycling in croplands because the organic waste in crop residues and cover crops recycles nitrogen back into the soils [6]. CA does not, however, conform to the smallholder agro-ecology outlined by Netting because it does not typically mix livestock with the cultivation of crops.

Foundations and development assistance agencies have recently tried to promote the adoption of CA practices among smallholders in Sub-Saharan Africa. The CA regimen does conserve soil moisture and thereby benefits plants in zones with erratic rainfall, but, aside from this advantage, CA has failed to produce yield gains consistently in Sub-Saharan Africa, which in turn has limited its appeal [22,23]. CA has been most widely adopted in countries like Canada with agricultural sectors in which large farms predominate [20].

While on a global scale some farms were expanding in size during the 20th century, other farms, particularly in South Asia and some regions of Sub-Saharan Africa, were declining in size. Large numbers of smallholders in South Asia and Sub-Saharan Africa have become proletarianized as a byproduct of a continuing subdivision of smallholdings through inheritance [24]. These smallholders began to work at least part-time as farm laborers because they could no longer earn a sufficient income from the small plot of land that they owned. In these circumstances landholdings in densely populated districts of, for example, western Kenya became so small (less than 0.4 hectares) that the owners had to maximize their revenue from the land, however they could, in order to survive. To generate income, they sold the crop residues and the manure from confined livestock rather than returning the manure to the land, so over time the soils on their landholdings degraded [24]. They were caught up in a natural resource degrading poverty trap [25].

A more socially benign version of this dynamic might apply to small farms in peri-urban districts where non-farm employment has reduced the family labor force to the point where no one has the time to perform the time-consuming, land-improving tasks of, for example, spreading manure. These dynamics involving the smallest landholdings, like those involving CA on the largest landholdings, depart from the theory of smallholder agro-ecology because they sever the connection between livestock and the land. They suggest that the bifurcation of global agriculture into large and very small farms has reduced the numbers of farms on which smallholder, mixed crop-livestock agro-ecology would be feasible. The meta-analysis presented below assesses this conjecture. It analyses the prevalence of sustainable agricultural practices across different types and sizes of farms, including some with crop-livestock mixes.

5. Methods and Materials

5.1. Case Selection and Farmer Categorization

To answer questions about the social and ecological conditions that foster sustainable agriculture, we reviewed and collected all of those studies that asked farmers about agricultural practices on their farms. The data in these studies make it possible to investigate the patterns of association between the characteristics of farms and sustainable agricultural practices. From this large set of studies, we selected for the meta-analysis those studies in which we could identify sets of farmers that practiced or did not practice sustainable agriculture. The authors of these studies created typologies of farmers, based on the size of their operations, the presence or absence of a crop-livestock product mix, and the prevalence of sustainable agricultural practices.

We wanted to identify the attributes of farms that had an effect on the probability that farmers would cultivate their lands in environmentally sustainable ways. Measurement problems precluded the measurement of the magnitude of these effects beyond the binary of statistically significant or insignificant. We just wanted to find out if a particular variable, the size of a farm or the presence of a mixed livestock-crop operation, made it significantly more likely that a farmer would employ practices that contribute to the environmental sustainability of agriculture on his/her farm. To answer this question, we looked for patterns across the case studies of farming in different geographical locales.

To assemble an exhaustive list of articles on types of farms and sustainable agricultural practices, we searched the scientific literature in the four different languages that we were able to read (English, French, Spanish, and Portuguese). We used one search engine (Google), the Web of Science database, and the International Livestock Research Institute's database. We employed a range of search terms including "agro-pastoral" and "crop-livestock." Personal knowledge of the relevant literatures also helped us to identify some case studies. These searches produced a list of 148 studies for possible inclusion in the meta-analysis. We examined all of these articles and, using the criteria outlined below, we ended up with 39 studies, including several books as well as articles. The 39 studies identified 141 types of farmers, some of whom practiced sustainable agriculture and some of whom did not. All of the case studies that we identified had publication dates after 1980.

To reduce the likelihood of selection bias in choosing case studies for analysis [26], we sampled on the independent variable of greatest interest, the presence or absence of a mixed crop-livestock operation on a farm. We searched for studies that reported variations from farm to farm in product mixes (crop-livestock, crop-only, livestock-only) and in sustainable agricultural practices like manuring, tillage frequency, crop rotations, and mulching (see Box 1). Using these criteria, we eliminated many studies from consideration. For example, some studies could not be included in the database because they presented models of sustainable and conventional farms instead of data on actual agricultural practices. Similarly, studies of conventional and alternative agricultural practices in experiment station settings did not represent the practices of actual farm households, so they could not be included in the study. Although a particular survey of farmers may have served as the basis for multiple publications, it only appears once in the database to prevent the double counting of data from particular farms.

The 141 different types of farmers in the 39 studies represent aggregated data from 6253 farms. The 39 studies are listed in the Supplementary Materials for this article. The immediate contexts of the studies varied tremendously. For example, a study of farmers in eastern Bolivia compared the agricultural practices of cattle ranchers (one type of farmer) with the practices of crop-livestock farmers (another type of farmer). Researchers in Kenya compared the agricultural practices of dairy farmers with crops and several cows (one type of farmer) with the agricultural practices of farmers who only grew corn (another type of farmer). In a study in the United States, researchers compared the agricultural practices of grain-only farmers (one type of farmer) with the agricultural practices of other farmers who grew grain and maintained livestock on their land (another type of farmer).

Given that researchers frequently selected the sites for case studies for idiosyncratic reasons, the cases in this meta-analysis cannot properly be considered to be anything other than a non-random sample from the universe of agricultural regions. The assembled case studies can, however, be evaluated in the same way that we would assess a purposive sample. It would be considered adequate for our analytic task if the 39 studies contained a sufficient range of variation from case to case to address the substantive hypothesis under consideration. There would have to be variations from farm type to farm type in sustainable agricultural practices and in crop-livestock, crop-only, and livestock-only product mixes. The 141 farm types contain variation along these dimensions. The numbers of farm types in different categories are listed below in parentheses. The assembled studies have crop-livestock types of operations (111) and more specialized crop-only or livestock-only farm types (30). The studies span the developed (44 farm types) and the developing world (97 farm types) with cases from North America (23), Latin America (15), Europe (17), Asia (19), Oceania (4), and Sub-Saharan Africa (63). The size of farms for each farm type was also reported. These variations in the data make it possible to assess the link between mixed crop-livestock operations and sustainable agricultural practices across a broad range of places.

Each of the 39 studies was read by multiple people, and the inevitable disagreements about how to code a particular case were resolved in discussions among the team of readers. Practices were coded as 'present' or 'absent' based on quantitative and qualitative data presented in the original article. In addition to the coded practices from the farms in each case study, the database contains data about the agricultural contexts of the case studies, in particular about the agricultural policies of the countries in which researchers did the case studies.

5.2. Quantitative Analyses

For our quantitative analyses, we built generalized linear models (GLM) of variations from farm type to farm type in sustainable agricultural practices. For these multivariate analyses, we supplemented data from the case studies with data from other sources (see Table 1). For our dependent variable, we employed an index that sums four common agricultural practices (minimum tillage, crop rotation, mulching crop residues, and manuring) that conserve soils and soil nutrients. To this sum, we added an additional variable that indicates whether or not the farmer knowingly practices sustainable agriculture or CA. When a series of items represent different, but still substantively related behaviors and their values co-vary from case to case, they represent a scale of a particular phenomenon, in this case the sustainability of agricultural practices. Testing for the appropriateness of treating these five items as a scale resulted in an acceptable Cronbach's alpha (a measure of how much items in a scale vary together) of 0.617. The relatively low alpha may reflect the differences in the substance of sustainability practices from the large-scale practitioners of CA to the smallholders who practice mixed crop-livestock agriculture.

Generalized linear models (GLM) are the preferable models for multivariate analyses when the dependent variable has a non-normal distribution, which is the case with an ordinal variable like our scale of sustainable agriculture. In addition it seemed heuristically useful to carry out, in addition to the global-scale GLM analysis, separate analyses of studies from the developed and developing worlds because the largest farms were in the developed countries and the smallest farms were in the developing world (see Table 2). There were influential cases in the developed and developing countries analyses—two cases in the developed country analysis and one case in the developing country analysis. These cases were removed. Their removal from the analyses did not, however, produce any substantively meaningful changes in the GLM coefficients for the predictor variables. This diagnostic test indicates a relatively robust set of findings from the multivariate analyses.

Table 1. Definitions and sources for variables in the multivariate analyses.

Variables	Values	Sources
Dependent Variable		
Sustainable Agriculture Index: Presence (1) or absence (0) of manure on fields + reduced tillage + crop rotation + mulching of stubble and litter + conscious practice of conservation or sustainable agriculture.	0–5	Case studies
Independent Variables		
Mixed Crop-Livestock Agriculture	0 = No, 1 = Yes	Case studies
Tree Crops on Farms	0 = No, 1 = Yes	Case studies
Farm Labor	0 = Some or all hired labor, 1 = Family labor	Case studies
Government Subsidies for Agricultural Products *	% of the value of agricultural products in the country containing the case study site.	[27]

* The subsidies vary in form from country to country. In some countries, like the United States, the government subsidizes the production of a crop like corn. In other countries, like Cameroon, the government subsidizes the acquisition of inputs like fertilizer for crops like cacao. Whatever the form of the subsidy, it can be converted into a proportion of the market value of the crop.

Table 2. Farm characteristics and conservation practices: developed and developing countries.

	Farm Characteristics			Conservation Practices (% of Farms)					
	Mean Size of Farms (hectares)	Mixed Crop-Livestock Farms (%)	Mean Number of Agricultural Products	Manuring	Less Tillage	Crop Rotation	Stubble Mulching	Little or No Chem. Fertilizer	Sustainable Agriculture Index: Mean
Developed Countries (Europe, N. America, Oceania)	464***	55***	1.98***	66*	41***	63*	54	10***	2.57
Developing Countries (Africa, Asia, Latin America)	31	89	2.51	79	9	76	45	40	2.21

Notes: Significance of Difference of Means: *** <0.001, ** <0.01, * <0.10. The number of agricultural products is an ordinal level variable with three categories ranging from 1 to 3: 1 = 1 or 2 crops, 2 = 3 or 4 crops, 3 = 5+ crops. Crop-livestock farms, manuring, less tillage, crop rotation, stubble mulching, little-no chemical fertilizer are all binary variables (1 = yes, 0 = no). The percentages refer to the percentage of farm types from these countries that exhibit the characteristic or engage in the particular conservation practice.

6. Results

We identified several important aggregate differences between richer and poorer countries in their farms and in their sustainable agriculture practices (see Table 2). The farms from the wealthier nations tended to be larger and specialized in either crops or livestock. Farmers from developed countries relied heavily on external (chemical) inputs. Among the sustainable agricultural practices, they emphasized minimum tillage. Farmers from the poorer countries had less land. They operated more mixed crop-livestock farms and produced a greater diversity of agricultural products than did farmers in the wealthier countries. Farmers in the developing world were more likely to choose, among the sustainable practices, to spread manure on their fields, rotate crops, and intercrop them.

The GLM analyses (Table 3) suggest explanations that would account for the differences between farm types in the frequency with which they practice sustainable agriculture. The most striking finding concerns the strong association between mixed crop-livestock operations and sustainable practices. It is the strongest effect in the global, developed, and developing country analyses. Farm size did not predict sustainable practices in any of the analyses, so it is not included among the independent variables in Table 3. There are also some substantive differences between the three analyses. In the global-scale analysis (column 1 of Table 3) the size of agricultural subsidies associates positively with sustainable practices. This effect, although weaker than the mixed crop-livestock effect, underscores how in the wealthier countries the subsidies, which encourage an expansion in the size of farms, are compatible with some sustainable agricultural practices like minimum tillage and stubble mulching. In the developed country equation (column 2), mixed crop-livestock operations associate positively with sustainable practices, as does family farming (where all of the labor comes from the family). For the farms in the developed countries, those that relied exclusively on family labor, while large in an international context (mean = 326 hectares), were smaller than farms that hired labor (mean = 624 hectares) and were more likely to use crop residues as mulch in their fields. In the developing country subsample (column 3 in Table 3), mixed crop-livestock operations made more use of sustainable practices, as did farms that harvested tree crops along with other agricultural products. The salience of tree crops reflects the widespread planting of fertilizer, fruit, and fodder trees on smallholdings, particularly in Sub-Saharan Africa and Central America.

Table 3. Variations in sustainable agriculture practices across farm types: Ɓ coefficients from generalized linear models †.

	(1) All Cases: Global Pattern	(2) Cases from Developed Countries	(3) Cases from Developing Countries
Mixed Crop-Livestock Farms	−1.577*** (0.282)	−1.815*** (0.521)	−1.376*** (0.319)
Tree Crops on Farms	−0.627* (0.260)		−0.657** (0.229)
Government Subsidies for Agriculture	0.076***(0.018)	−0.037(0.079)	−0.037(0.091)
Labor Source (Family = 1)	−0.127(0.205)	−0.956*(0.513)	0.155(0.205)
Farm Size	0.019(0.028)	−0.035(0.057)	0.023(0.081)
Likelihood Ratio	39.05	13.10	30.20
Chi Square	(sig. 0.000)	(sig. 0.011)	(sig. 0.000)
N of Cases	133	36	97

Notes: *** <0.001, ** <0.01, * <0.10. † The equations for the different geographic groupings of farms describe the efficacy of different factors, analyzed together, in predicting the score of a particular farm type on the index of sustainable agricultural practices. The Ɓ coefficients (unstandardized regression coefficients) indicate the change in the response variable when the predictor variable is set to 0 (not present in the case of the mixed crop-livestock and tree crop variables). The numbers in parentheses are the standard errors of the Ɓs. There were no cases of farm types with tree crops in the developed country subsample. Given the use of the maximum likelihood estimation framework, the significance tests for the partial regression coefficients are for Wald Chi Squares. Missing data for the labor variable, along with the removal of the three influential cases, reduced the number of cases in the analysis to 133.

7. Discussion

The importance of mixed crop-livestock operations for the practice of sustainable agriculture is the clearest finding to emerge from the meta-analysis. The presence of animals on farms where people also cultivate crops encourages cultivators to recycle animal wastes onto croplands, and these practices seem to encourage farmers to adopt other practices like the rotation of cover crops and minimum tillage that also enhance agricultural sustainability. This tight integration of the recycling of animal wastes with crop production is the hallmark of the sustainable production practices in Netting's [5] theory about the ecology of small-scale, sustainable agricultural livelihoods. For this reason the association of mixed crop-livestock farming with sustainable agricultural practices in the meta-analysis provides broad confirmation for Netting's approach to understanding the sources for sustainable agriculture.

This finding takes on added significance given the projections of more variable and more extreme climatic conditions in the future. An acceleration in climate change would probably reinforce the attractiveness of mixed crop-livestock practices to farmers, largely because the greater diversification of these mixed operations might reduce the risk of complete crop failures when climatic conditions fluctuate [2]. For example, adverse climactic conditions for a row crop might not limit animal production on a farm. The same logic of diversification in the face of climate change would appear to favor the spread of tree crops on farms with row crops. Because the greater agro-biodiversity of mixed farming would presumably enable mixed-crop livestock farmers to survive climatic extremes more easily, they should exhibit more resilience during periods of recovery [6]. The superiority of the sustainable practices used by mixed crop-livestock farmers may however depend on the types of extreme weather encountered. A recent global meta-analysis of differences in crop yields between conventional and conservation agriculture concluded that the practice of conservation agriculture led to yield losses in humid growing conditions and to yield gains in arid conditions, presumably because CA practices increased moisture retention in soils during droughts [28].

Other findings from the meta-analysis point indirectly to the social and ecological limits of Netting's theory. The argument that small farmers practice more sustainable agriculture does not receive support. The ongoing divergence in the agrarian sector, outlined above, may explain the absence of a relationship between farm size and sustainable agricultural practices across the studies. First, the spread of conservation agriculture (CA) across the agricultural sectors of the grain-exporting nations has made sustainable practices more common among large growers [6,20], but, because it does not mix crops with livestock and does not involve manuring, CA represents a somewhat different form of sustainability than the smallholder agro-ecologies described by Netting. Second, changes at the other end of the size distribution of farms have weakened the association between smallholdings and sustainability. In densely settled, impoverished districts of South Asia and Sub-Saharan Africa, the number of agriculturally unsustainable smallholdings appears to be increasing [24].

In sum it would seem that the growing worldwide bifurcation in agricultural livelihoods between large, highly capitalized agricultural operations and poorly capitalized smallholdings has begun to characterize sustainable agricultural practices as well. Conservation agriculture (CA), with its less invasive treatments of soils and redesigned machinery represents ecological modernization [29], a process of change that depends on technological innovations and characterizes large-scale, highly capitalized agricultural economies. The natural resource degrading poverty traps afflicting the poorest landowners represent substantively a type of proletarianization in which smallholders are over time reduced to selling their own labor as the productive value of their land falls to zero. The dynamics that characterize both of these processes depart dramatically from the smallholder agro-ecologies described by Netting. The increasing numbers of these very large and very small farms suggest that the mixed crop-livestock agro-ecologies found on small to middle sized farms may be declining in number.

8. Conclusion: Limitations and Implications for Policy

The limits of the analysis presented here temper our conclusions in important ways. First, we have focused only on the environmental sustainability of agricultural practices, so important factors in the overall acceptance of these practices by cultivators, in particular their economic return, have not been systematically considered here. Second, there are shifting cultivators, probably declining in number, whose practices are not fairly represented in any of the case studies, so generalizations about synergies between crops and livestock in their livelihoods are beyond the scope of this analysis [30]. Similarly, the sample does not contain a large enough number of cases of some agro-ecological practices like silvo-pastoralism and establishment of pollination corridors to warrant generalizations about the determinants of these practices. With these caveats in mind, we explore the policy implications of the preceding analysis.

Two conclusions with policy implications would seem to follow from the foregoing analysis. First, as emphasized elsewhere [6], the substance of sustainable agricultural practices appears to vary with the size of farms. Larger farms seem more frequently to specialize their operations, focusing in particular on crop-only operations. For these farms the path towards more environmentally sustainable operations may involve the introduction of new machines and new seeds that increase the ecosystem services and economic returns associated with conservation agriculture. For the usually, but not invariably, middle-sized farms with product mixes that include livestock as well as crops, innovations in sustainability would try to capitalize on the synergies present at the interface between crops and livestock on farms. These innovations might apply on even the smallest crop-livestock operations in poor districts of the rural South. In sum, policymakers would want to encourage environmentally sustainable agriculture in different ways depending on the size of the participating farmers' landholdings [31].

A second, more structural policy implication acknowledges and tries to build upon the primary finding from this analysis, the strong relationship between mixed crop-livestock operations and agricultural practices that promote environmental sustainability. Given the historical role that increases in the size of agricultural operations played in the segregation of crop from livestock production and the loss of environmental synergies that has come with the disappearance of crop-livestock operations, land reforms [32] that reduce the size of the larger operations might provide some environmental synergies by making the more diverse, climate resilient crop-livestock operations more appealing to farmers who, after the reform, would cultivate middle-sized farms. The appeal of mixed crop-livestock operations grows when the land base for farms is limited because crops plus livestock generate multiple streams of income from fairly small tracts of land. Compared to larger farms, a higher proportion of the total land area on these middle-sized farms would be under cultivation, so farmers might bring sub-marginal lands into cultivation after a land reform, as they did in Bolivia in the 1950s [33]. In sum, there would be environmental trade-offs with a land reform, increases in sustainable procedures on fields occupied by livestock at the cost of increases in the cultivation of sub-marginal lands. These structural reforms will most likely only work if they are accompanied by familiar government and NGO initiatives like lines of credit for innovations in sustainable agriculture and certification schemes that promise higher returns for growers who produce crops in sustainable ways.

Supplementary Materials: The following are available online at www.mdpi.com/2073-445X/5/1/6/s1, Supplementary Materials: Case Studies Included in the Meta-analysis.

Acknowledgments: Holly Berman provided valuable assistance in the conduct of this research. Funds from National Science Foundation Grant CNH10009499 facilitated this research. Researchers from CIAT received funds from two CGIAR Research Programs: (1) Livestock and Fish, and (2) Climate Change, Agriculture, and Food Security (CCAFS).

Author Contributions: TKR and OJK engaged in case selection, coding, and data analysis. The other authors made contributions to the search for relevant case studies.

Conflicts of Interest: The authors declare no conflict of interest.

References

1. Montpellier Panel. *Sustainable Intensification: A New Paradigm for African Agriculture*; Agriculture for Impact: London, UK, 2013.
2. Eakin, H. *Weathering Risk in Rural Mexico: Climatic, Economic and Institutional Change*; University of Arizona Press: Tucson, AZ, USA, 2006.
3. Intergovernmental Panel on Climate Change (IPCC). *Climate Change 2014: Mitigation of Climate Change*; Cambridge University Press: Cambridge, UK, 2014.
4. Palm, C.; Blanco-Canqui, H.; DeClerck, F.; Gatere, L.; Grace, P. Conservation agriculture and ecosystem services: An overview. *Agric. Ecosyst. Environ.* **2014**, *187*, 87–105. [CrossRef]

5. Netting, R.M. *Smallholders, Householders: The Ecology of Small Scale, Sustainable Agriculture*; Stanford University Press: Stanford, CA, USA, 1993.
6. National Research Council. *Towards a Sustainable Agriculture for the 21st Century*; National Academies Press: Washington, DC, USA, 2010.
7. Coughenour, C.M.; Chamala, S. *Conservation Tillage and Cropping Innovation. Constructing the New Culture of Agriculture*; Iowa State University Press: Ames, IA, USA, 2000.
8. Edmeades, D.C. The long-term effects of manures and fertilizers on soil productivity and quality: A review. *Nutr. Cycl. Agroecosyst.* **2003**, *66*, 165–180. [CrossRef]
9. Rusinamhodzi, L.; Corbeels, M.; Zingore, S.; Nyamangara, J.; Giller, K. Pushing the envelope? Maize production intensification and the role of cattle manure in recovery of degraded soils in smallholder farming areas of Zimbabwe. *Field Crop. Res.* **2013**, *147*, 40–53. [CrossRef]
10. Sommerfeldt, T.G.; Chang, C; Entz, T. Long-term annual manure applications increase soil organic matter and nitrogen and decrease the carbon to nitrogen ratio. *Soil Sci. Soc. Am. J.* **1988**, *52*, 1668–1672. [CrossRef]
11. Wilk, R.R.; Stone, P. Introduction to a very human ecology: Celebrating the work of Robert McC. *Netting. Hum. Ecol.* **1998**, *26*, 175–188. [CrossRef]
12. Ostrom, E. Crossing the great divide: Coproduction, synergy, and development. *World Dev.* **1996**, *24*, 1073–1086. [CrossRef]
13. Peters, P. Inequality and conflict over land in Africa. *J. Agrar. Change* **2004**, *4*, 269–314. [CrossRef]
14. Sulc, R.M.; Tracy, B.F. Integrated crop-livestock systems in the US Corn Belt. *Agron. J.* **2007**, *99*, 335–345. [CrossRef]
15. Hendrickson, J.; Sassenrath, G.; Archer, D.; Hansen, J.; Halloran, J. Interactions in integrated U.S. agricultural systems: The past, the present, and the future. *Renew. Agric. Food Syst.* **2008**, *23*, 314–324. [CrossRef]
16. Mintz, S. *Sweetness and Power: The Place of Sugar in Modern History*; Viking: New York, NY, USA, 1986.
17. Hart, J.F. *The Changing Scale of American Agriculture*; University of Virginia Press: Charlottesville, VA, USA, 2004.
18. Sharma, R. *Food Export Restrictions: Review of the 2007–2010 Experience and Considerations for Disciplining Restrictive Measures*; FAO Commodity and Trade Policy Research Working Paper, No. 32; Food and Agricultural Organization of the United Nations: Rome, Italy, 2011.
19. Borras, S.; Hall, R.; Scoones, I.; White, B.; Wolford, W. Towards a better understanding of global land grabbing: An editorial introduction. *J. Peasant Stud.* **2011**, *38*, 209–216. [CrossRef]
20. Hobbs, P.R. Conservation agriculture: What is it and why is it important for future sustainable food production. *J. Agric. Sci.* **2007**, *145*, 127–137. [CrossRef]
21. Hurt, R.D. *The Dust Bowl: A Cultural and Social History*; Nelson-Hall: Chicago, IL, USA, 1981.
22. Giller, K.; Witter, E.; Corbeels, M.; Tittonell, P. Conservation agriculture and smallholder agriculture in Africa: The Heretics view. *Field Crops Res.* **2009**, *114*, 23–34. [CrossRef]
23. Corbeels, M.; de Graaff, J.; Nash, T.; Penot, E.; Baudron, F.; Naudin, K.; Andrieu, N.; Chirot, G.; Schuler, J.; Nyagumbo, I.; *et al.* Understanding the impact and adoption of conservation agriculture in Africa: A multi-scale analysis. *Agric. Ecosyst. Environ.* **2014**, *187*, 155–170. [CrossRef]
24. Valbuena, D.; Erenstein, O.; Tui, S.H.-K.; Abdoulaye, T.; Claessens, L.; Duncan, A.; Gerard, B.; Rufino, M.C.; Teufel, N.; van Rooyen, A.; *et al.* Scoping crop residue trade-offs in Sub-Saharan Africa and South Asia. *Field Crops Res.* **2012**, *132*, 175–184. [CrossRef]
25. McPeak, J.; Barrett, C. Differential risk exposure and stochastic poverty traps among East African pastoralists. *Am. J. Agric. Econ.* **2001**, *83*, 674–679. [CrossRef]
26. Berk, R.A. An introduction to sample selection bias in sociological data. *Am. Sociol. Rev.* **1983**, *48*, 386–398. [CrossRef]
27. Anderson, K.; Martin, W.; Valenzuela, E. The relative importance of global agricultural subsidies and market access. *World Trade Rev.* **2006**, *5*, 357–376. [CrossRef]
28. Pittelkow, C.; Liang, X.; Linquist, B.; van Groenigen, K.; Lee, J.; Lundy, M.; van Gestel, N.; Six, J.; Venterea, R.; van Kessel, C. Productivity limits and potentials of the principles of conservation agriculture. *Nature* **2015**, *517*, 365–368. [CrossRef] [PubMed]
29. Mol, A.J.; Sonnenfeld, D. *Ecological Modernization around the World: Perspectives and Critical Debates*; Frank Cass: London, UK, 2000.

30. Van Vliet, N.; Mertz, O.; Heinimann, A.; Langanke, T.; Pascual, U.; Schmook, B.; Adams, C.; Schmidt-Vogt, D.; Messerli, P.; Leisz, S.; *et al.* Trends, drivers and impacts of changes in swidden cultivation in tropical forest-agriculture frontiers: A global assessment. *Glob. Environ. Change* **2012**, *22*, 418–429. [CrossRef]
31. Hayami, Y.; Ruttan, V. *Agricultural Development: An International Perspective*; Johns Hopkins University Press: Baltimore, MD, USA, 1971.
32. Lipton, M. *Land Reform in Developing Countries: Property Rights and Property Wrongs*; Routledge: London, UK, 2009.
33. Preston, D. The revolutionary landscape of highland Bolivia. *Geogr. J.* **1969**, *135*, 1–17. [CrossRef]

land

MDPI

Article

Toward the Integrated Framework Analysis of Linkages among Agrobiodiversity, Livelihood Diversification, Ecological Systems, and Sustainability amid Global Change

Karl S. Zimmerer [1,2,*] and Steven J. Vanek [2]

1 David Rockefeller Center, Harvard University, Cambridge, MA 02138, USA
2 Department of Geography and GeoSyntheSES Laboratory, Pennsylvania State University, 302 Walker Building, State College, PA 16802, USA; sjv13@psu.edu
* Correspondence: ksz2@psu.edu; Tel.: +1-617-495-3366

Academic Editors: Claudia A. Radel and Jacqueline M. Vadjunec
Received: 12 August 2015; Accepted: 28 March 2016; Published: 21 April 2016

Abstract: Scientific and policy interest in the biological diversity of agriculture (agrobiodiversity) is expanding amid global socioeconomic and environmental changes and sustainability interests. The majority of global agrobiodiversity is produced in smallholder food-growing. We use meta-analyses in an integrated framework to examine the interactions of smallholder agrobiodiversity with: (1) livelihood processes, especially migration, including impacts on agrobiodiversity as well as the interconnected resource systems of soil, water, and uncultivated habitats; and (2) plant-soil ecological systems. We hypothesize these interactions depend on: (1) scope of livelihood diversification and type resource system; and (2) plant residues and above-/belowground component ecological specificity. Findings show: (1) livelihood diversification is linked to varied environmental factors that range from rampant degradation to enhancing sustainability; and (2) significant ecological coupling of aboveground and soil agrobiodiversity (AGSOBIO assemblages). The environmental impacts of livelihood interactions correspond to variation of diversification (migration, on-farm diversification) and resource system (*i.e.*, agrobiodiversity *per se*, soil, water). Our findings also reveal mutually dependent interactions of aboveground and soil agrobiodiversity. Results identify livelihood diversification-induced reduction of environmental resource quality with lagged agrobiodiversity declines as a potentially major avenue of global change. Our contribution re-frames livelihood interactions to include both agrobiodiversity and ecological systems. We discuss this integrated social-environmental re-framing through the proposed spatial geographic schema of regional agri-food spaces with distinctive matrices of livelihood strategies and relations to biodiversity and resources. This re-framing can be used to integrate livelihood, agrobiodiversity, and ecological analysis and to guide policy and scientific approaches for sustainability in agriculture and food-growing.

Keywords: livelihood diversification; migration; environment; development; agrobiodiversity; agroecology; water resources; soil microbial diversity; ecosystem services; global change; meta-analysis; social-ecological systems; ecological intensification

---

1. Introduction: Integrating the Analysis of Livelihoods, Agrobiodiversity, and Ecological Interactions

Science and policy increasingly recognize biodiversity as central to land-use and agri-food systems amid global-scale changes that are socioeconomic (e.g., trade policy) and environmental (e.g., climate change) [1,2]. Biodiversity is also being appreciated as embedded in the context of human-environment interactions involving the use of ecological systems comprising agricultural resources such as soil and water [3]. Agrobiodiversity is particularly interdependent on human-environment interactions. It is defined broadly as the biodiversity of food-producing organisms and their landscapes and ecosystems, wild counterparts in natural areas, and the realms of knowledge, skills, management, access, and the related socioeconomic and cultural factors that are integral to these human-environment systems. Previously, agrobiodiversity was considered chiefly as a source of the stocks of genetic resources and corresponding raw material for "improved" crops and livestock [4]. By contrast, we adopt and advance a new understanding of agrobiodiversity that places emphasis on its relations to current livelihoods and human well-being [5–8], including the function of agrobiodiversity in diet quality and health. Ecological interactions [2,9–11] including vegetation, soil biodiversity, water resources, and the diversity of landscape and agroecosystem components are also integral to agrobiodiversity.

The wide-ranging interactions of agrobiodiversity amid global change encompass: (a) the management, quality, and access to livelihood-related resource inputs (e.g., skills, knowledge, and labor engaged directly with agrobiodiversity and supporting resources, such as soils and water) amid the livelihood diversification of food-growers [12–14]; (b) ecosystem services and specifically interactions within agrobiodiversity-supporting ecological systems (plant-soil interactions) [10,15]; (c) crop and food-growing strategies using agrobiodiversity for combined economic, environmental, and cultural rationales [16–21] including agrodiversity, which refers to management of environmental variation in agriculture [5,22]; (d) adaptation, resilience, and mitigation in response to climate change [23–27]; (e) biodiversity use and conservation [9,28–32]; (f) market opportunities [33,34]; and (g) food security and sovereignty together with nutrition and dietary diversity and human well-being [35–39]. Each of the above linkages is receiving increased research and policy interest that includes synthesis treatments of multiple linkage types [40,41].

Our analysis is directed specifically at the linkages of agrobiodiversity to livelihood diversification and the ecological system of plant-soil interactions. These linkages are represented as the first ones in our aforementioned list. Our analysis thus addresses an important gap in current research on global changes since these factors are increasingly influential in sustainability challenges and potential opportunities. Focusing on this pair of core linkages requires a selective integration across thematic and conceptual fields. First, the concept of social-ecological systems highlights causal linkages in global change and the relationship between resources and governance [42–44]. Second, the concept of human-environment interactions draws attention to the complex coupling of these systems [45–49]. Third, political ecology has examined the driving forces as well as the "inner workings" (e.g., gender relations) of rapidly expanding livelihood diversification [50–54]. Political ecology also includes a focus on the impacts of labor migration indicative of the "new rurality" [55,56]. Fourth, the ecology of soil-plant systems is needed to analyze agrobiodiversity interactions and environmental sustainability [9,57–59], in order to include the important non-human interactions of these systems. The selective integration of these four fields is needed to address the multi-faceted issues that arise in the livelihood-environmental interactions of smallholders. Such issues range from poverty alleviation, resilience, and food security [60,61] to sustainability and the intensification of land use [4,52,62–66].

Topically, this study engages with agri-food systems and global socioeconomic and environmental changes of small-scale agri-food producers that are also agricultural workers (smallholder) [52,66–70]. We focus specifically on two types of interactions of agrobiodiversity: (i) with the diversification of smallholder livelihoods whose scope includes migration and cash cropping that reflect global socioeconomic change; and (ii) within the context of the ecological systems of plant-soil biotic interactions that range from the cultivated and managed to the partly "wild" (e.g., forests and woodlands). These ecological systems strongly influence the resilience capacity of food growers and their fields and agricultural landscapes [11,71–73]. We recognize that many works to date in the global change sciences have focused on forest ecosystems and deforestation. Our research intends to incorporate agri-food systems and environments as central to integrated scientific frameworks of global changes. Indeed, the major drivers of global change on forest cover, such as neoliberal political economy, urbanization, trade liberalization, and climate change, also influence agri-food systems, agrobiodiversity, and livelihood diversification. More broadly, our approach contributes to the scientific and policy analysis of smallholder agrobiodiversity and thereby advances environmental sustainability and social issues of food and equity.

Our conceptual model is focused on the linkages of livelihood diversification to the biodiversity of agriculture and food-growing in particular (Figure 1). This model highlights the direct livelihood linkages to related yet distinct sub-components comprised of above- and belowground agrobiodiversity (lower left of Figure 1) and the components of soil and water resources (lower right of Figure 1). Our model offers a re-framing of livelihood analysis to include the linkages to both food-growing and the ecology of resource systems. The increasingly complex roles of livelihood interactions are pronounced in the agri-food systems of smallholders both locally and globally (Sections 2 and 3).

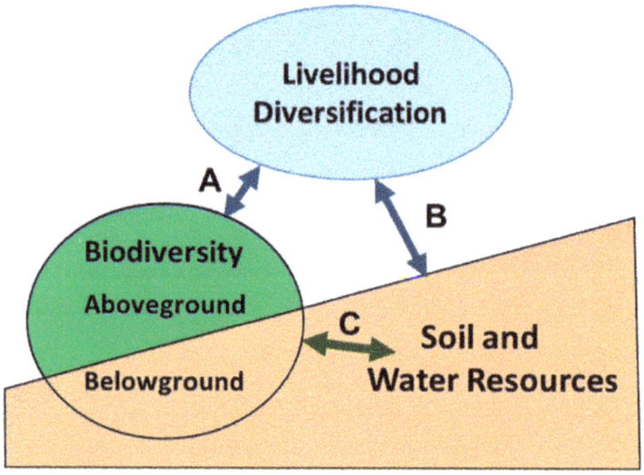

Figure 1. Conceptual diagram of the linkages of livelihood diversification to agrobiodiversity, as both aboveground and belowground sub-components, and to soil and water resources. Arrows A and B represent principal linkages of livelihoods to natural systems; C represents coupling between natural resources and agrobiodiversity.

Current smallholder livelihoods rely on a mix of economic, social, and cultural activities undertaken as individuals and as household groups, as well as in extra-household networks and spaces in their communities and beyond. In the context of expanding globalization and urbanization, migration is increasingly vital to the livelihood diversification of smallholders worldwide. A majority of the world's 500–800 million migrants between 1990 and 2010 belong to smallholder households (see also below, Section 2). Smallholders have thus extensively diversified their off- and non-farm incomes while continuing to derive livelihoods partially from the use of local land and resources. Ecological systems are integral to these changing local agri-food spaces. The local social-ecological systems of large populations of smallholders are also essential to global agrobiodiversity (see current demographic estimate in the next section). Seen in the context of smallholder livelihoods and ecological interactions, agrobiodiversity is recognized as fulfilling a range of functions ranging from agroecosystem dynamics to nutritional and dietary diversity, food supply, and cultural roles.

This paper's methodological goal is to examine the case study-based documentation of the linkages of agrobiodiversity and related resource use to global change consisting of livelihood diversification (Section 3) and ecological systems with an emphasis on soil-plant interactions and biodiversity (Section 4). We employ the methodology of meta-analysis to research these linkages [74–76]. First, we conduct an analysis of the interactions of recent livelihood diversification with agrobiodiversity use and the related management of soils, water, and uncultivated biodiversity (Section 3). We place emphasis on migration and on-farm cash cropping as two common forms of smallholder livelihood diversification. Second, we analyze the ecological interactions of aboveground biodiversity to belowground, soil-level biotic functions and diversity (Section 4).

Meta-analysis has become significant in research on global change and social-ecological systems in order to synthesize expanded fields of study as well as to advance new understandings and hypotheses [74–76]. Design of this study's meta-analyses draws also on our extensive personal experience in field-based research on the linkages of biodiversity and soil and water management to livelihoods and livelihood changes. This occurs in three ways: (a) our backgrounds as an interdisciplinary environmental scientist and geographer (ksz) focused on human-environment, ecological, and livelihoods research and as a soil scientist and ecologist (sjv) focused on agroecosystems helped to generate the specific ideas about linkages being tested and evaluated in this meta-analysis; (b) our field experiences suggested that the widely spanning human-environmental linkages examined here are scientifically important, policy-relevant, and often escape consideration in the way we formulate them here; and (c) more generally our field experiences motivate our collaboration with one another and other teams to undertake the meta-analysis research and synthesis undertaken in this article.

In keeping with our emphasis on geographic linkages, we have incorporated our meta-analysis findings into a spatially explicit conceptual framework of livelihood-diversification and agricultural intensification regimes that we relate to expected or hypothesized interactions and outcomes with regard to agrobiodiversity (Section 5). Relating agrobiodiversity to regional spaces of agri-food systems with characteristic livelihood-diversification regimes enables us to offer hypotheses for future scientific study and insights for policy analysis. Our discussion of livelihood interactions with agrobiodiversity and ecological systems using this spatial framework reflects the need for sustainability research with a meso-level focus. Focus on meso-level geographic and thematically specified social-ecological systems has gained importance in light of the growing number of high-profile and influential yet general summary-style calls in science and policy for sustainable intensification, ecological intensification, and smallholder sustainability ([3,4,11,62,64,65,77]; see example of related meso-level research in [52,66]).

2. Empirical Focus and Research Methodology: Global Smallholders and Nested Meta-Analysis

Globally, smallholders comprise a population of 2.0–2.5 billion persons whose lands are significantly smaller in areal extent and endowed with fewer capital- and technology-intensive resources than mean-size agricultural enterprises [77]. Livelihoods and socioeconomic capacity demonstrate a high level of heterogeneity within this group. Most smallholders live in geographic areas of Asia, Africa, and Latin America that range from remote rural to intermediate-distance and peri-urban locales [52]. Smallholder livelihood strategies involve production for agri-food systems that are local, national, and global in scope. Non- and off-farm labor, including agricultural work and migration at multiple scales (*i.e.*, local, national, global), are livelihood activities that link to their agri-food spaces and systems. This lived complexity of smallholders is far from an ideal socioeconomic reality since the large majority are poor by local and national standards, and many also suffer food insecurity. In addition, our perspective on global smallholders is non-teleological in the sense of not presupposing either smallholder disappearance or persistence. Most smallholders live a grueling, marginal existence underscoring the importance of livelihood issues. This reality is fused with the economic dominance of corporate, industrial agri-food systems globally as well as land grabs displacing smallholder populations [64,73,78]. These circumstances do not, however, detract from importance of the world's present-day population of 2.0–2.5 billion smallholders and their access to food and resources, combined with non- and off-farm linkages, wrought through livelihoods that include the use of agriculture and ecological systems.

Smallholder agri-food systems are distinguished through the varied use of land, water, and other resources, including agrobiodiversity. Across a wide range of global contexts, smallholders generally manage marginal land that is vulnerable to degradation [5]. Their access to soil and water varies widely, with smallholder resources often threatened through input-intensive agri-food production. Nonetheless significant global agrobiodiversity is concentrated in the fields and gardens of smallholders in the context of these limitations and constraints [52]. Agrobiodiversity is furnished through locally accessible seed systems and it reduces the risk of crop loss in low-input agroecological conditions [1,17,27], in addition to furnishing locally valued foods. The adequate management of soil and water resources is integral to these functions of agrobiodiversity that, in turn, provides nutritional and health benefits to smallholder populations.

To address the complexity of this human-environment system, we designed a meta-analysis on smallholder livelihoods, migration, agrobiodiversity and environmental resources. We used the Web of Science database to identify relevant publications since 1985. Forty-two research publications were chosen for our analysis in the first area of meta-analysis using the combinations of specific search terms (Table 1). Each of these studies explicitly addresses the household- and local-level mechanisms of livelihood-environment interaction, thus guiding us with regard to the issue of scale. Publications were chosen that provide case studies combining results on *both* livelihood diversification *and* environmental impacts and interactions. Our analysis is focused on studies of the environments (*i.e.*, agrobiodiversity, soils, water, uncultivated biodiversity) of smallholder *home* communities (referred to a "source" in the migration literature), while it recognizes that an important yet quite distinct analysis would focus on environments in migrant *destinations*. (The latter type of analysis would characterize much of migration's role in environment–development interactions in frontier settings in countries such as Brazil.) The 42 research publications chosen were found to contain 56 descriptions of specific livelihood diversification–environment interactions. Many research works (n = 32) are based on case studies undertaken after 1999. Case studies cover examples in Latin America, Africa, Asia, and Europe. Using a coding approach, we created quantitative estimates of: (a) change of agrobiodiversity (decrease, increase, and stasis) typically described at the taxonomic level of species or varieties (often termed landraces in the case of traditional crops); (b) soil quality and degradation (e.g., soil erosion); (c) water quantity and quality; and (d) uncultivated biodiversity. The studies were also stratified by the type of livelihood diversification depending on the predominance of migration or on-farm livelihood diversification. In addition, studies were coded according to changes in the gendered roles of resource use.

Table 1. Search terms and published sources (N = 42) in the meta-analysis of the interactions of livelihood diversification and resource and environmental systems (*i.e.*, agrobiodiversity, soils, water, wild biodiversity). Several sources address more than one resource or environmental system.

Search terms consulted (Web of Science database)
("Livelihood diversification" OR "migration" OR "development") AND ("agrobiodiversity" OR "crop diversity" OR "soil" OR "water" OR "biodiversity" OR "secondary forest transition")
Authors and publication year of sources with numbered reference
Aide, and Grau 2004 [79], Batterbury 2001 [80], Bilsborrow 1992 [81], Garcia-Barrio and García-Barrios 1990 [82], Bilsborrow and DeLargy 1990 [83], Carr 2009 [84], Chambers and Momsen 2007 [85], Chen *et al.* 2012 [86], Collins 1987, 1988 [87,88], Davis and López-Carr 2010, 2014 [89,90], European Commission 2015 [91], Grau and Aide 2007 [92], Gray 2009a [93] Gray and Bilsborrow 2014 [94], Gray 2005 [95], Heartherington 2014 [96], Hecht *et al.*, 2006 [97], Huijun *et al.* 2002 [98], Iniesta-Arandia *et al.*, 2015 [99], Jewitt 2000 [100], Jiménez Olivencia 2010 [101], Jiménez Olivencia *et al.*, 2015 [102], Jokisch 2002 [103], Lerner *et al.*, 2013 [104], Li *et al.* 2014 [105], López 2006 [106], McCord *et al.* 2015 [27], Mutersbaugh 2006 [107]; Nuijten 2010 [108], Preston, D 1997 [109], Qin 2010 [110], Radel and Schmook 2008 [111], Radel *et al.* 2012 [112], Reardon *et al.* 2001 [113], Rudel *et al.* 2002 [114], Schmook *et al.* 2013 [115], Turner 1999 [116], Zimmerer 1991 [117], 1993 [118], 1996 [7], 2014 [119].

The second area of meta-analysis focused on the interactions between above- and belowground components of agrobiodiversity. We use the term AGSOBIO (AGricultural and SOil BIOdiversity system) to denominate these coupled environmental components. For the AGSOBIO meta-analysis, 50 research publications (Table 2) were coded. Ecological systems in these publications range from croplands to perennial landscape components (*i.e.*, agroforestry, managed forests, fallows, and pastures). We gathered quantitative estimates and overall interpretations about the impact of increasing aboveground biodiversity on (a) belowground biotic community structure, regardless of diversity; (b) taxonomic diversity or functional diversity, using the measures of diversity given in the reviewed papers (e.g., the Shannon-Wiener index, or belowground species richness depending on the paper); and (c) changes in ecological function of soils for crop production that characterize the supporting role of soil biota as coupled to aboveground biodiversity, e.g., impacts on mycorrhizal colonization or activities of soil enzymes that catalyze biogeochemical cycling. The studies were also stratified by the type of change in vegetation diversity, between studies that examined land-use level change, e.g., forest conversion, secondary succession, adoption of agroforestry, and those studies that compared plant diversity within only agricultural land uses (e.g., crop rotation, fallow type, and introduction of short-term pasture). We also stratified the studies by functional type of soil biota: whole community metagenomic studies of bacteria, archaea, and fungi; arbuscular mycorrhizal symbioses in land-use systems, for which there were a particularly large number of studies; and macrofauna, which are increasingly valued as a soil quality indicator. The AGSOBIO meta-analysis is conceptualized as the lower-left vertex of our conceptual model (Figure 1) and thus nested within the triad of interactions that includes livelihood diversification.

Table 2. Search terms and published sources (N = 50) used to construct the meta-analysis of relationships between above-ground biodiversity and soil biotic communities. Sources are divided according to the domain of soil biota each research paper addressed. In addition to the search terms, backwards and forwards reference searches were used to more fully explore the database.

Search terms consulted (Web of Science database)	
("Land use" OR "crop rotation" OR "crop diversity" OR "cropping system") AND ("microbial diversity" OR "microbial functional diversity" OR "macrofaunal diversity" OR "macrofauna" OR "mycorrhiza" OR "fungal diversity" OR "bacterial diversity")	
Domain of soil biota	**Authors and publication year of sources with numbered reference**
Whole community: e.g., soil bacteria, archaea, and/or fungal communities	Acosta-Martínez *et al.* 2010 [120], Alvey *et al.* 2001 [121], Asgharipour *et al.* 2011 [122], Bach *et al.* 2012 [123], Berthrong *et al.* 2009 [124], Buckley and Schmidt 2003 [125], Duchicela 2013 [126], Garbeva *et al.* 2006 [127], Garcia *et al.* 2012 [128], Gomez *et al.* 2004 [129], Gonzalez- Chavez *et al.* 2010 [130], Jiang *et al.* 2014 [131], Johnson *et al.* 2010 [132], Larkin and Honeycutt 2006 [133], Lienhard *et al.* 2014 [134], McDaniel *et al.* 2014 [135], Meriles *et al.* 2009 [136], Millard and Singh 2010 [137], Nayyar *et al.* 2009c[138], Nurulita *et al.* 2015 [139], Shen *et al.* 2008 [140], Sheng *et al.* 2013 [141], Vasileiadis 2013 [142], Yao *et al.* 2006 [143], Yu *et al.* 2012 [144]
Mycorrhizae	Almaca and Ortas 2010 [145], Bedini *et al.* 2013 [146], Carpenter *et al.* 2001 [147], Chaturvedi *et al.* 2009 [148], Dai *et al.* 2013 [149], Gavito and Miller 1998 [150], Hendrix *et al.* 1995 [151], Higo *et al.* 2014 [152], Hijri *et al.* 2006 [153], Lehman *et al.* 2012 [154], Lekberg *et al.* 2008 [155], Melo *et al.* 2014 [156], Muchane 2012 [157], Oehl *et al.* 2003 [158], Rao *et al.* 1995 [159], Säle *et al.* 2015 [160], Sturmer *et al.* 2011 [161], Tchabi *et al.* 2008 [162], Tian *et al.* 2013 [163], Verbruggen *et al.* 2010 [164], Vestberg *et al.* 2011 [165]
Macrofauna	Dominguez *et al.* 2010 [166], Hulugalle *et al.* 1997 [167], Maria De Aquino *et al.* 2008 [168], Sileshi *et al.* 2008 [169]

3. Results: Livelihood Diversification and Environmental Linkages

The linkages among smallholder livelihood diversification and environmental impacts are increasingly known to be complex [170]. Nonetheless, our findings demonstrate there is ample evidence to identify and assess the role of specific linkage mechanisms and, more generally, to advance the understanding of these impacts (Table 3). The studies surveyed here and the perspective of our research urge that the increased amount of attention to livelihood-diversification processes view these phenomena as both cause and effect in relation to environmental change. Likewise, they reveal complex links to socioeconomic changes, such as globalization in the form of international labor market integration that is, in turn, an important driver of migration. At least three hypothesized mechanisms underlie the linkages of livelihood diversification to environmental impacts (Figure 2a). First, conventional understanding asserts the reasonable expectation that migration lessens environmental pressures in the home communities since livelihoods are expected to become less dependent on cropping and other land use including grazing and forestry [93,171]. Second and by contrast, various studies beginning in the late 1980s and early 1990s suggest that livelihood diversification reduces the supply of labor needed for the sustainable management of agroecosystems, including terracing of agricultural fields on sloped soils [87,88] and the cultivation of high-agrobiodiversity crops with staggered or extended growing seasons [117]. Third, the role of migration remittances is complex and may be either detrimental or beneficial with regard to environmental impacts in sending communities [89,90], which has seemed to motivate the growing number of case studies of livelihood diversification and environmental impacts in development. Gendered social relations and their role in agriculture and land use are an increasingly important focus of these studies [112,172–178]. Yet research has not focused as systematically as we do here across the geographic and interdisciplinary integration of multiple forms of livelihood diversification (e.g., migration, on-farm diversification), environmental resource- and environment-specific systems (e.g., agrobiodiversity, soil, water, uncultivated biodiversity), and related gender relations.

Results of our first meta-analysis demonstrate significant variation in the interactions of smallholder livelihood diversification with agrobiodiversity levels (Table 3). Agrobiodiversity increase, decline, and stasis are all demonstrated. Decreases in agrobiodiversity with livelihood diversification are significant in these results (33%) and exceed the cases of increased agrobiodiversity (8%) (Table 3.2). Overall, studies showing no change or mixed outcomes, combining increases and decreases of agrobiodiversity that were locally or regionally differentiated, are predominant (58%). These latter results are significant since they evidence the partial conservation of agrobiodiversity. They support calls for on-farm, *in situ* conservation in the context of current smallholder livelihoods and diversification trends. The meta-analysis also shows that livelihood diversification via migration is more favorable to agrobiodiversity than diversification using non-migration portfolios (Table 3). A key factor influencing these outcomes is whether smallholders had the inclination and capacity to commit resources (e.g., labor, land, water) to agrobiodiversity amid other demands. Such capacity can be increased non-traditionally (e.g., high levels of internal migration combined with local in-migration; 119) so that flexible household-level land use and the corresponding versatility of resource allocation enables the use of agrobiodiversity more effectively than in the high-intensity agriculture of prime areas of commercial cropping [2,5,52,104,119]. These findings illustrate so-called emergent social-ecological properties of smallholder agrobiodiversity, illustrating through either its disappearance or continued production the increase of linkages to global change.

Livelihood diversification frequently results in a decline in the quality of smallholder soil resources (59%) and water resources (53%) (Table 3.2). At the same time, occurrence of quality increases is shown to be significant albeit less common (26% and 35%, respectively). In general, a greater level of research has been focused on the interactions of livelihood diversification with soils (n = 27 studies) and water (n = 17 studies) compared to agrobiodiversity. This research emphasis owes to the immediate impacts of soil and water management on human well-being as well as the potential reverse causation, namely, that soil or water degradation, or "agro-ecological drivers" more generally, contribute to the migration decisions of individuals and households [93,176–179]. The importance of soil degradation in smallholder land use, for example, can be symptomatic of resource-poverty traps associated with livelihood diversification that is detrimental to both the environment and human well-being [61,173,174]. Water resources and management show a discernible trend toward increased quality (60%) under changes associated with on-farm livelihood diversification in contrast to migration, since the former often incorporates small-scale irrigation systems (Table 3.3b). Variation in outcomes with regard to smallholder soil and water resources depends principally on the viability of investing labor, capital, and land and the availability of these factors. This viability hinges on household capacities and endowments amid livelihood diversification that occasionally incorporates their access to the inputs of institutions working in the realm of environment and development [180].

Table 3. Results of the meta-analysis of research studies (N = 42) on the interactions of livelihood diversification with resource and ecological systems. This analysis drew from 42 published research papers that related the diversification of livelihood activities, with emphasis on migration, non-migration diversification of livelihoods, and the conditions of agrobiodiversity, soil and water resources, and uncultivated biodiversity.

3.1. All studies: association of livelihood diversification with change *vs.* no change in agrobiodiversity, soil, and water resource factors					
	***N* (Studies Reporting)**	**Change in One or More Factors**		**No Change (in Any Factor)**	
	42	93%		7%	
3.2. All studies: changes in food plant agrobiodiversity richness (species or varieties) and the environmental quality of soil and water resources					
	***N* (Number of Comparisons)**	**Increase**	**No Change or Mixed**	**Decrease**	**Observations and Main Tendencies**
Changes in agrobiodiversity	N = 12	8%	58%	33%	Change interpreted as a mixed outcome shows both increase and decrease occurring as local variation among households and local places as well as regionally.
Changes in soil quality	N = 27	26%	15%	59%	
Changes in water quality	N = 17	35%	12%	53%	
Changes in wild biodiversity	N = 18	39%	22%	39%	
3.3. Studies grouped by predominant type of livelihood diversification					
(a) Studies that focus primarily on migration as the major form of livelihood diversification					
Changes in agrobiodiversity	N = 6	17%	50%	33%	Livelihood diversification studies with a migration emphasis tend to incorporate a focus on the abandonment of soil and water infrastructure (e.g., historic irrigation systems)
Changes in soil quality	N = 20	25%	15%	65%	
Changes in water quality	N = 12	25%	8%	67%	
Changes in wild biodiversity	N = 12	42%	25%	33%	
(b) Studies with emphasis primarily on-farm livelihood diversification (without migration emphasis)					
Changes in agrobiodiversity	N = 6	0%	67%	33%	Livelihood diversification studies (without migration emphasis) tend to focus on local land-use capacity, including increased infrastructure for soil and water management (e.g., irrigation)
Changes in soil quality	N = 7	33%	17%	50%	
Changes in water quality	N = 5	60%	20%	20%	
Changes in wild biodiversity	N = 6	33%	16%	50%	
3.4. All studies: gender analysis in diversified livelihood interactions with food plant agrobiodiversity, soil, and water resources					
	***N* (Number of Comparisons)**		**Frequency**		
Contains analysis of gendering	N = 23		55%		Level of gender analysis varies widely in these studies
Women's increased role in resource use	N = 22		96%		
Men's increased role in resource use	N = 8		35%		Associated principally with livelihood diversification and relatively minor role of migration and, in some cases, recent increased migration among rural women (e.g., Spain and Andean countries)

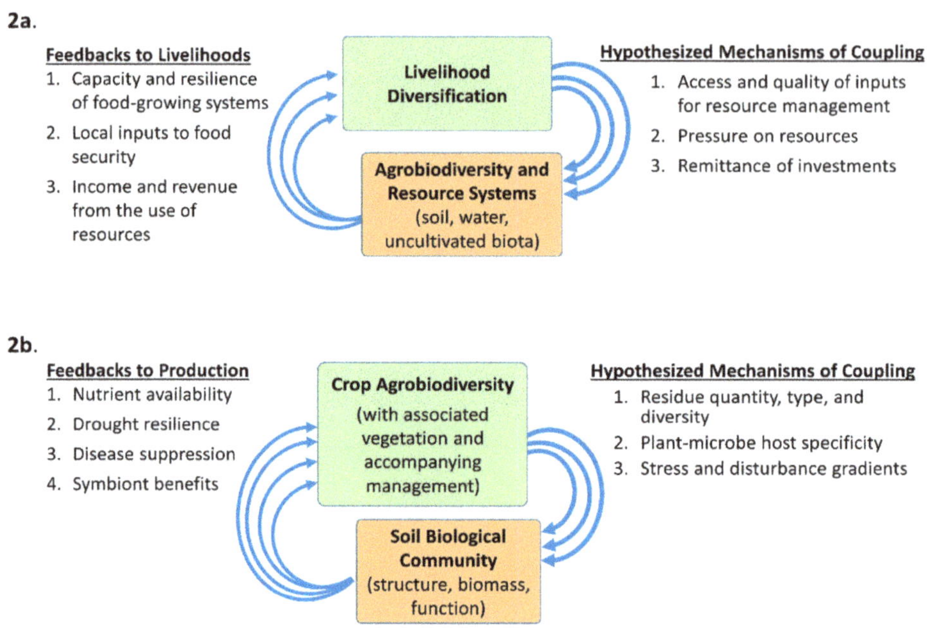

Figure 2. (**a**) Coupling of the interactions of livelihood diversification and agrobiodiversity and (**b**) the above- and belowground components of agrobiodiversity.

Uncultivated or "wild" biodiversity associated with forest cover is incorporated as a focus in several studies related to livelihood diversification (n = 18; Table 3). A significant portion of these studies (39%) report favorable interactions with forest and woodland biodiversity resulting from forest regrowth, or so-called secondary forest transitions, occurring through the disintensification of agriculture linked to household-level livelihood diversification (in addition to the case studies in Table 1, see also [181,182]). These findings of favorable impacts on uncultivated biodiversity through forest regrowth resulting from livelihood diversification are found more often in conjunction with predominant migration (42%) than on-farm activities (33%) (Table 3.3a,b). Conversely, livelihood diversification can be a significant driver of deforestation and resulting loss of wild biodiversity through land use (39%), with the remainder reporting mixed outcomes or no change (22%) (Table 3.2). These declines of forest and woodland biodiversity reflect outcomes discernible within the group of studies characterized as indicating the prominence of migration (33%) and within the contexts of predominantly on-farm activities (50%) (see also [183]). Whether forest and woodland biodiversity is increased or decreased depends on such factors as household-level decision-making with regard to the expansion of agriculture and land use, including the grazing of livestock, and feasibility of access to forested lands. In sum, our meta-analysis of case studies shows that the impacts of livelihood diversification on wild biodiversity are mixed and dependent on factors in the local milieu of smallholder and, in the case of migration, in the destination context. These findings resemble the findings of our analysis with regard to agrobiodiversity, soils, and water resources.

Dynamics of gendering and land-use intensification are two additional themes increasingly common within the suite of research dealing with livelihood diversification and migration related to the environment. Interests in and the analysis of gendering in livelihood diversification are expanding significantly and build upon vibrant existing research traditions in household-level analysis in agricultural, resource, and development economics and sociology and also in such fields as geography and anthropology (e.g., [172,184–192]), as well as interdisciplinary approaches such as feminist political ecology (FPE; [53,54,193]). Gendered resource access and knowledge systems are integral to the resource use and agri-food systems of smallholders amid livelihood diversification and global changes. These interactions may include unanticipated outcomes. For example, gendered changes have been found to be influential in soil management and the potential of resource-poverty traps [194]. The gendered livelihood changes associated with purportedly beneficial land use, such as conservation agriculture (CA) in African smallholder systems, have been found to undermine nutrient-management goals due to an inequitable shift of labor requirements to women [195]. Our meta-analysis results show the gendering of work activities related to livelihood diversification and resources to involve an expanded role of women's efforts in nearly all studies (96%; Table 3.4). Men's work activities were expanded less commonly in resource use (35%), which owes partly to the expansion of off- and non-farm livelihood activities.

Environmental impacts of the intensification or disintensification of land use vary significantly as a result of livelihood diversification. Since these impacts are often not measured in these studies, we noted only qualitative information in our meta-analysis. Generally, the process of disintensification is associated with livelihood diversification, although several studies mention it, and a few are able to demonstrate the increase of intensification, while others suggest the occurrence of no change in intensification level. Gender dynamics are critical to these trends, since in many cases women become the main farm managers, and this "feminization" of resource use exerts a major influence on the specifics of agricultural change (e.g., crop choice) and general outcomes (e.g., intensification or disintensification) [14,48,51,89,90,97,172–177,196–198]. For example, the decline of chili pepper markets led to disintensification in the southern Yucatán region of Mexico through processes of livelihood diversification based on international migration and the influence of women's role in subsequent resource use [173,196,197]. In other cases, where women have migrated significantly, such as the case of remote-rural regions of Spain and elsewhere in Europe, including the former German Democratic Republic [99,198], new arrangements combine the "masculinization" of continued land use with social innovations in women's continued albeit reduced involvement.

In summary, our results combine with other new findings to build upon earlier demographic approaches [199]. These findings demonstrate how the influences of livelihood diversification on agricultural intensification vary significantly and are conditioned on the existence of geographically specific opportunities and constraints in both sending areas and receiving areas.

4. Results: Ecological Linkages within Agrobiodiversity Systems

We turn now to the results of meta-analysis of the ecological linkages within agrobiodiversity systems (Table 4). Our analysis focuses on the interactions between soil biotic diversity and the biodiversity of crop species, including local varieties of food crops known as landraces, along with associated management as well as managed wild species (bottom of Figure 1). The component of soil biota is a key form of "associated biodiversity" [1,11] that provides supporting ecosystem services key to food production [200–203]. Considering AGSOBIOs as coupled above- and belowground agrobiodiversity components provides the explicitly ecological framing to complement our recognition that each of these components may be impacted through livelihood diversification, as discussed above (Section 3) and illustrated graphically (Figure 1). Livelihood changes that affect either the above- or belowground component will also potentially impact the other component. The ecology of these plant-soil interactions is the focus of meta-analysis in this section.

Table 4. AGSOBIO meta-analysis. This analysis drew from 50 published research papers with 53 land-use and crop rotation comparisons of community composition, total biomass of target biota in the study, and soil ecological function under increasing levels of aboveground plant diversity. Sections 4.1 and 4.2 show the change of community structure, and any increase, lack of change, or decrease in soil biological diversity with increases in aboveground diversity. Section 4.3 divides the analysis into studies that assessed increasing diversity via land use change (e.g., forest conversion to crops, secondary succession from crops, and adoption of agroforestry) *versus* those that assessed increasing diversity in comparisons of crop rotations and associated short-term pasture or fallow management. Section 4.4 groups the studies by soil biota functional type for which there were sufficient research papers.

4.1. All studies: changes in soil biotic community structure					
	N (Number of Comparisons)	Change	No change		
	N = 53	92%	8%		
4.2. All studies: changes in soil biological diversity, abundance, and biological function under increasing aboveground biodiversity					
	N (Number of Comparisons)	Increase	No Change	Decrease	Observations and Main Tendencies
Changes in soil biotic diversity	*N* = 53	55%	34%	9%	
Changes in biotic abundance or biomass	*N* = 26	69%	23%	4%	
Changes in soil biological function (e.g., enzyme activity, mycorrhizal colonization)	*N* = 28	79%	14%	7%	
4.3. Studies grouped by land management types that alter aboveground biodiversity:					
Studies that examined changes in land use (forest conversion, succession, agroforestry *vs.* cropping)					
Changes in soil biotic diversity	*N* = 25	64%	28%	8%	Differences in soil organic matter accumulation, disturbance, and species type that drive soil biotic differences are especially strong between long-term perennial uses as a group (e.g., forest) versus cropping systems as a group.
Changes in biotic abundance or biomass	*N* = 9	78%	11%	11%	
Changes in soil biological function	*N* = 8	100%	0%	0%	
Studies that examined changes in agricultural and livestock management within agricultural land uses (among crop rotations, fallows, pastures)					
Changes in soil biotic diversity	*N* = 34	41%	47%	9%	Management accompanying crop diversity is important, e.g., increased tillage under continuous monoculture *vs.* diverse rotations with fallows and variation in tillage regimes.
Changes in biotic abundance or biomass	*N* = 16	69%	31%	0%	
Changes in soil biological function	*N* = 17	65%	24%	12%	
4.4. Studies grouped by soil biota type impacted by increasing aboveground biodiversity:					
	N (Number of Comparisons)	Increase	No Change	Decrease	Observations and Main Tendencies
Analyses of soil whole microbial communities (using e.g., bacteria, archaea, fungi using whole metagenome analysis)					
Increases in soil biotic diversity	*N* = 20	50%	40%	10%	Long lag times: decadal "imprint" of management based on previous land use. High degree of functional diversity and redundancy in bacterial and archaeal communities that creates temporal stability/hysteresis in community composition. Less well-understood aboveground/belowground relationships than for symbiont communities or macrofauna. In addition the fungal: bacterial ratio of various biomass measures (e.g., PLFA) often increases with the amount of recalcitrant residues in systems and decreases under soil disturbance (e.g., grassland has a higher fungal: bacterial ratio than an intensified vegetable rotation.)
Increases in biotic abundance or biomass	*N* = 9	78%	22%	0%	
Analyses of arbuscular mycorrhizal communities (using spore classification and counts and fingerprinting using mycorrhizal DNA primers)					
Increases in soil biotic diversity	*N* = 34	59%	35%	6%	Strong influence of whether aboveground species that add biodiversity are mycorrhizal or non-mycorrhizal, and specialist *vs.* generalist mycorrhizal fungal species.
Increases in biotic abundance or biomass	*N* = 12	67%	25%	8%	
Analyses of macrofaunal communities (using field methods and visual classification of macrofauna)					
Increases in soil biotic diversity	*N* = 4	75%	25%	0%	Mediation by plant residue quantity and quality accompanying aboveground species diversity and associated management is especially strong.
Increases in biotic abundance or biomass	*N* = 3	100%	0%	0%	

Several hypothesized mechanisms underlie the coupling of above- and belowground biodiversity in AGSOBIOs (Figure 2b). First, various research studies suggest that soil biodiversity responds to niche differentiation corresponding to plant diversity as proposed by Reynolds *et al.* [204]. In particular, soil biotic diversity could arise from different amounts and temporal dynamics of nutrients and labile carbon for microbes in the root systems of different species [205], or based on plant species-dependent quality and quantity of residues from dead above- and belowground biomass (e.g., [137,206]). Second, plants often host symbiotic microbes such as mycorrhizal fungi and N-fixing rhizobia and associative bacteria. Host-specificity of these symbioses may affect the microbial community that inhabits plant root systems. Third, the agricultural management associated with crop and associated plant communities, such as tillage, fertilization, irrigation, and weed management, create different levels of nutrient and water stress and physical disturbance (e.g., tillage, harvest) for soil biotic communities that favor higher or lower diversity and the ecological function of microbes, with the highest levels of diversity theorized to occur at intermediate levels of stress and disturbance [141,207].

These varied potential mechanisms illustrate that the linkages between the aboveground and belowground components of biodiversity are undoubtedly complex [59]. Furthermore, there are differences in magnitude between above- and belowground biodiversity: Aboveground agrobiodiversity incorporates crop varietal differences in phenology and growth habit as well as suites of perennial and annual species that embody plant family and functional type differences. By contrast, the soil component of this biodiversity is far more varied, ranging from thousands of species of soil bacteria and archaea that mediate nutrient cycling and interact with plant roots, to fungal symbionts and pathogens, to fauna including macrofauna that serve as so-called soil ecosystem engineers and provide the physico-chemical scaffolding for both root and microbial activities that support food production [200,208].

In spite of this complexity, our meta-analysis yielded sufficient evidence to contribute to understanding how and under what conditions aboveground plant agrobiodiversity influences soil biological communities (Table 4). Over 90% of the 53 comparisons included in the analysis found that changes in plant biodiversity changed the microbial or faunal community structure of soils, including cases when some measure of community structuring was found to change even if diversity indices (e.g., richness, Shannon-Wiener index, functional diversity) of the soil communities did not change.

In addition, just over half of studies show increasing soil biodiversity with aboveground diversity, while only 9% showed declines in soil agrobiodiversity. Moreover, the plausibility of feedback effects of soil biota for contributing to food production is verified: For studies that measured aspects of soil biological function that contribute to aboveground productivity, a majority (79%) show that these functions were improved under higher diversity plantings (e.g., more complex crop rotations with legumes, or reforested areas compare to intensive agriculture). Increased soil biodiversity and improved soil properties relate to aboveground biodiversity were relatively common when comparisons were made between non-agricultural land uses and agriculture (Table 4.3). Nevertheless, comparisons of vegetation diversity made only within agriculture and pasture land uses also showed a preponderance of increases over decreases in soil biotic diversity and biological soil function as plant communities became more diverse. Very few studies tested the impact of varietal diversity within a crop species on soil biological communities. However, a few studies demonstrated that crop varietal interactions with soil biota were plausible, for example, those that found higher response of local maize landraces and older varieties to mycorrhizal symbionts than from modern varieties [209–211].

5. Discussion: Hypothesizing Pathways and the Spatial-Geographic Dynamics of Livelihoods and Agrobiodiversity amid Global Change

These results offer a new framing of the study of the interactions of livelihoods (Figure 1, top) with agrobiodiversity (Figure 1, lower left) and closely related resource systems (Figure 1, lower right). This framing is focused on globally predominant modes of diversification that include off-farm activities (migration) and on-farm shifts (toward more mixed portfolios), gender dimensions, and land-use intensification or disintensification. Focus of our framing also encompasses the elements of biodiversity in agriculture (agrobiodiversity), resource use, and the large global population of socioeconomically vulnerable smallholders (Section 2). Our framing needs to be understood as contributing to the interdisciplinary analysis of global change that links major trends, such as migration, urbanization, and commodity trade, with the environment [212–214]. Finally, our framing engages and contributes to growing interdisciplinary research on migration-dependent livelihoods and the environment [215,216]. Here our intent is also to offer explicit re-framing that integrates multiple ecological components rather than treating the environment as a simple, single sub-system.

This re-framing is evident in the design and results of our meta-analysis to encompass multiple, related environmental resources (e.g., agrobiodiversity, water, soils): the increased links of livelihood diversification to agrobiodiversity are associated with minor or mixed changes while this increase shows pronounced negative impacts on the environmental quality of soils and water, albeit with some exceptions (Section 3). The noted absence of significant change in agrobiodiversity systems may be influenced by observational challenges. For example, agrobiodiversity loss tends to be less visually apparent than erosion or other prevalent forms of soil degradation. Noteworthy too is that primarily negative consequences for soil systems, in particular soil biotic functioning and diversity, may subsequently incur deleterious impacts on biodiversity of crops as well as uncultivated vegetation.

The results of our AGSOBIO meta-analysis (Section 4) demonstrate the mutually dependent interactions between aboveground and soil biodiversity in agricultural landscapes, and suggest that crop/land-use diversity and soil biodiversity act in a coupled way to support local food production. Understanding the linkages between vegetation and soil biodiversity in a way that is more sensitive to both biophysical context and the management techniques driven by types of livelihood diversification will help to better understand sustainability linkages to livelihoods. Methodologically the coupling of soil and crop diversity suggests that remote sensing could in the future be helpful in first characterizing aboveground biodiversity and then stratifying and orienting subsequent ground-based research to understand AGSOBIOs. Crop-level distinctions will require the use of medium- to high-resolution imagery to allow distinguishing crops, pasture, and fallow areas within rotations. Lower resolution phenology-based greenness data (e.g., MODIS) with high repeat acquisition rates may be useful if periods of greenness and seasonality of productivity can be associated with the phenologies of different crop varieties and species or perennial landscape components.

The interactions between above- and belowground components may potentially serve as a conduit transferring impacts from deteriorating soil and water systems to biodiversity. Without an explicit ecological framework, as detailed in the results of Sections 3 and 4 the consideration of livelihood impacts tends to be directed at separate, isolated environmental elements. By designing a framework of the interactions of livelihoods with above- and belowground biodiversity components, rather than either one in dyadic isolation with livelihood diversification, our research is able to identify and distinguish these crucial functional interdependencies.

Delimitation of the principal interactions of livelihood diversification are highlighted throughout this study (Interaction A and Interaction B in Figure 1). These bi-directional interactions represent the coupling of human-environment systems of resource use. These systems are also social-ecological in the sense that spatial units of agroecosystems, their governance, and resource properties are all considered (for human-environment and social-ecological approaches see Section 1). In order to both expand and strengthen our analysis of livelihood-agrobiodiversity interactions in our first meta-analysis (Section 3), we focused our second meta-analysis on the ecological coupling of above- and belowground agrobiodiversity assemblages, referred to as AGSOBIOs (Section 4). Results from the studies we analyzed suggest the functional interdependence of above- and belowground AGSOBIO components, with plant-host microbe specificity and primary production of residues as major coupling drivers and the ecological services of soil biota as dominant feedbacks. In addition, responses of whole soil microbial communities to management in these systems are characterized by significant lag times. This finding, combined with other results, suggests the potentially influential yet indirect pathway of agrobiodiversity change through degrading soil and water resources that influence plant productivity. In the case of soil degradation, which is probably most common, the linkages are mediated by processes such as erosion and nutrient depletion and resulting changes in the amount, quality, and diversity of crop residue (Figure 2). Current and future decline of agrobiodiversity—also referred to as genetic erosion—may thus be widely, albeit partially, linked to soil degradation [2].

The comprehensive framing and careful analysis of ecological interactions within agrobiodiversity systems is an important and novel addition to understandings of livelihood—environment interactions. It promises new insights well-suited to integration with livelihood analysis that together provide a significantly fuller perspective on the suite of agrobiodiversity interactions with human-social and ecological factors. Here, our findings highlight the importance and interconnectedness of the interactions specified in the framework we have developed and implemented in this research (Interactions A, B, and C in Figure 1).

We propose that a productive way to integrate further the above results into a new spatial analytic approach is through the concept of large-area regional spaces comprised of characteristic agri-food systems with corresponding livelihood diversification strategies and agrobiodiversity. Here, our discussion uses the hypothesized role of distinct Regional Agri-Food and Livelihood Diversification Spaces (RALDS) (Table 5, left column). Factors determining these RALDS include distance to markets and resource functions (e.g., access, availability, quality). While suggesting the possible influence of a distance-based model, we emphasize the predicted patterning of RALDS and agrobiodiversity outcomes as spatial mosaics with the occurrence of non-linear and even irregular relations to distance rather than linear or regular gradients [52,104,119]. Proposed examples in the United States and Peru (Table 5) reflect the non-linear, mosaic patterning of RALDS [217].

Globally, the land use of smallholders is relatively common in two of the four RALDS (Table 5, second column). Elsewhere, we have used the recent estimate of global smallholder population as 2.0–2.5 billion to offer the conceptual model of these geographic spaces [52]. It focuses on the hypothesized roles of gender and agrobiodiversity in the context of agricultural intensification and disintensification. Here, we are able to use the results of the above section to extend this model (Table 5). We focus on how each of the RALDS is anticipated to correspond not only to an anticipated agricultural intensity level (third column) but also environmental change pathway (fifth column) and agrobiodiversity (sixth and seventh columns). The latter is hypothesized to incorporate distinct agrobiodiversity in the aboveground and belowground components.

This type of conceptual categorization is well suited to elevated policy interest in the intensification and sustainability of food-growing [3,4,52,62,65]. For example, it predicts how RALDS with less significant livelihood diversification may turn out to be associated with either low or high agrobiodiversity levels (Table 5). Similar potential of bimodal outcomes is predicted in spaces of high livelihood diversification. In sum, the features of each of the RALDS are hypothesized to reflect the intersection of agri-food system properties, livelihood diversification strategies, and distinct agrobiodiversity interactions. These delineations are predicted in advance of rigorous empirical study, models, and testing.

Table 5. Predicted spatial-geographic patterning and the predominant processes of agrobiodiversity and Regional Agri-Food and Livelihood Diversification Spaces (RALDS is discussed in Section 5. Impacts in this table refer to both above- and belowground agrobiodiversity per Section 4.)

Regional Agri-Food and Livelihood Diversification Space	Predicted Social-Ecological Characterization	Predicted Agricultural Intensification Level	Predicted Livelihood Diversification Level	Anticipated Environmental Change Pathway [1]	Above-ground Agrobiodiversity	Belowground Agrobiodiversity	Examples (United States and Peru)
Prime Agricultural	Productivist, corporate industrial; smallholders nominal	High	Low	Predominant agrochemical substitution (LR)	Low, non-local seed and breed systems	Low, limited by low soil organic matter and residue returns	Central Valley, California, US; major valleys, Peru
Marginal-Land and Remote Rural Food Growing	Low-input land use for local and regional food; smallholder significance high	Low-Medium	Low	Locally varied; includes ecological disintensification (UL) and intensification (UR)	High, labor-saving local seed and breed systems	High/varied, both crop and "wild" managed ecosystem components	California coastal range, US; Andean uplands
Remote Rural Specialized Livestock or Cropping	Transition from food-growing; smallholder significance medium	Low-Medium	Low	Specialized intensification (LR) and disintensification (LL)	Low, non-local seed and breed systems	Medium/varied, limited by low residue inputs.	Great Plains, US; Altiplano, Peru [2]
Peri-Urban, Ex-Urban	Multi-purpose; smallholder significance medium	Medium	Low	Locally varied; includes ecological intensification (UR)	High, local seed and breed systems	High, managed to foster agroecosystem services	Bay Area, US; Huancayo, Peru

[1] Summaries provided in the column on Anticipated Environmental Change Pathway combine a first descriptor associated with disintensification vs. intensification and a second descriptor associated with ecological vs. specialized agrochemical management (Additional information that appears parenthetically in this column refers to the quadrants of Figure 3 found on page 52 of Reference [52]: LR = lower-right; UL = upper-left; LL = lower-left; UR = upper-right). [2] Includes processes of landscape polarization. (See also Reference [52], including Figure 2 on page 51).

Finally, by focusing on AGSOBIOs, this study highlights the importance of the interactions among non-human components as being advanced in sociocultural anthropology, human geography, and related fields of the social sciences and humanities [218,219]. Similarly, this incorporation of ecological analysis contributes to understanding other dimensions of livelihood diversification (Figure 3). For example, influential interactions occur within the human system influencing livelihood diversification. These include the interactions of macro-scale neoliberal political economy with important demographic and social factors (Figure 3). The latter especially include the productive analysis of gender in these interactions as discussed in Section 3. We hasten to underscore that the literature consulted in Sections 1–4 is comprehensive to the best of our knowledge and could be made even more extensive if we were to broaden further our research goals. In addition, while interdisciplinary to a considerable extent, the literature here is not exhaustive across all possible points of reference with regard to the other factors of both the social-ecological and ecological systems of agrobiodiversity.

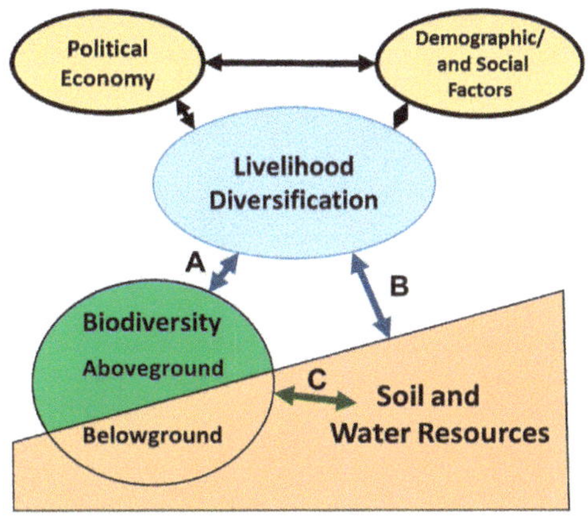

Figure 3. Expanded conceptual diagram of the linkages of livelihood diversification to human-social factors and environmental factors. Arrows A and B represent principal linkages of livelihoods to natural systems; C represents coupling between natural resources and agrobiodiversity.

6. Conclusion: Interactions of Agrobiodiversity and Smallholder Livelihood Diversification in the Context of Global Change

Analysis of the interactions of smallholder livelihoods with agrobiodiversity and interactions (including the ecological inner-actions within AGSOBIO assemblages) yield a total of more than ten synthesis-based statistical summaries. The meta-analysis of empirical studies contributes scientific research and policy insights. The past 25 years have witnessed a significant evolution from the view of smallholder or peasant livelihoods and their environments as significantly separate from global systems, such as the once common concept of disarticulated economies ([220]; pp. 32–40). Currently, concepts and conditions highlight the relevance of smallholder connectedness with national and global systems including so-called planetary urbanism. Correspondingly, these interactions of smallholder livelihoods are producing emergent qualities in agrobiodiversity and other land, water, and biotic resource systems. This perspective highlighting the potential social-ecological emergence of agrobiodiversity [8,14,52,119] is replacing the earlier blanket assumption of purely traditional resource management and, therefore, the inference of agrobiodiversity as a vestige or relict.

Our study's meta-analysis demonstrates how expanding research on livelihood diversification is evidencing causal influences along with noticeable variation in regard to local outcomes and, additionally, in regard to the type of resource or environmental system (agrobiodiversity, soil resources, water resources, uncultivated biodiversity). This finding can be used to nest our interpretation of interactions in agrobiodiversity systems. Combined in this way, the two meta-analyses confirm the conceptual importance and empirical validity of interactions among and within the dynamics of agrobiodiversity, livelihood diversification, and environmental resources.

The meta-analysis of agrobiodiversity components furnishes an important additional dimension to existing livelihood-environment frameworks. Our approach re-frames the predominantly dyadic concepts used to understand the linkages of livelihoods to the environment where the latter are single elements or environmental sub-systems isolated from ecological interactions. The expanded conceptual framework and analysis is essential to understanding the interactions of agrobiodiversity. Livelihood diversification commonly exerts immediate impacts on either the above- or belowground components of agrobiodiversity, thus leading to the need for the framework developed here. The interactions of above- and belowground components must be seen as critical to the health and vitality of agrobiodiversity in the context of smallholder livelihoods and diversification.

This research offers insights and contributions to the major scientific and policy approaches of Sustainable Intensification and Ecological Intensification. It potentially strengthens the bridging of this pair of approaches—two goals that are among the highest priority issues of global change facing societies and science worldwide. The foundations of Sustainable Intensification and Ecological Intensification suggest a productive tension [65] insofar as the former is focused on such issues as social equity and food security [3,4] while the latter prioritizes ecosystem services and ecological sustainability [62]. Our framework of livelihood diversification—environment interactions and the coupled system of AGSOBIO interactions can contribute to the expanded integration of the goals associated with both Sustainable Intensification and Ecological Intensification. It illustrates an integration of themes by carefully combining the concerns of livelihood sustainability (prioritized in Sustainable Intensification) with the analysis of agrobiodiversity and ecological functions *per se* (prioritized in Ecological Intensification). The potential for this selective fusion using the re-framing of livelihood-environment interactions we have developed here can aid in the practical application of combined Sustainable Intensification and Ecological Intensification in future scientific research and policy formulations.

Acknowledgments: We gratefully acknowledge the support of the College of Earth and Mineral Sciences, the Department of Geography, and the GeoSyntheSES Lab at Pennsylvania State University. Connie Almekinders, Judith Carney, Carolyn Sachs, Stef de Haan, Steven Brush, Wendy Wolford, Bradford Barham, Michael Carter, Louise Jackson, Susan Paulson, Clark Gray, Chris Barrett, Erica Smithwick, and Dave Mortensen provided excellent discussions on livelihood diversification, gender, related environment–development interactions, and plant-soil ecological interactions at various times. Table 5 was presented in a symposium at the 2015 AAAS meeting in San Jose, California where helpful inputs were received from Bill Turner, Bill Easterling, Louise Jackson, Darla Munroe, and Tom Rudel. Jerry Glover and Sieglinde Snapp commented on sustainable and ecological intensification. Oliver Chadwick, Peter Vitousek, and Patrick Kirch stimulated ideas through earlier participation in their "Roots of Conflict" project. Portions of the paper were presented at the University of Texas at Austin in October 2014 and at the 2015 AAG meeting in Chicago, with helpful comments from Tim Beach, Sheryl Luzzadder-Beach, Ken Young, Oliver Coomes, and Bill Doolittle. Helpful comments were provided by anonymous reviews and the Guest Editors. Remaining errors are those of the authors.

Author Contributions: K.S.Z. and S.J.V. contributed to writing and research including data compilation and analysis.

Conflicts of Interest: The authors declare no conflict of interest.

References

1. Vandermeer, J.; van Noordwijk, M.; Anderson, J.; Ong, C.; Perfecto, I. Global change and multi-species agroecosystems: Concepts and issues. *Agric. Ecosyst. Environ.* **1998**, *67*, 1–22. [CrossRef]
2. Zimmerer, K.S. Biological diversity in agriculture and global change. *Annu. Rev. Environ. Resour.* **2010**, *35*, 137–166. [CrossRef]
3. Garnett, T.; Appleby, M.C.; Balmford, A.; Bateman, I.J.; Benton, T.G.; Bloomer, P.; Godfray, H.C.J. Sustainable intensification in agriculture: Premises and policies. *Science* **2013**, *341*, 33–34. [CrossRef] [PubMed]

4. Godfray, H.C.J.; Beddington, J.R.; Crute, I.R.; Haddad, L.; Lawrence, D.; Muir, J.F.; Pretty, J.; Robinson, S.; Thomas, S.M.; Toulmin, C. Food security: The challenge of feeding 9 billion people. *Science* **2010**, *327*, 812–818. [CrossRef] [PubMed]
5. Brookfield, H.C. *Exploring Agrodiversity*; Columbia University Press: New York, NY, USA, 2001.
6. Brush, S.B. *Farmers' Bounty: Locating Crop Diversity in the Contemporary World*; Yale University Press: New Haven, CT, USA, 2004.
7. Zimmerer, K.S. *Changing Fortunes: Peasant Livelihood and Biodiversity in the Peruvian Andes*; University of California Press: Los Angeles, Berkeley, CA, USA, 1996.
8. Zimmerer, K.S. Understanding agrobiodiversity and the rise of resilience: Analytic category, boundary concept, or meta-level transition? *Resil. Int. Policies Pract. Discourses* **2015**, *3*, 1–16. [CrossRef]
9. Jackson, L.E.; Pascual, U.; Hodgkin, T. Utilizing and conserving agrobiodiversity in agricultural landscapes. *Agric. Ecosyst. Environ.* **2007**, *121*, 196–210. [CrossRef]
10. Jarvis, D.I.; Padoch, C.; Cooper, H.D. *Managing Biodiversity in Agricultural Ecosystems*; Columbia University Press: New York, NY, USA, 2007.
11. Tscharntke, T.; Clough, Y.; Wanger, T.C.; Jackson, L.; Motzke, I.; Perfecto, I.; Vandermeer, J.; Whitbread, A. Global food security, biodiversity conservation and the future of agricultural intensification. *Biol. Conserv.* **2012**, *151*, 53–59. [CrossRef]
12. Nabhan, G.P. *Where Our Food Comes from: Retracing Nikolay Vavilov's Quest to End Famine*; Island Press: Washington, DC, USA, 2008.
13. Zimmerer, K.S. Woodlands and agrobiodiversity in irrigation landscapes amidst global change: Bolivia, 1990–2002. *Prof. Geogr.* **2010**, *62*, 335–356. [CrossRef]
14. Zimmerer, K.S. The compatibility of agricultural intensification in a global hotspot of smallholder agrobiodiversity (Bolivia). *Proc. Natl. Acad. Sci. USA* **2013**, *110*, 2769–2774. [CrossRef] [PubMed]
15. Altieri, M.A. The ecological role of biodiversity in agroecosystems. *Agric. Ecosyst. Environ.* **1999**, *74*, 19–31. [CrossRef]
16. Brush, S.B. *In situ* conservation of landraces in centers of crop diversity. *Crop Sci.* **1995**, *35*, 346–354. [CrossRef]
17. Brush, S.B.; Perales, H.R. A maize landscape: Ethnicity and agro-biodiversity in Chiapas Mexico. *Agric. Ecosyst. Environ.* **2007**, *121*, 211–221. [CrossRef]
18. Benin, S.; Smale, M.; Pender, J.; Gebremedhin, B.; Ehui, S. The economic determinants of cereal crop diversity on farms in the Ethiopian highlands. *Agric. Econ.* **2004**, *31*, 197–208. [CrossRef]
19. Bezabih, M.; Sarr, M. Risk preferences and environmental uncertainty: Implications for crop diversification decisions in Ethiopia. *Environ. Resour. Econ.* **2012**, *53*, 483–505. [CrossRef]
20. Chavas, J.P.; Di Falco, S. On the productive value of crop biodiversity: Evidence from the highlands of Ethiopia. *Land Econ.* **2012**, *88*, 58–74. [CrossRef]
21. Samberg, L.H.; Shennan, C.; Zavaleta, E.S. Human and environmental factors affect patterns of crop diversity in an Ethiopian Highland agroecosystem. *Prof. Geogr.* **2010**, *62*, 395–408. [CrossRef]
22. Almekinders, C.J.; Fresco, L.O.; Struik, P.C. The need to study and manage variation in agro-ecosystems. *NJAS Wagening J. Life Sci.* **1995**, *43*, 127–142.
23. Bellon, M.R.; Hodson, D.; Hellin, J. Assessing the vulnerability of traditional maize seed systems in Mexico to climate change. *Proc. Natl. Acad. Sci. USA* **2011**, *108*, 13432–13437. [CrossRef] [PubMed]
24. Hellin, J.; Bellon, M.R.; Hearne, S.J. Maize landraces and adaptation to climate change in Mexico. *J. Crop Improv.* **2014**, *28*, 484–501. [CrossRef]
25. Lin, B.B.; Perfecto, I.; Vandermeer, J. Synergies between agricultural intensification and climate change could create surprising vulnerabilities for crops. *BioScience* **2008**, *58*, 847–854. [CrossRef]
26. Lin, B.B. Resilience in agriculture through crop diversification: Adaptive management for environmental change. *BioScience* **2011**, *61*, 183–193. [CrossRef]
27. McCord, P.F.; Cox, M.; Schmitt-Harsh, M.; Evans, T. Crop diversification as a smallholder livelihood strategy within semi-arid agricultural systems near Mount Kenya. *Land Use Policy* **2015**, *42*, 738–750. [CrossRef]
28. Brussaard, L.; Caron, P.; Campbell, B.; Lipper, L.; Mainka, S.; Rabbinge, R.; Babin, D.; Pulleman, M. Reconciling biodiversity conservation and food security: Scientific challenges for a new agriculture. *Curr. Opin. Environ. Sustain.* **2010**, *2*, 34–42. [CrossRef]

29. De Haan, S.; Núñez, J.; Bonierbale, M.; Ghislain, M. Multilevel agrobiodiversity and conservation of Andean potatoes in Central Peru: Species, morphological, genetic, and spatial diversity. *Mt. Res. Dev.* **2010**, *30*, 222–231. [CrossRef]
30. Jackson, L.E.; Pulleman, M.M.; Brussaard, L.; Bawa, K.S.; Brown, G.G.; Cardoso, I.M.; De Ruiter, P.C.; García-Barrios, L.; Hollander, A.D.; Lavelle, P.; *et al.* Social-ecological and regional adaptation of agrobiodiversity management across a global set of research regions. *Glob. Environ. Chang.* **2012**, *22*, 623–639. [CrossRef]
31. Bellon, M.R. Conceptualizing interventions to support on-farm genetic resource conservation. *World Dev.* **2004**, *32*, 159–172. [CrossRef]
32. Smale, M.; Bellon, M.R.; Jarvis, D.; Sthapit, B. Economic concepts for designing policies to conserve crop genetic resources on farms. *Genet. Res. Crop Evol.* **2004**, *51*, 121–135. [CrossRef]
33. Keleman, A.; Hellin, J. Specialty maize varieties in Mexico: A case study in market-driven agro-biodiversity conservation. *J. Lat. Am. Geogr.* **2009**, *8*, 147–174. [CrossRef]
34. Keleman, A.; Hellin, J.; Flores, D. Diverse varieties and diverse markets: Scale-related maize "profitability crossover" in the central Mexican highlands. *Hum. Ecol.* **2013**, *41*, 683–705. [CrossRef]
35. Jones, A.D.; Shrinivas, A.; Bezner-Kerr, R. Farm production diversity is associated with greater household dietary diversity in Malawi: Findings from nationally representative data. *Food Policy* **2014**, *46*, 1–12. [CrossRef]
36. Skarbø, K. The cooked is the kept: Factors shaping the maintenance of agrobiodiversity in the Andes. *Hum. Ecol.* **2014**, *42*, 711–726. [CrossRef]
37. Johns, T. Agrobiodiversity, diet, and human health. In *Managing Biodiversity in Agricultural Ecosystems*; Jarvis, D.I., Padoch, C., Cooper, H.D., Eds.; Columbia University Press: New York, NY, USA, 2007; pp. 382–406.
38. Johns, T.; Eyzaguirre, P.B. Linking biodiversity, diet and health in policy and practice. *Proc. Nutr. Soc.* **2006**, *65*, 182–189. [CrossRef] [PubMed]
39. Oyarzun, P.J.; Borja, R.M.; Sherwood, S.; Parra, V. Making sense of agrobiodiversity, diet, and intensification of smallholder family farming in the Highland Andes of Ecuador. *Ecol. Food Nutr.* **2013**, *52*, 515–541. [CrossRef] [PubMed]
40. Jarvis, D.I.; Brown, A.H.D.; Cuong, P.H.; Collado-Panduro, L.; Latournerie-Moreno, L.; Gyawali, S.; Tanto, T.; Sawadogo, M.; Mar, I.; Sadiki, M.; *et al.* A global perspective of the richness and evenness of traditional crop-variety diversity maintained by farming communities. *Proc. Natl. Acad. Sci. USA* **2008**, *105*, 5326–5331. [CrossRef] [PubMed]
41. Zimmerer, K.S. Cultural ecology: Placing households in human-environment studies—The cases of tropical forest transitions and agrobiodiversity change. *Prog. Hum. Geogr.* **2004**, *28*, 795–806. [CrossRef]
42. Folke, C. Resilience: The emergence of a perspective for social-ecological systems analyses. *Glob. Environ. Chang.* **2006**, *16*, 253–267. [CrossRef]
43. Ostrom, E. A general framework for analyzing sustainability of social-ecological systems. *Science* **2009**, *325*, 419–422. [CrossRef] [PubMed]
44. Zimmerer, K.S. Retrospective on nature-society geography: Tracing trajectories (1911–2010) and reflecting on translations. *Ann. Assoc. Am. Geogr.* **2010**, *100*, 1076–1094. [CrossRef]
45. Moran, E. *Environmental Social Science: Human-Environment Interactions and Sustainability*; John Wiley & Sons: New York, NY, USA, 2010.
46. Moran, E.F.; Ostrom, E. *Seeing the Forest and the Trees. Human-Environment Interactions in Forest Ecosystems*; MIT Press: Cambridge, MA, USA, 2005.
47. Liu, J.; Dietz, T.; Carpenter, S.R.; Alberti, M.; Folke, C.; Moran, E.; Pell, A.N.; Deadman, P.; Kratz, T.; Lubchenko, J.; *et al.* Complexity of coupled human and natural systems. *Science* **2007**, *317*, 1513–1516. [CrossRef] [PubMed]
48. Brondízio, E.S.; Chowdhury, R.R. Human-environment research: Past trends, current challenges, and future directions. In *Human-Environment Interactions: Current and Future Directions*; Brondizio, E.S., Moran, E.F., Eds.; Springer: Dordrecht, The Netherlands, 2013; pp. 391–400.
49. Zimmerer, K.S. Geographic approaches to LTSER: Principal themes and concepts with a case study of Andes-Amazon watersheds. In *Long Term Socio-Ecological Research*; Singh, S.J., Haberl, H., Chertow, M., Mirtl, M., Schmid, M., Eds.; Springer: Dordrecht, The Netherlands, 2013; pp. 163–187.

50. Brannstrom, C.; Vadjunec, J.M. *Land Change Science, Political Ecology and Sustainability: Synergies and Divergences*; Routledge: London, UK, 2014.
51. Watts, N.; Scales, I.R. Seeds, agricultural systems and socio-natures: Towards an actor–network theory informed political ecology of agriculture. *Geogr. Compass* **2015**, *9*, 225–236. [CrossRef]
52. Zimmerer, K.; Carney, J.A.; Vanek, S.J. Sustainable smallholder intensification in global change? Pivotal spatial interactions, gendered livelihoods, and agrobiodiversity. *Curr. Opin. Environ. Sustain.* **2015**, *14*, 49–60. [CrossRef]
53. Rocheleau, D.E. Gender and biodiversity: A feminist political ecology perspective. *IDS Bull.* **1995**, *26*, 9–16. [CrossRef]
54. Rocheleau, D.; Thomas-Slayter, B.; Wangari, E. *Feminist Political Ecology: Global Issues and Local Experience*; Routledge: New York, NY, USA, 1996.
55. Hecht, S. The new rurality: Globalization, peasants and the paradoxes of landscapes. *Land Use Policy* **2010**, *27*, 161–169. [CrossRef]
56. Kay, C. Reflections on Latin American rural studies in the neoliberal globalization period: A new rurality? *Dev. Chang.* **2008**, *39*, 915–943. [CrossRef]
57. Postma-Blaauw, M.B.; de Goede, R.G.M.; Bloem, J.; Faber, J.H.; Brussaard, L. Soil biota community structure and abundance under agricultural intensification and extensification. *Ecology* **2010**, *91*, 460–473. [CrossRef] [PubMed]
58. Brussaard, L.; de Ruiter, P.C.; Brown, G.G. Soil biodiversity for agricultural sustainability. *Agric. Ecosyst. Environ.* **2007**, *121*, 233–244. [CrossRef]
59. Hooper, D.U.; Bignell, D.E.; Brown, V.K.; Brussaard, L.; Dangerfield, J.M.; Wall, D.H.; Wardle, D.A.; Coleman, D.C.; Giller, K.E.; Lavelle, P.; *et al.* Interactions between aboveground and belowground biodiversity in terrestrial ecosystems: Patterns, mechanisms, and feedbacks. *BioScience* **2000**, *50*, 1049–1061. [CrossRef]
60. Barrett, C.; Constas, M. Toward a theory of resilience for international development applications. *Proc. Natl. Acad. Sci. USA* **2014**, *111*, 14625–14630. [CrossRef] [PubMed]
61. Reardon, T.; Vosti, S.A. Links between rural poverty and the environment in developing countries: Asset categories and investment poverty. *World Dev.* **1995**, *23*, 1495–1506. [CrossRef]
62. Cassman, K.G. Ecological intensification of cereal production systems: Yield potential, soil quality, and precision agriculture. *Proc. Natl. Acad. Sci. USA* **1999**, *96*, 5952–5959. [CrossRef] [PubMed]
63. Omer, A.; Pascual, U.; Russell, N. A theoretical model of agrobiodiversity as a supporting service for sustainable agricultural intensification. *Ecol. Econ.* **2010**, *69*, 1926–1933. [CrossRef]
64. Pretty, J. Agricultural sustainability: Concepts, principles and evidence. *Philos. Trans. R. Soc. Lond. B: Biol. Sci.* **2008**, *363*, 447–465. [CrossRef] [PubMed]
65. Snapp, S.S.; Blackie, M.J.; Gilbert, R.A.; Bezner-Kerr, R.; Kanyama-Phiri, G.Y. Biodiversity can support a greener revolution in Africa. *Proc. Natl. Acad. Sci. USA* **2010**, *107*, 20840–20845. [CrossRef] [PubMed]
66. Zimmerer, K.S.; Rojas Vaca, H.L. Fine-grain spatial patterning and dynamics of land use and agrobiodiversity amid global changes in the Bolivian Andes. *Reg. Environ. Change* **2016**. [CrossRef]
67. McMichael, P. Food system sustainability: Questions of environmental governance in the new world (dis) order. *Glob. Environ. Chang.* **2011**, *21*, 804–812. [CrossRef]
68. Thompson, J.; Scoones, I. Addressing the dynamics of agri-food systems: An emerging agenda for social science research. *Environ. Sci. Policy* **2009**, *12*, 386–397. [CrossRef]
69. Turner, B.L. The sustainability principle in global agendas: Implications for understanding land-use/cover change. *Geogr. J.* **1997**, *163*, 133–140.
70. Turner, B.L.; Lambin, E.F.; Reenberg, A. The emergence of land change science for global environmental change and sustainability. *Proc. Natl. Acad. Sci. USA* **2007**, *104*, 20666–20671. [CrossRef] [PubMed]
71. Scherr, S.J.; McNeely, J.A. Biodiversity conservation and agricultural sustainability: Towards a new paradigm of "ecoagriculture" landscapes. *Philos. Trans. R. Soc. Lond. B: Biol. Sci.* **2008**, *363*, 477–494. [CrossRef] [PubMed]
72. International Fund for Agricultural Development (IFAD). *Smallholders, Food Security, and the Environment*; International Fund for Agricultural Development, U.N. Environment Programme: Rome, Italy, 2013.
73. DeFries, R.; Rosenzweig, C. Toward a whole-landscape approach for sustainable land use in the tropics. *Proc. Natl. Acad. Sci. USA* **2010**, *107*, 19627–19632. [CrossRef] [PubMed]

74. Keys, E.; McConnell, W.J. Global change and the intensification of agriculture in the tropics. *Glob. Environ. Chang.* **2005**, *15*, 320–337. [CrossRef]
75. Rudel, T.K. Meta-analyses of case studies: A method for studying regional and global environmental change. *Glob. Environ. Chang.* **2008**, *18*, 18–25. [CrossRef]
76. Magliocca, N.R.; Rudel, T.K.; Verburg, P.H.; McConnell, W.J.; Mertz, O.; Gerstner, K.; Ellis, E.C. Synthesis in land change science: Methodological patterns, challenges, and guidelines. *Reg. Environ. Chang.* **2015**, *15*, 211–226. [CrossRef] [PubMed]
77. High Level Panel of Experts on Food Security and Nutrition (HLPE). *Investing in Smallholder Agriculture for Food Security*; High Level Panel of Experts on Food Security and Nutrition of the Committee on World Food Security: Rome, Italy, 2013; pp. 25–28.
78. DeFries, R.S.; Foley, J.A.; Asner, G.P. Land-use choices: Balancing human needs and ecosystem function. *Front. Ecol. Environ.* **2004**, *2*, 249–257. [CrossRef]
79. Aide, T.M.; Grau, H.R. Globalization, migration, and Latin American ecosystems. *Science* **2004**, *305*, 1915–1916. [CrossRef] [PubMed]
80. Batterbury, S. Landscapes of diversity: A local political ecology of livelihood diversification in south-western Niger. *Cult. Geogr.* **2001**, *8*, 437–464. [CrossRef]
81. Bilsborrow, R.E. *Rural Poverty, Migration, and the Environment in Developing Countries: Three Case Studies*; World Bank Publications: Washington, DC, USA, 1992; Volume 1017.
82. Garcia-Barrio, R.; García-Barrios, L. Environmental and technological degradation in peasant agriculture: A consequence of development in Mexico. *World Dev.* **1990**, *18*, 1569–1585. [CrossRef]
83. Bilsborrow, R.E.; DeLargy, P.F. Land use, migration, and natural resource deterioration: The experience of Guatemala and the Sudan. *Popul. Dev. Rev.* **1990**, *16*, 125–147. [CrossRef]
84. Carr, D. Rural migration: The driving force behind tropical deforestation on the settlement frontier. *Prog. Hum. Geogr.* **2009**, *33*, 355–378. [CrossRef] [PubMed]
85. Chambers, K.J.; Momsen, J.H. From the kitchen and the field: Gender and maize diversity in the Bajío region of Mexico. *Singap. J. Trop. Geogr.* **2007**, *28*, 39–56. [CrossRef]
86. Chen, X.; Frank, K.A.; Dietz, T.; Liu, J. Weak ties, labor migration, and environmental impacts: Toward a sociology of sustainability. *Organ. Environ.* **2012**, *25*, 3–24. [CrossRef]
87. Collins, J.L. Labor scarcity and ecological change. In *Lands at Risk in the Third World: Local-Level Perspectives*; Little, P.D., Horowitz, M.M., Eds.; Westview: Boulder, CO, USA, 1987; pp. 19–37.
88. Collins, J.L. *Unseasonal Migrations: The Effects of Rural Labor Scarcity in Peru*; Princeton University Press: Princeton, NJ, USA, 1988.
89. Davis, J.; Lopez-Carr, D. The effects of migrant remittances on population–environment dynamics in migrant origin areas: International migration, fertility, and consumption in highland Guatemala. *Popul. Environ.* **2010**, *32*, 216–237. [CrossRef] [PubMed]
90. Davis, J.; Lopez-Carr, D. Migration, remittances and smallholder decision-making: Implications for land use and livelihood change in Central America. *Land Use Policy* **2014**, *36*, 319–329. [CrossRef] [PubMed]
91. European Commission. *The Impact of European Water Policy on the Water Cultural Heritage*; European Policy Brief; Mediterranean Mountainous Landscapes (MEMOLA): Granada, Spain, 2015.
92. Grau, H.R.; Aide, T.M. Are rural-urban migration and sustainable development compatible in mountain systems? *Mt. Res. Dev.* **2007**, *27*, 119–123. [CrossRef]
93. Gray, C.L. Rural out-migration and smallholder agriculture in the southern Ecuadorian Andes. *Popul. Environ.* **2009**, *30*, 193–217. [CrossRef]
94. Gray, C.L.; Bilsborrow, R.E. Consequences of out-migration for land use in rural Ecuador. *Land Use Policy* **2014**, *36*, 182–191. [CrossRef] [PubMed]
95. Gray, L.C. What kind of intensification? Agricultural practice, soil fertility and socioeconomic differentiation in rural Burkina Faso. *Geogr. J.* **2005**, *171*, 70–82. [CrossRef]
96. Heartherington, T. Tasting cultural ecology: Foodscapes of sustainability in the Mediterranean. *Gastr. J. Crit. Food Stud.* **2014**, *14*, 16–26. [CrossRef]
97. Hecht, S.B.; Kandel, S.; Gomes, I.; Cuellar, N.; Rosa, H. Globalization, forest resurgence, and environmental politics in El Salvador. *World Dev.* **2006**, *34*, 308–323. [CrossRef]

98. Huijun, G.; Padoch, C.; Coffey, K.; Aiguo, C.; Yongneng, F. Economic development, land use and biodiversity change in the tropical mountains of Xishuangbanna, Yunnan, Southwest China. *Environ. Sci. Policy* **2002**, *5*, 471–479. [CrossRef]
99. Iniesta-Arandia, I.; del Amo, D.G.; García-Nieto, A.P.; Piñeiro, C.; Montes, C.; Martín-López, B. Factors influencing local ecological knowledge maintenance in Mediterranean watersheds: Insights for environmental policies. *AMBIO* **2015**, *44*, 285–296. [CrossRef] [PubMed]
100. Jewitt, S. Unequal knowledges in Jharkhand, India: De-romanticizing women's agroecological expertise. *Dev. Chang.* **2000**, *31*, 961–985. [CrossRef]
101. Jiménez Olivencia, Y. Consecuencias del abandono del regadío den la montaña mediterránea. In *El Agua Domesticada, Los Paisajes del Regadíos de Montaña en Andalucía*; Guzmán Álvarez, J.R., Navarro Cerrillo, R.M., Eds.; Consejería de Medio Ambiente de la Junta de Andalucía: Sevilla, Spain, 2010; pp. 508–513.
102. Jiménez Olivencia, Y.; Porcel Rodríguez, L.; Caballero Calvo, A. A half-century of landscape evolution in the Sierra Nevada (Spain). *Bol. Asoc. Geógr. Españoles* **2015**, *68*, 497–502.
103. Jokisch, B.D. Migration and agricultural change: The case of smallholder agriculture in Highland Ecuador. *Hum. Ecol.* **2002**, *30*, 523–550. [CrossRef]
104. Lerner, A.M.; Eakin, H.; Sweeney, S. Understanding peri-urban maize production through an examination of household livelihoods in the Toluca Metropolitan Area, Mexico. *J. Rural Stud.* **2013**, *30*, 52–63. [CrossRef]
105. Li, Y.; López-Carr, D.; Chen, W. Factors affecting migration intentions in ecological restoration areas and their implications for the sustainability of ecological migration policy in arid northwest China. *Sustainability* **2014**, *6*, 8639–8660. [CrossRef]
106. López, E.; Bocco, G.; Mendoza, M.; Velázquez, A.; Aguirre-Rivera, J.R. Peasant emigration and land-use change at the watershed level: A GIS-based approach in Central Mexico. *Agric. Syst.* **2006**, *90*, 62–78. [CrossRef]
107. Mutersbaugh, T. Certifying biodiversity: Conservation networks, landscape connectivity, and certified agriculture in Southern Mexico. In *Globalization and New Geographies of Conservation*; Zimmerer, K.S., Ed.; University of Chicago Press: Chicago, IL, USA, 2006; pp. 49–70.
108. Nuijten, E. Gender and management of crop diversity in The Gambia. *J. Political Ecol.* **2010**, *17*, 42–58.
109. Preston, D.; Macklin, M.; Warburton, J. Fewer people, less erosion: The twentieth century in southern Bolivia. *Geogr. J.* **1997**, 198–205. [CrossRef]
110. Qin, H. Rural-to-urban labor migration, household livelihoods, and the rural environment in Chongqing Municipality, Southwest China. *Hum. Ecol.* **2010**, *38*, 675–690. [CrossRef] [PubMed]
111. Radel, C.; Schmook, B. Male transnational migration and its linkages to land-use change in a southern Campeche ejido. *J. Lat. Am. Geogr.* **2008**, *7*, 59–84. [CrossRef]
112. Radel, C.; Schmook, B.; McEvoy, J.; Mendez, C.; Petrzelka, P. Labour migration and gendered agricultural relations: The feminization of agriculture in the Ejidal sector of Calakmul, Mexico. *J. Agrar. Chang.* **2012**, *12*, 98–119. [CrossRef]
113. Reardon, T.; Berdegue, J.; Escobar, G. Rural nonfarm employment and incomes in Latin America: Overview and policy implications. *World Dev.* **2001**, *29*, 395–409. [CrossRef]
114. Rudel, T.K.; Bates, D.; Machinguiashi, R. A tropical forest transition? Agricultural change, out-migration, and secondary forests in the Ecuadorian Amazon. *Ann. Assoc. Am. Geogr.* **2002**, *92*, 87–102. [CrossRef]
115. Schmook, B.; van Vliet, N.; Radel, C.; de Jesus Manzon-Che, M.; McCandless, S. Persistence of Swidden cultivation in the face of globalization: A case study from communities in Calakmul, Mexico. *Hum. Ecol.* **2013**, *41*, 93–107. [CrossRef]
116. Turner, M.D. Labor process and the environment: The effects of labor availability and compensation on the quality of herding in the Sahel. *Hum. Ecol.* **1999**, *27*, 267–296. [CrossRef]
117. Zimmerer, K.S. Labor shortages and crop diversity in the southern Peruvian sierra. *Geogr. Rev.* **1991**, *81*, 414–432. [CrossRef]
118. Zimmerer, K.S. Soil erosion and labor shortages in the Andes with special reference to Bolivia, 1953–1991: Implications for "conservation-with-development". *World Dev.* **1993**, *21*, 1659–1675. [CrossRef]
119. Zimmerer, K.S. Conserving agrobiodiversity amid global change, migration, and nontraditional livelihood networks: The dynamic uses of cultural landscape knowledge. *Ecol. Soc.* **2014**, *19*, 1.

120. Acosta-Martínez, V.; Bell, C.W.; Morris, B.E.L.; Zak, J.; Allen, V.G. Long-term soil microbial community and enzyme activity responses to an integrated cropping-livestock system in a semi-arid region. *Agric. Ecosyst. Environ.* **2010**, *137*, 231–240. [CrossRef]
121. Alvey, S.; Bagayoko, M.; Neumann, G.; Buerkert, A. Cereal/legume rotations affect chemical properties and biological activities in two West African soils. *Plant Soil* **2001**, *231*, 45–54. [CrossRef]
122. Asgharipour, M.R.; Rafiei, M. The effects of land use on biomass and catabolic diversity of soil microbial communities. *Afr. J. Agric. Res.* **2011**, *6*, 4607–4612.
123. Bach, E.M.; Baer, S.G.; Six, J. Plant and soil responses to high and low diversity grassland restoration practices. *Environ. Manag.* **2012**, *49*, 412–424. [CrossRef] [PubMed]
124. Berthrong, S.T.; Schadt, C.W.; Pineiro, G.; Jackson, R.B. Afforestation alters the composition of functional genes in soil and biogeochemical processes in South American grasslands. *Appl. Environ. Microbiol.* **2009**, *75*, 6240–6248. [CrossRef] [PubMed]
125. Buckley, D.H.; Schmidt, T.M. The structure of microbial communities in soil and the lasting impact of cultivation. *Microb. Ecol.* **2001**, *42*, 11–21. [PubMed]
126. Duchicela, J.; Sullivan, T.S.; Bontti, E.; Bever, J.D. Soil aggregate stability increase is strongly related to fungal community succession along an abandoned agricultural field chronosequence in the Bolivian Altiplano. *J. Appl. Ecol.* **2013**, *50*, 1266–1273. [CrossRef]
127. Garbeva, P.; Postma, J.; van Veen, J.A.; van Elsas, J.D. Effect of aboveground plant species on soil microbial community structure and its impact on suppression of Rhizoctonia solani AG3. *Environ. Microbiol.* **2006**, *8*, 233–246. [CrossRef] [PubMed]
128. Garcia, M.R.L.; Nahas, E. Microbial populations and the activity of the soil under agricultural and agricultural–pastoral systems. *Arch. Agron. Soil Sci.* **2012**, *58*, 511–525. [CrossRef]
129. Gomez, E.V.; Garland, J.L.; Roberts, M.S. Microbial structural diversity estimated by dilution–extinction of phenotypic traits and T-RFLP analysis along a land-use intensification gradient. *FEMS Microbiol. Ecol.* **2004**, *49*, 253–259. [CrossRef] [PubMed]
130. González-Chávez, M.A.; Aitkenhead-Peterson, J.A.; Gentry, T.J.; Zuberer, D.; Hons, F.; Loeppert, R. Soil microbial community, C, N, and P responses to long-term tillage and crop rotation. *Soil Tillage Res.* **2010**, *106*, 285–293. [CrossRef]
131. Jiang, D.; Cao, C.; Zhang, Y.; Cui, Z.; Han, X. Plantations of native shrub species restore soil microbial diversity in the Horqin Sandy Land, northeastern China. *J. Arid Land* **2014**, *6*, 445–453. [CrossRef]
132. Johnson, D.; Anderson, I.C.; Williams, A.; Whitlock, R.; Grime, J.P. Plant genotypic diversity does not beget root-fungal species diversity. *Plant Soil* **2010**, *336*, 107–111. [CrossRef]
133. Larkin, R.P.; Honeycutt, C.W. Effects of different 3-year cropping systems on soil microbial communities and *Rhizoctonia* diseases of potato. *Phytopathology* **2006**, *96*, 68–79. [CrossRef] [PubMed]
134. Lienhard, P.; Terrat, S.; Prévost-Bouré, N.C.; Nowak, V.; Régnier, T.; Sayphoummie, S.; Panyasiri, K.; Tivet, F.; Mathieu, O.; Levêque, J.; *et al.* Pyrosequencing evidences the impact of cropping on soil bacterial and fungal diversity in Laos tropical grassland. *Agron. Sustain. Dev.* **2014**, *34*, 525–533. [CrossRef]
135. McDaniel, M.D.; Tiemann, L.K.; Grandy, A.S. Does agricultural crop diversity enhance soil microbial biomass and organic matter dynamics? A meta-analysis. *Ecol. Appl.* **2014**, *24*, 560–570. [CrossRef] [PubMed]
136. Meriles, J.M.; Vargas Gil, S.; Conforto, C.; Figoni, G.; Lovera, E.; March, G.J.; Guzmán, C.A. Soil microbial communities under different soybean cropping systems: Characterization of microbial population dynamics, soil microbial activity, microbial biomass, and fatty acid profiles. *Soil Tillage Res.* **2009**, *103*, 271–281. [CrossRef]
137. Millard, P.; Singh, B.K. Does grassland vegetation drive soil microbial diversity? *Nutr. Cycl. Agroecosyst.* **2010**, *88*, 147–158. [CrossRef]
138. Nayyar, A.; Hamel, C.; Lafond, G.; Gossen, B.D.; Hanson, K.; Germida, J. Soil microbial quality associated with yield reduction in continuous-pea. *Appl. Soil Ecol.* **2009**, *43*, 115–121. [CrossRef]
139. Nurulita, Y.; Adetutu, E.M.; Kadali, K.K.; Zul, D.; Mansur, A.A.; Ball, A.S. The assessment of the impact of oil palm and rubber plantations on the biotic and abiotic properties of tropical peat swamp soil in Indonesia. *Int. J. Agric. Sustain.* **2015**, *13*, 150–166. [CrossRef]
140. Shen, W.; Lin, X.; Gao, N.; Zhang, H.; Yin, R.; Shi, W.; Duan, Z. Land use intensification affects soil microbial populations, functional diversity and related suppressiveness of cucumber *Fusarium* wilt in China's Yangtze River Delta. *Plant Soil* **2008**, *306*, 117–127. [CrossRef]

141. Sheng, R.; Meng, D.; Wu, M.; Di, H.; Qin, H.; Wei, W. Effect of agricultural land use change on community composition of bacteria and ammonia oxidizers. *J. Soil Sediments* **2013**, *13*, 1246–1256. [CrossRef]
142. Vasileiadis, S.; Puglisi, E.; Arena, M.; Cappa, F.; van Veen, J.A.; Cocconcelli, P.S.; Trevisan, M. Soil microbial diversity patterns of a lowland spring environment. *FEMS Microbiol. Ecol.* **2013**, *86*, 172–184. [CrossRef] [PubMed]
143. Yao, H.; Bowman, D.; Shi, W. Soil microbial community structure and diversity in a turfgrass chronosequence: Land-use change *versus* turfgrass management. *Appl. Soil Ecol.* **2006**, *34*, 209–218. [CrossRef]
144. Yu, Y.; Shen, W.; Yin, Y.; Zhang, J.; Cai, Z.; Zhong, W. Response of soil microbial diversity to land-use conversion of natural forests to plantations in a subtropical mountainous area of southern China. *Soil Sci. Plant Nutr.* **2012**, *58*, 450–461. [CrossRef]
145. Almaca, A.; Ortas, I. Growth response of maize plants (*Zea mays* L.) to wheat and lentil pre-cropping and to indigenous mycorrhizae in field soil. *Span. J. Agric. Res.* **2010**, *8*, 131–136. [CrossRef]
146. Bedini, S.; Avio, L.; Sbrana, C.; Turrini, A.; Migliorini, P.; Vazzana, C.; Giovannetti, M. Mycorrhizal activity and diversity in a long-term organic Mediterranean agroecosystem. *Biol. Fertil. Soils* **2013**, *49*, 781–790. [CrossRef]
147. Carpenter, F.L.; Mayorga, S.P.; Quintero, E.G.; Schroeder, M. Land-use and erosion of a Costa Rican Ultisol affect soil chemistry, mycorrhizal fungi and early regeneration. *For. Ecol. Manag.* **2001**, *144*, 1–17. [CrossRef]
148. Chaturvedi, S.; Mehta, C.M.; Singh, S.; Sharma, A.K. Host influences Arbuscular mycorrhizal fungal diversity. *J. Mycol. Plant Pathol.* **2009**, *39*, 124.
149. Dai, M.; Bainard, L.D.; Hamel, C.; Gan, Y.; Lynch, D. Impact of land use on arbuscular mycorrhizal fungal communities in Rural Canada. *Appl. Environ. Microbiol.* **2013**, *79*, 6719–6729. [CrossRef] [PubMed]
150. Gavito, M.E.; Miller, M.H. Changes in mycorrhiza development in maize induced by crop management practices. *Plant Soil* **1998**, *198*, 185–192. [CrossRef]
151. Hendrix, J.W.; Guo, B.Z.; An, Z.Q. Divergence of mycorrhizal fungal communities in crop production systems. *Plant Soil* **1995**, *170*, 131–140. [CrossRef]
152. Higo, M.; Isobe, K.; Drijber, R.A.; Kondo, T.; Yamaguchi, M.; Takeyama, S.; Suzuki, Y.; Niijima, D.; Matsuda, Y.; Ishii, R.; *et al.* Impact of a 5-year winter cover crop rotational system on the molecular diversity of arbuscular mycorrhizal fungi colonizing roots of subsequent soybean. *Biol. Fertil. Soils* **2014**, *50*, 913–926. [CrossRef]
153. Hijri, I.; SýKorová, Z.; Oehl, F.; Ineichen, K.; Mäder, P.; Wiemken, A.; Redecker, D. Communities of arbuscular mycorrhizal fungi in arable soils are not necessarily low in diversity. *Mol. Ecol.* **2006**, *15*, 2277–2289. [CrossRef] [PubMed]
154. Lehman, R.M.; Taheri, W.I.; Osborne, S.L.; Buyer, J.S.; Douds, D.D. Fall cover cropping can increase arbuscular mycorrhizae in soils supporting intensive agricultural production. *Appl. Soil Ecol.* **2012**, *61*, 300–304. [CrossRef]
155. Lekberg, Y.; Koide, R.T.; Twomlow, S.J. Effect of agricultural management practices on arbuscular mycorrhizal fungal abundance in low-input cropping systems of southern Africa: A case study from Zimbabwe. *Biol. Fertil. Soils* **2008**, *44*, 917–923. [CrossRef]
156. Melo, C.D.; Walker, C.; Rodríguez-Echeverría, S.; Borges, P.A.; Freitas, H. Species composition of arbuscular mycorrhizal fungi differ in semi-natural and intensively managed pastures in an isolated oceanic island (Terceira, Azores). *Symbiosis* **2014**, *64*, 73–85. [CrossRef]
157. Muchane, M.N. Effect of land use system on Arbuscular Mycorrhiza fungi in Maasai Mara ecosystem, Kenya. *Afr. J. Microbiol. Res.* **2012**, *6*, 3904–3916.
158. Oehl, F.; Sieverding, E.; Ineichen, K.; Mader, P.; Boller, T.; Wiemken, A. Impact of land use intensity on the species diversity of arbuscular mycorrhizal fungi in agroecosystems of Central Europe. *Appl. Environ. Microbiol.* **2003**, *69*, 2816–2824. [CrossRef] [PubMed]
159. Rao, A.V.; Tarafdar, J.C.; Sharma, S.K.; Kumar, P.; Aggarwal, R.K. Influence of cropping systems on soil biochemical properties in an arid rain-fed environment. *J. Arid Environ.* **1995**, *31*, 237–244. [CrossRef]
160. Säle, V.; Aguilera, P.; Laczko, E.; Mäder, P.; Berner, A.; Zihlmann, U.; van der Heijden, M.G.A.; Oehl, F. Impact of conservation tillage and organic farming on the diversity of arbuscular mycorrhizal fungi. *Soil Biol. Biochem.* **2015**, *84*, 38–52. [CrossRef]
161. Stürmer, S.L.; Siqueira, J.O. Species richness and spore abundance of arbuscular mycorrhizal fungi across distinct land uses in Western Brazilian Amazon. *Mycorrhiza* **2011**, *21*, 255–267. [CrossRef] [PubMed]

162. Tchabi, A.; Coyne, D.; Hountondji, F.; Lawouin, L.; Wiemken, A.; Oehl, F. Arbuscular mycorrhizal fungal communities in sub-Saharan Savannas of Benin, West Africa, as affected by agricultural land use intensity and ecological zone. *Mycorrhiza* **2008**, *18*, 181–195. [CrossRef] [PubMed]
163. Tian, H.; Drijber, R.A.; Zhang, J.L.; Li, X.L. Impact of long-term nitrogen fertilization and rotation with soybean on the diversity and phosphorus metabolism of indigenous arbuscular mycorrhizal fungi within the roots of maize (*Zea mays* L.). *Agric. Ecosyst. Environ.* **2013**, *164*, 53–61. [CrossRef]
164. Verbruggen, E.; Röling, W.F.M.; Gamper, H.A.; Kowalchuk, G.A.; Verhoef, H.A.; van der Heijden, M.G.A. Positive effects of organic farming on belowground mutualists: Large-scale comparison of mycorrhizal fungal communities in agricultural soils. *New Phytol.* **2010**, *186*, 968–979. [CrossRef] [PubMed]
165. Vestberg, M.; Kahiluoto, H.; Wallius, E. Arbuscular mycorrhizal fungal diversity and species dominance in a temperate soil with long-term conventional and low-input cropping systems. *Mycorrhiza* **2011**, *21*, 351–361. [CrossRef] [PubMed]
166. Domínguez, A.; Bedano, J.C.; Becker, A.R. Negative effects of no-till on soil macrofauna and litter decomposition in Argentina as compared with natural grasslands. *Soil Tillage Res.* **2010**, *110*, 51–59. [CrossRef]
167. Hulugalle, N.R.; de Bruyn, L.A.L.; Entwistle, P. Residual effects of tillage and crop rotation on soil properties, soil invertebrate numbers and nutrient uptake in an irrigated Vertisol sown to cotton. *Appl. Soil Ecol.* **1997**, *7*, 11–30. [CrossRef]
168. Maria de Aquino, A.; Ferreira da Silva, R.; Mercante, F.M.; Fernandes Correia, M.E.; de Fátima Guimarães, M.; Lavelle, P. Invertebrate soil macrofauna under different ground cover plants in the no-till system in the Cerrado. *Eur. J. Soil Biol.* **2008**, *44*, 191–197. [CrossRef]
169. Sileshi, G.; Mafongoya, P.; Chintu, R.; Akinnifesi, F. Mixed-species legume fallows affect faunal abundance and richness and N cycling compared to single species in maize-fallow rotations. *Soil Biol. Biochem.* **2008**, *40*, 3065–3075. [CrossRef]
170. Hunter, L.M.; Luna, J.K.; Norton, R.M. Environmental dimensions of migration. *Ann. Rev. Sociol.* **2015**, *41*, 377–397. [CrossRef]
171. Carr, D.L.; Lopez, A.C.; Bilsborrow, R.E. The population, agriculture, and environment nexus in Latin America: Country-level evidence from the latter half of the twentieth century. *Popul. Environ.* **2009**, *30*, 222–246. [CrossRef]
172. Paulson, S. Gendered practices and landscapes in the Andes: The shape of asymmetrical exchanges. *Hum. Organ.* **2003**, *62*, 242–254. [CrossRef]
173. Radel, C.; Schmook, B.; McCandless, S. Environment, transnational labor migration, and gender: Case studies from southern Yucatan, Mexico and Vermont, USA. *Popul. Environ.* **2010**, *32*, 177–197. [CrossRef]
174. Barrett, C.B.; Reardon, T.; Webb, P. Nonfarm income diversification and household livelihood strategies in rural Africa: Concepts, dynamics, and policy implications. *Food Policy* **2001**, *26*, 315–331. [CrossRef]
175. Barrett, C.B.; Travis, A.J.; Dasgupta, P. On biodiversity conservation and poverty traps. *Proc. Natl. Acad. Sci. USA* **2011**, *108*, 13907–13912. [CrossRef] [PubMed]
176. Gray, C.L. Environment, land, and rural out-migration in the Southern Ecuadorian Andes. *World Dev.* **2009**, *37*, 457–468. [CrossRef]
177. Gray, C.L. Soil quality and human migration in Kenya and Uganda. *Glob. Environ. Chang.* **2011**, *21*, 421–430. [CrossRef] [PubMed]
178. Gray, C.L.; Bilsborrow, R.E. Environmental influences on human migration in rural Ecuador. *Demography* **2013**, *50*, 1217–1241. [CrossRef] [PubMed]
179. López-Carr, D. Agro-ecological drivers of rural out-migration to the Maya Biosphere Reserve, Guatemala. *Environ. Res. Lett.* **2012**, *7*, 045603. [CrossRef] [PubMed]
180. Zimmerer, K.S. The landscape technology of spate irrigation amid development changes: Assembling the links to resources, livelihoods, and agrobiodiversity-food in the Bolivian Andes. *Glob. Environ. Chang.* **2011**, *21*, 917–934. [CrossRef]
181. Rudel, T.K.; Schneider, L.; Uriarte, M.; Turner, B.L.; DeFries, R.; Lawrence, D.; Geoghegan, J.; Hecht, S.; Ickowitz, A.; Lambin, E.F.; *et al.* Agricultural intensification and changes in cultivated areas, 1970–2005. *Proc. Natl. Acad. Sci. USA* **2009**, *106*, 20675–20680. [CrossRef] [PubMed]
182. Rudel, T.K.; Defries, R.; Asner, G.P.; Laurance, W.F. Changing drivers of deforestation and new opportunities for conservation. *Conserv. Biol.* **2009**, *23*, 1396–1405. [CrossRef] [PubMed]

183. López-Carr, D.; Burgdorfer, J. Deforestation drivers: Population, migration, and tropical land use. *Environment* **2013**, *55*, 3–11. [CrossRef] [PubMed]
184. Carney, J.A. Peasant women and economic transformation in The Gambia. *Dev. Chang.* **1992**, *23*, 67–90. [CrossRef]
185. Carney, J.A. The bitter harvest of Gambian rice policies. *Globalizations* **2008**, *5*, 129–142. [CrossRef]
186. Deere, C.D. The division of labor by sex in agriculture: A Peruvian case study. *Econ. Dev. Cult. Chang.* **1982**, *30*, 795–811. [CrossRef]
187. Deere, C.D.; León de Leal, M. *Women in Andean Agriculture: Peasant Production and Rural Wage Employment in Colombia and Peru*; International Labor Office: Geneva, Switzerland, 1982.
188. Momsen, J.H. Gender and agrobiodiversity: Introduction to the Special Issue. *Singap. J. Trop. Geogr.* **2007**, *28*, 1–6. [CrossRef]
189. Sachs, C. Reconsidering diversity in agriculture and food systems: An ecofeminist approach. *Agric. Hum. Values* **1992**, *9*, 4–10. [CrossRef]
190. Sachs, C. *Gendered Fields: Rural Women, Agriculture, and Environment*; Westview Press: Boulder, CO, USA, 1996.
191. Sachs, C. *Women Working in the Environment: Resourceful Natures*; Taylor & Francis: New York, NY, USA, 2014.
192. Sachs, C.; Gajurel, K.; Bianco, M. Gender, seeds, and biodiversity. In *Women Working in the Environment*; Sachs, C., Ed.; Taylor and Francis: Washington, DC, USA, 1996; pp. 177–192.
193. Kerr, R.B. Lost and found crops: Agrobiodiversity, indigenous knowledge, and a feminist political ecology of sorghum and finger millet in Northern Malawi. *Ann. Assoc. Am. Geogr.* **2014**, *104*, 577–593. [CrossRef]
194. Giller, K.E.; Tittonell, P.; Rufino, M.C.; van Wijk, M.T.; Zingore, S.; Mapfumo, P.; Adjei-Nsiah, S.; Herrero, M.; Chikowo, R.; Corbeels, M.; *et al.* Communicating complexity: Integrated assessment of trade-offs concerning soil fertility management within African farming systems to support innovation and development. *Agric. Syst.* **2011**, *104*, 191–203. [CrossRef]
195. Giller, K.E.; Witter, E.; Corbeels, M.; Tittonell, P. Conservation agriculture and smallholder farming in Africa: The heretics' view. *Field Crops Res.* **2009**, *114*, 23–34. [CrossRef]
196. Radel, C. Gendered livelihoods and the politics of socio-environmental identity: Women's participation in conservation projects in Calakmul, Mexico. *Gend. Place Cult.* **2012**, *19*, 61–82. [CrossRef]
197. Schmook, B.; Radel, C. International labor migration from a tropical development frontier: Globalizing households and an incipient forest transition. *Hum. Ecol.* **2008**, *36*, 891–908. [CrossRef]
198. Kröhnert, S.; Klingholz, R. *Not am Mann: Von Helden der Arbeit zur Neuen Unterschicht?*; Berlin-Institut für Bevölkerung und Entwicklung: Berlin, Germany, 2007.
199. Bilsborrow, R.E.; Ogendo, H.W. Population-driven changes in land use in developing countries. *AMBIO* **1992**, *21*, 37–45.
200. Barrios, E. Soil biota, ecosystem services and land productivity. *Ecol. Econ.* **2007**, *64*, 269–285. [CrossRef]
201. Culman, S.W.; Young-Mathews, A.; Hollander, A.D.; Ferris, H.; Sánchez-Moreno, S.; O'Geen, A.T.; Jackson, L.E. Biodiversity is associated with indicators of soil ecosystem functions over a landscape gradient of agricultural intensification. *Landsc. Ecol.* **2010**, *25*, 1333–1348. [CrossRef]
202. Millennium Ecosystem Assessment. *Ecosystems and Human Well-Being: Biodiversity Synthesis*; World Resources Institute: Washington, DC, USA, 2005.
203. Moreira, F.M.S.; Siqueira, J.O.; Brussaard, L. Soil organisms in tropical ecosystems: A key role for Brazil in the global quest for the conservation and sustainable use of biodiversity. In *Soil Biodiversity in Amazonian and Other Brazilian Ecosystems*; Moreira, F.M.S., Siqueira, J.O., Brussaard, L., Eds.; CABI: Wallingford, UK, 2006; pp. 1–12.
204. Reynolds, H.L.; Packer, A.; Bever, J.D.; Clay, K. Grassroots ecology: Plant-microbe-soil interactions as drivers of plant community structure and dynamics. *Ecology* **2003**, *84*, 2281–2291. [CrossRef]
205. Kowalchuk, G.A.; Buma, D.S.; de Boer, W.; Klinkhamer, P.G.; van Veen, J.A. Effects of aboveground plant species composition and diversity on the diversity of soil-borne microorganisms. *Antonie Van Leeuwenhoek* **2002**, *81*, 509–520. [CrossRef] [PubMed]
206. Berthrong, S.T.; Buckley, D.H.; Drinkwater, L.E. Agricultural management and labile carbon additions affect soil microbial community structure and interact with carbon and nitrogen cycling. *Microb. Ecol.* **2013**, *66*, 158–170. [CrossRef] [PubMed]
207. Giller, K.E.; Witter, E.; Mcgrath, S.P. Toxicity of heavy metals to microorganisms and microbial processes in agricultural soils: A review. *Soil Biol. Biochem.* **1998**, *30*, 1389–1414. [CrossRef]

208. Lavelle, P.; Bignell, D.E.; Austen, M.C.; Brown, V.K.; Behan-Pelletier, V.; Garey, J.R.; Giller, P.S.; Hawkins, S.J.; Brown, G.G.; St. John, M.; *et al.* Connecting soil and sediment biodiversity: The role of scale and implications for management. In *Sustaining Biodiversity and Ecosystem Services in Soils and Sediments*; Wall, D.H., Ed.; Island Press: Washington, DC, USA, 2004; pp. 193–224.
209. Liu, A.; Hamel, C.; Begna, S.H.; Ma, B.L.; Smith, D.L. Soil phosphorus depletion capacity of arbuscular mycorrhizae formed by maize hybrids. *Can. J. Soil Sci.* **2003**, *83*, 337–342. [CrossRef]
210. Martinez, T.N.; Johnson, N.C. Agricultural management influences propagule densities and functioning of arbuscular mycorrhizas in low- and high-input agroecosystems in arid environments. *Appl. Soil Ecol.* **2010**, *46*, 300–306. [CrossRef]
211. Evans, L.T. *Crop Evolution, Adaptation and Yield*; Cambridge University Press: Cambridge, UK, 1996.
212. Kates, R.W.; Parris, T.M. Long-term trends and a sustainability transition. *Proc. Natl. Acad. Sci. USA* **2003**, *100*, 8062–8067. [CrossRef] [PubMed]
213. Lambin, E.F.; Meyfroidt, P. Global land use change, economic globalization, and the looming land scarcity. *Proc. Natl. Acad. Sci. USA* **2011**, *108*, 3465–3472. [CrossRef] [PubMed]
214. Seto, K.C.; Reenberg, A.; Boone, C.G.; Fragkias, M.; Haase, D.; Langanke, T.; Simon, D. Urban land teleconnections and sustainability. *Proc. Natl. Acad. Sci. USA* **2012**, *109*, 7687–7692. [CrossRef] [PubMed]
215. Greiner, C.; Sakdapolrak, P. Rural–urban migration, agrarian change, and the environment in Kenya: A critical review of the literature. *Popul. Environ.* **2013**, *34*, 524–553. [CrossRef]
216. Neumann, K.; Hilderink, H. Opportunities and challenges for investigating the environment-migration nexus. *Hum. Ecol.* **2015**, *43*, 309–322. [CrossRef] [PubMed]
217. Aguilar, J.; Gramig, G.G.; Hendrickson, J.R.; Archer, D.W.; Forcella, F.; Liebig, M.A. Crop species diversity changes in the United States: 1978–2012. *PLoS ONE* **2015**, *10*, e0136580. [CrossRef] [PubMed]
218. Tsing, A.L. *Friction: An Ethnography of Global Connection*; Princeton University Press: Princeton, NJ, USA, 2005.
219. Sundberg, J. Decolonizing posthumanist geographies. *Cult. Geogr.* **2014**, *21*, 33–47. [CrossRef]
220. De Janvry, A. *The Agrarian Question and Reformism in Latin America*; Johns Hopkins University Press: Baltimore, MD, USA, 1981.

land

MDPI

Article

Poverty and Environmental Degradation in Southern Burkina Faso: An Assessment Based on Participatory Methods

Daniel Etongo [1,*], Ida Nadia S. Djenontin [2] and Markku Kanninen [1,3]

1 Viikki Tropical Resources Institute, Department of Forest Sciences, University of Helsinki, Latokartanonkaari 7, P.O. Box 27, Helsinki 00014, Finland; markku.kanninen@helsinki.fi or m.kanninen@cgiar.org
2 West Africa Regional Office (WARO), Center for International Forestry Research (CIFOR), 06 P.O. Box 9478, Ouagadougou 06, Burkina Faso; I.Djenontin@cgiar.org
3 Center for International Forestry Research (CIFOR), Bogor 16115, Indonesia
* Correspondence: daniel.etongobau@helsinki.fi; Tel.: +358-449-151-781; Fax: +358-919-158-100

Academic Editors: Claudia A. Radel and Jacqueline M. Vadjunec
Received: 31 July 2015; Accepted: 4 June 2016; Published: 24 June 2016

Abstract: The poverty and environmental degradation vicious circle hypothesis considers the poor as agents and victims of environmentally degrading activities. Despite some studies, however, there still has not been a sufficient empirical examination of the poverty-environment nexus. Based on participatory poverty assessment (PPA) methods with two hundred farm households categorized by wealth status in southern Burkina Faso, six indicators of environmental degradation and a set of land management practices were examined to answer the following questions: (i) Which households (non-poor, fairly-poor, or poorest) are responsible for environmental degradation? (ii) Does poverty constrain adoption of land management practices considered to improve the land? Results indicate deforestation is highest for non-poor farmers, and non-poor and fairly-poor farmers have higher rates of overgrazing. In addition, the entire non-poor group, mainly recent migrants to the area, occupy borrowed lands with tenure perceived as insecure, considered by farmers to be a disincentive for assisted natural regeneration of vegetation. Thus, non-poor and fairly-poor farmers participate most in activities locally identified as environmentally degrading, and the former contribute more than the latter. On the other hand, adoption of land management practices considered to improve the land is relatively low amongst the poorest farmers.

Keywords: poverty; tenure security; deforestation; land management practices; Burkina Faso

1. Introduction

Until the last decade, the concept of a vicious circle, fostered by the World Bank [1,2] and related international agencies [3], dominated the debate on poverty–environment linkages. This concept first appeared in the Brundtland Report launched by the World Commission on Environment and Development (WCED) [4]. According to this report, "Many parts of the world are caught in a vicious downward spiral: poor people are forced to overuse environmental resources to survive on a daily basis, and their impoverishment of their environment further impoverishes them, making their survival more uncertain and difficult" ([4]; p. 27). Contrary to the assumptions of WCED, later studies [5–7] show that poverty–environment linkages do not conform to a simple downward-spiraling two-way relationship. According to these findings, existing social networks and institutions that govern access to natural resources mediate the relationships between poverty and environmental degradation and improvement. In addition, the popular portrayal of the vicious circle has treated poverty in uniform and generic ways and thereby caused the broader context behind this poverty to be under-represented. Given that the natural resource base constitutes an important source of income in rural developing countries, especially for the poorer groups in society, a better understanding of the poverty–environment relationship can lead to better policies that support the natural resource assets of the poor.

Poverty itself is contextualized differently across academic disciplines and regions of the world [8]. It is now generally accepted that human poverty has many dimensions, and it is not just poverty of income or not having things necessary for material well-being. Human poverty also means the deprivation that people suffer throughout their lives which differs among nations. Furthermore, differences occur within developing countries when rural areas are compared to their urban counterparts. The result has been the emergence of different ways of approaching poverty–environmental linkages. On one hand, the state of the environment is increasingly a focus of development practitioners when determining the magnitude of poverty. On the other hand, a major concern for environmentalists is the role of poverty in resource degradation [9]. While the former focuses on the poor state of the environment as a factor in pushing rural households that are dependent on natural resources into poverty, the latter concerns itself with the poor as agents of environmental degradation [10]. Can the potential reinforcing interactions between environmental degradation and poverty provide explanations as to why this relationship is far from being linear? Additional empirical research is needed to further specify the relationship.

In addition, the poverty–environment vicious circle hypothesis suggests that economic growth is needed to break the poverty-environment downward spiral. Economic growth policies are expected to reduce poverty while providing incentives for investment in the land [10]. However, a simple generalization of the relationship between poverty and environmental degradation is misleading [9]. This is because economic growth is not always inclusive of the poor and policy instruments often fail to address inequality. On the other hand, previous studies have limited environmental degradation to soil erosion and deforestation at the expense of attention being paid to other environmentally degrading activities such as overgrazing, use of pesticides, etc. [6]. Understanding the environmental entitlements or resource rights of farmers provides valuable information for understanding land-use decision making [11,12].

In Burkina Faso, approximately 70 percent of the population is rural and depends on farming and livestock for their livelihoods [13]. The rural population depends on the natural resource base and adopts different resource management strategies that can either improve or degrade the environment. The study reported here examines the role of different rural wealth groups in contributing to environmental degradation in four rural communities in southern Burkina Faso. Given that many farmers in the Sahel manage complex crop and livestock portfolios [8], expanding the focus to include other environmentally degrading activities aside from soil erosion and deforestation is invaluable for further understanding of the poverty–environment nexus in the Burkina Faso context. Previous Burkina Faso studies identify the cutting and selling of fuel wood [13,14], cotton cultivation [15], and the conversion of forests to croplands [16] as the main drivers of environmental degradation. However, studies on the relationship between individual wealth status and environmentally degrading activities are lacking.

For the purposes of this study, environmental degradation is defined as the deterioration of the environment through depletion of resources such as air, water and soil, leading to the destruction of ecosystems and the extinction of wildlife [17]. This study focuses on the depletion and degradation of land and forest resources as examples of environmental degradation. In addition to the above cited activities, overgrazing resulting from livestock stocking densities that exceed the available fodder supply is also considered to be a cause of degradation of land and forest resources [18]. Agro-pastoral systems dominate in the Sahel, and livestock numbers have increased significantly during the last 30 years in southwestern Burkina Faso [19]. Such increases inevitably lead to increases in rangeland and fodder demand that is likely to affect the management of natural resources and especially forests. Understanding the livelihood activities of farming households characterized by different economic means will provide insight into the natural resource management problems implicit in poverty–environment linkages. These problems are considered to be largely related to agriculture [20] and may be responsible for the depletion of both individual and common-pool resources.

Southern Burkina Faso offers greater opportunities for rain-fed agriculture, fuel wood supply, forest and tree products, fodder supply, etc., compared to the country's central and northern regions, which suffer from periodic drought. The favorable climate and soil conditions attract migrants and agribusiness investors in search of arable land. Increased demand for land has caused land scarcity [21] and resulted in higher competition among different land uses [22], which not only threatens environmental sustainability [16] but also causes deforestation [23,24]. Consequently, previous studies in Burkina Faso [16,23,24] focus on population–environment rather than poverty–environment interactions. Although these studies are important for assessing land degradation, they address only some of the potential causes. This study tests the poverty–environment relationship through the following questions: (i) When households are categorized based on poverty and wealth, which groups are more responsible for environmental degradation? (ii) Does poverty constrain the adoption of land management practices that are considered to improve the land? Understanding the relationship between poverty, wealth, and natural resource management activities leading to environmental degradation is important for prescribing policy measures to mitigate these problems.

2. Materials and Methods

2.1. Description of Study Area

This paper is based on field research conducted in four adjacent community forest villages: Cassou, Vrassan, Dao, and Kou, all in the Ziro province, southern Burkina Faso (Figure 1). These villages were chosen under the framework of the Building Biocarbon and Rural Development (BIODEV) project in West Africa. This project was financed by Finland's Ministry of Foreign Affairs as an initiative to achieve developmental benefits by building biological and natural carbon resources through improved agroforestry and forest management practices. This area lies within the South-Sudanian climate zone, with annual precipitation of 800–1000 mm. Rains consisting of short intense storms fall over a single wet season lasting for approximately four months from June to September [25]. During the hot season, the average daily temperature stands at 30 °C, with peaks of 40 °C as a result of hot dry air that blows from the Sahara Desert. The area is characterized by low relief and homogenous soil types including silt-clay cambisols, sandy lixisols, and loamy ferric luvisols [26].

The average population density in the Ziro province was estimated at 28 persons/km^2 in 2006 [27], but this figure is increasing due to rural–rural in-migration [16]. The population consists of three main ethnic groups: Gourounsi (indigenous), Mossi (originating from the central plateau) and Fulani (originating from the north of Burkina Faso).

The farming system is dominated by crops grown under a discontinuous cover of scattered trees that constitutes the so-called parklands. Parklands are considered to be agroforestry systems, but their biodiversity depends on the original vegetation cover, the number and type of trees and shrubs spared during conversion to farmland, the needs of farmers, etc. [20]. Subsistence production includes the cultivation of cereals (such as sorghum, sesame, maize and millet) and tubers (yam and sweet potatoes) and animal husbandry. In addition to the above, a more complex and lucrative production system exists that involves the extraction of fuel wood and non-timber forest products, the cultivation of cash crops (cotton and fruit-tree plantations) and ranching [20]. The natural flora is dominated by perennial grasses such as *Andropogon gayanus* Kunth, *A. ascinodis* C.B. Clarke, and *Schizachyrium sanguineum* (Retz.) Alston [28]. Tree species commonly found in the parklands include *Vitellaria paradoxa* C. F. Gaertn, *Parkia biglobosa* (Jacq.) R. Br. ex G. Don., and *Tamarindus indica* L., amongst others. Forests in the study villages are under one of two management regimes: protected and classified forest. The classified forests, or national parks, (25 percent) are strictly protected from livestock and farming activities, while the protected forests (*chantiers d'aménagements forestiers—CAF* and *forêts villageoises*) are subject to field expansion and managed by local communities in collaboration with the government [14].

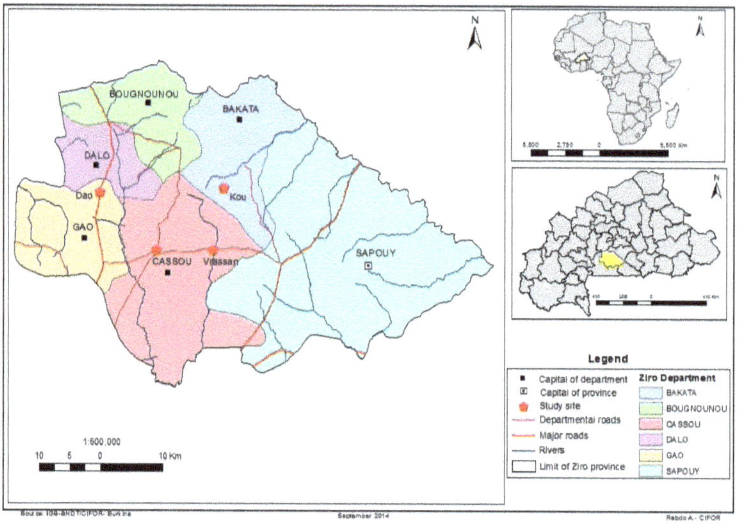

Figure 1. Study area.

2.2. *Sampling and Data Collection*

The first step of field data collection consisted of constructing a participatory poverty assessment (PPA) based on local indicators developed during a focus group discussion (FGD) in each of the four villages. Although approximately 70 percent of the population in Burkina Faso is rural and supported by the informal economy, national poverty assessment is based on income criteria [21]. Use of the money metric criterion [29,30] in such rural communities as in Burkina Faso is misleading and is not the most adequate and applicable means for assessing poverty. Such a criterion is more applicable in urban areas where it is possible to assess income and expenditures using money as the unit of assessment. This is not the case in rural areas where banks are not available for money to be deposited, and which are dominated by an informal economy where wealth is stored in the form of assets. In rural areas, the informal economy sustains the livelihoods of households through natural resource and land-based economic activities such as farming, logging, trade, etc. This rural informal sector is highly complex and often rooted in traditional resources and land rights [31]. Therefore, the livelihood approach for assessing a household's wealth in relation to its asset holdings has been widely applied in rural developing countries [32]. In this approach, the real assets of a household are the unit of measurement, not money.

A total of five participants, including two women and three men from each of the three ethnic groups, participated in each FGD. The aim of the FGDs was to develop a poverty profile for the study area through a participatory exercise and to identify indicators of environmental degradation. The participants of the FGDs had to satisfy two conditions. Participants had to have lived in the community in or before 2003, which coincides with the start of the period used by the study for assessing deforestation. This time period is considered to be sufficiently long enough for the participants to know the level of well-being of other households. Participants were selected to represent a cross-section of the community in characteristics such as gender, ethnicity, wealth status, and neighborhood. Based on these criteria, participants were selected from among all the 'sub-chiefs' and women leaders of farmer management groups. This is because in Cassou, for example, the Mossi and Fulani ethnic groups live in separate quarters from the indigenous ethnic group (Gourounsi).

Each of the ethnic groups has a leader, or sub-chief, who knows all the families in the village within that group. The women leaders were additionally selected because all the sub-chiefs were men.

Following Narayan et al. [33], the participants of each FGD were guided during the discussion to: (i) list local indicators used to assess wealth status in the community; and (ii) describe the specifications of each local indicator and its corresponding wealth category (see Table 1 for indicators). The resulting indicators differed slightly across the villages; therefore, the 20 participants of the four FGDs participated in a final meeting in Cassou to agree upon a common list (see the PPA below for more details). Participants also identified the following as environmentally degrading activities or indications of environmental degradation: field expansion (leading to deforestation), cotton cultivation, fuel wood exploitation, overgrazing, soil fertility loss, and local perceptions of tenure insecurity. Local participants identified tenure security as a factor playing a role in the adoption land management practices such as assisted natural regeneration. Tenure insecurity is likely to act as a disincentive for land management practices with long time horizons [25] compared to the use of practices such as composting, which are more immediately effective [34].

These outputs from the FGDs were reinforced with a literature review to locate corroborating scholarship on the activities identified by the FGDs as environmentally degrading in the region. These activities were found to be consistent with those identified in previous studies in Africa and Latin America [6,15,35,36]. A detailed questionnaire was then designed to collect both quantitative and qualitative data related to the environmentally degrading activities and which also addressed a specific set of land management practices (use of fallows, planting pits, composting, stone bunds, and live hedges) as well as household and farm characteristics (to be used for wealth categorization).

With the assistance of local youth leaders, 200 households from the study villages were randomly selected from a list of all households considered to fall within the different wealth status groups. Furthermore, 10 farms from each proposed wealth group were randomly selected for an additional farm survey, for a total of 30 farms. During the farm surveys, field sizes were estimated alongside other specific features of the farming systems such as fallows, evidence of assisted natural regeneration, live hedges, planting pits, stone bunds, use of compost, etc. The purpose of the farm survey was to confirm the interview questionnaire data.

2.3. Analytical Methods: Categorizing Households Based on Local Indicators Derived from Participatory Poverty Assessment (PPA)

As described above, participatory research methods were applied in which community members defined the wealth criteria based on local indicators [6]. Twelve wealth status indicators and their descriptions (Table 1) were identified during the FGDs. As a part of this process, participants selected a schema of wealth groups, with households categorized based on these indicators. A different set of wealth groups was identified in each study village, and three wealth groups were adopted for a common list. In Cassou, participants identified the groups: rich, fairly rich, poor, and poorest; in Dao and Kou: rich, fairly rich, and poor; and in Vrassan: rich and poor. To correct for these differences in income groups across locations, Ravnborg et al. [37] use a mean value, which was adopted in our study as follows: $(4 + 3 + 3 + 2)/4 = 3$. Thus, the three income groups of non-poor, fairly poor and poorest were numbered 1, 2 and 3, respectively.

The next stage in this method was to transform the numbers to scores representing poverty levels. The qualitative rankings were quantified using the following equation adapted from Ravnborg et al. [37] as follows:

$$S = (A - 1)/(P - 1) \times 100 \quad (1)$$

where S = Well-being score; A = Income group of the household based on local perception of well-being indicators (Table 1); and P = the total number of wealth groups. The result was multiplied by 100 to avoid operating with decimals, resulting in:

Level 1—$[(1 - 1)/(3 - 1)] \times 100 = 0$, where 0 implies **non-poor** household

Level 2—[(2 − 1)/(3 − 1)] × 100 = 50, where 50 implies **fairly poor** household
Level 3—[(3 − 1)/(3 − 1)] × 100 = 100, where 100 implies **poorest** household

During the interviews, data were collected from each household based on the local indicators agreed upon in the common list (Table 1). In addition, data were collected on household resource management strategies self-reported by farmers, based on their perceptions, to capture activities considered environmentally degrading. Each household was assigned a corresponding score for all 12 indicators, which was later averaged by the first author to classify the household into its corresponding wealth group. The threshold values were then calculated to define the range of each wealth group based on the MEAN of all 200 household as follows: 0 to 50 (non-poor), 50 to 75 (fairly poor) and 75 to 100 for the poorest. Within these threshold values, it was possible to assign each individual mean to a category.

Table 1. Household poverty indicators and scoring system in southern Burkina Faso.

Indicator	Score	Description
Access to land	0	Owns more than 10 ha of land
	50	Owns between 4 and 10 ha of land
	100	Owns less than 4 ha of land
Food security	0	Household without a period of food shortage in the last 3 years
	50	Experienced a food shortage in the last 3 years that lasted <3 months
	100	Experienced a food shortage in the last 3 years that lasted >3 months
Healthcare	0	Capable of paying for the services of a doctor in the district hospital and beyond
	50	Capable of paying for doctors' services limited to the district hospital
	100	Household is unable to pay for a doctor's service and relies on herbal medicine
Nonagricultural sources of income	0	Receives income from the sale of livestock, household shops, owns a truck for transportation
	50	Uses cart to transport crops for income, sells food and non-timber forest products
	100	Household does not have any other source of nonagricultural income
Sale of crops	0	Sells more than half of cereals produced while satisfying household needs
	50	Selling up to half of cereal produced will lead to a food shortage
	100	Does not sell cereals and is not self-sufficient, depends heavily on non-timber forest products (NTFPs)
Agricultural equipment	0	Cultivates the land with tractor and draught ox, owns compost production facilities
	50	Cultivates with donkeys and is capable of buying compost to use on farm
	100	Cultivates the land with hand hoes and cutlasses
Tree resources	0	Owns tree plantations (fruit trees, poles for construction, etc.)
	50	Has a few trees on farm and around compound for subsistence and commercial use
	100	Does not own trees on farm and compound but depends on the forest for NTFPs
Livestock ownership	0	Owns three or more herds of cattle (a herd is 10 cows)
	50	Owns less than three herds of cattle
	100	Does not own any cattle
Ownership of other animals	0	Owns three droves of donkeys, goats, and sheep (a drove is 10 animals)
	50	Owns less than three droves of donkeys, goats, and sheep
	100	Does not own any donkeys, goats, or sheep
Household gadgets	0	Owns TV/solar panel, radio/radio-cassette player and Yamaha generator
	50	Owns radio/radio-cassette player, uses motor battery to generate electricity
	100	Does not own electrical appliances but uses kerosene lamp
Transportation	0	Owns $\geqslant$ 1 car and $\geqslant$ 1 motorcycle
	50	Owns a motorcycle and a cart
	100	Owns a bicycle and other members of the household often go on foot
Institutional credit	0	Has the required collateral security for credit and is capable of paying back
	50	Limited collateral for credit and might be unable to pay if externalities arises
	100	Lacks collateral for credit and also lacks the potential for repayment

Note: NTFPs are goods obtained from the forests without harvesting whole trees.

2.4. *Study Variables Indicating Environmentally Degrading Activities*

The collected activities-related data were used to create variables of two kinds: numeric measurement-based variables and categorical variables (often self-reported). The former variables include mean annual deforestation (2003–2013), cotton cultivation, cutting and selling of fuel wood, and overgrazing based on cattle numbers. Categorical data were gathered through farmers' self-reported assessments on the following: overgrazing, soil fertility loss, and tenure security as an incentive for assisting natural regeneration. Overgrazing occurs in both categories because both numerical data and self-reported assessments of respondents were recorded. Several variables based on numeric measurement-based data were also converted to additional categorical variables for analyses (see below).

2.4.1. Deforestation

Through the interviews, data were collected on changes in farm area between 2003 and 2013. Farm size included areas under shifting cultivation with fallows. A recent study in Burkina Faso finds that farmers without fallows are more likely to expand their fields, thereby causing deforestation [38]. The lack of strict monitoring of community forest areas exposes them to encroachment. As such, field expansion into protected forest areas has been identified as the dominant proximate driver of deforestation in Burkina Faso [39,40]. In this study, the cultivation of fallows was not considered to represent deforestation because they represent potential areas for reuse when the rotation cycle is completed or as need arises. Therefore, deforestation in the current study focuses on the expansion of fields into protected forest.

$$\text{Mean Annual Deforestation} = (\text{Farm Area in 2013 ha} - \text{Farm Area in 2003 ha})/10 \qquad (2)$$

The difference in farm area was calculated as the change in farm area, excluding fallow, between 2013 and 2003. The result was divided by 10 to get the mean annual change value for the 10-year-period. A categorical variable was created in addition to the continuous variable. Households that did not experience a change in farm area were assigned the value 0, while those that cleared forests were assigned the value 1. Jones et al. [41], in an assessment of deforestation driven by farming systems, apply a mean annual area of forest cleared using this methodology. For additional details, see Etongo et al. [38]. Land is not for sale due to customary rules that prevail in these communities, thereby reducing the options for farm expansion.

2.4.2. Cotton Cultivation

Aside from the role of cotton as a driver of deforestation in Burkina Faso, where a threefold increase in cultivated area (ha) occurred between 1992 and 2007 [15], the use of pesticides also constitutes a threat to the environment [42]. An increase in the annual rate of pesticide consumption over the last two decades is attributed to the treatment of cotton fields [43]. The term pesticide covers a wide range of compounds including insecticides, fungicides, herbicides, etc. A pilot study in Burkina Faso on agricultural pesticide poisoning indicates its effect on the environment, livestock and human health [44]. This study employs both a continuous variable of cotton produced (in kg) and a categorical variable for presence or absence of cotton cultivation at the farm level.

2.4.3. Cutting and Selling of Fuel Wood

Fuel wood is a major source of household energy in Burkina Faso and is collected from fields, fallows, forests, and plantations. The traditional measuring unit is a cart-driven system called a *charet*. A *charet* full of fuel wood is estimated at three m^3. The estimated monthly fuel wood consumption per household in 2013 (for both subsistence use and sale) was recorded in *charets* and converted to cubic meters. A recent study in Burkina Faso estimates the average rural household's daily fuelwood consumption to be 0.04 m^3 [45], which is roughly the same as estimated in Kenya [46]. Based on this daily estimate, monthly and annual consumption should stand at 1.24 m^3 and 14.88 m^3, respectively. These estimates provide a guide on household fuel consumption. In addition, information was collected on quantity of fuel wood sales and places of collection. This study employs both a continuous variable of reported monthly fuel wood consumption and a derived categorical variable which assumes fuel wood sales when reported monthly consumption was greater than 1.24 m^3.

2.4.4. Overgrazing

Average cattle herd sizes were collected for the last five years (2009–2013) to assess overgrazing in rangelands and farmlands. Studies on total livestock units (TLUs) in relation to stocking density consider overgrazing to occur when demand for fodder exceeds supply [18,47]. Niemeijer and Mazzucato [47] study total livestock units (TLUs) and arrive at a similar conclusion that overgrazing occurs when stocking density increases without an increase in fodder. Studies in the West African Sahel [19] and eastern Mediterranean [18] find that increases in livestock numbers are not consistent with fodder supply, thereby causing overgrazing. Furthermore, another study in Peru reports losses in range species as a consequence of high stocking rates, drought, and the absence of fallow areas [7]. In this study, the number of cattle was used as a continuous variable indicating overgrazing. In addition, a categorical variable was created using the mean value of 9.05 calculated among the 200 respondents, with two categories formulated as follows: households with $\leqslant$9 cattle and households with $\geqslant$10 cattle. Thus, owning more than the mean number of cattle was defined as indicative of overgrazing. To reinforce the above analysis, farmers also self-reported practices of overgrazing on their own farms in terms of environmental degradation as either low, moderate, and high.

2.4.5. Farmers' Assessment of Tenure Insecurity on FMNR

The role of indigenous tree species in remediating land degradation through farmer-managed natural regeneration (FMNR) is widely acknowledged in the Sahel [48,49]. FMNR fosters tree ownership and land tenure security for farmers. Also known as assisted natural regeneration, FMNR is the protection of indigenous tree species in the Sahel so that they can regenerate naturally to maturity [49]. It provides environmental benefits such as restoration of tree cover, increased biodiversity, climate change adaptation and mitigation, etc. [50]. In the interviews, farmers were asked if they participate in assisted natural regeneration. However, due to the sensitive nature of this line of questioning (as indigenous species are protected by law), this question was asked in the context of other questions to determine farmers' self-reported assessments of the state of natural regeneration of indigenous trees on their farms, who owns the trees, and how tenure security influences this practice. For this study, farmers' assessments of the role of tenure security for FMNR was used because these data were assessed to be the most accurate proxies for determining FMNR practice due to the sensitivity of the issue. Additionally, two studies in the Sahel indicated the importance of FMNR as a land management practice that improve tenure security and also tree cover on farms, thereby reducing environmental degradation [48,49].

2.4.6. Farmers' Assessment of Soil Fertility Loss

A study in Rwanda finds that local perceptions of soil fertility align well with soil fertility measured using local indicators such as crop yield, soil softness, indicator plants, and soil color [51]. Karltun et al. [52] also find consensus between farmers' indigenous knowledge and scientifically validated indicators of soil fertility. Using locally reported soil fertility thus is a viable alternative to direct or indirect measurement. Farmers' perceptions of soil fertility loss were adopted for this study and assessed at three levels: high, moderate and low. Three levels of soil fertility loss were adopted in the current study because this approach was found in Ethiopia to be effective when using perception-based methods [36].

2.5. Statistical Analysis

In order to compare the indicators of environmental degradation across the households in different wealth categories, we use descriptive statistics and non-parametric tests to analyze the data collected. Specifically, we applied Chi square to check the relationships between categorical variables. We use one-way ANOVA to compare the variation within the continuous variables across a single factor and determine whether significant differences existed among their means [53].

To run the ANOVA, Levene's test was used as a prior test for homogeneity of variance within the continuous variables. When a significant difference was found in the one-way ANOVA test at the 5 percent level, the Scheffe pairwise multiple comparison test was conducted for cases exhibiting homogenous variance [54]. For variables that exhibited non-homogenous variance, a Welch test was performed to correct the violation; then, Dunnett's C pairwise multiple comparison test was applied [55]. The multiple comparison tests highlighted where mean differences varied between pairs of wealth categories for the selected indicators.

3. Results and Discussion

3.1. Poverty Level and Households' Characteristics

Based on the results of the participatory poverty assessment (PPA) indicators, developed on the basis of local perceptions of poverty, 51.5 percent of the respondents were fairly poor, 32.5 percent were the poorest and 16 percent were non-poor (Figure 2).

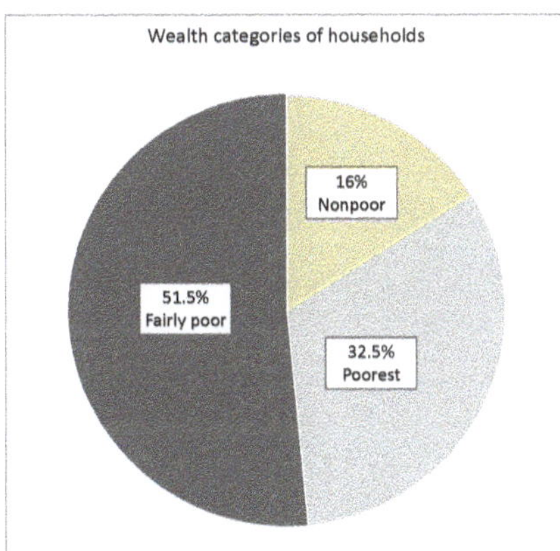

Figure 2. Percentage of farmers in each wealth category in the study area, across poverty levels (N= 200).

The two hundred farm households interviewed were distributed among the Gourounsi, Mossi and Fulani ethnic groups as 46 percent, 41.5 percent and 12.5 percent, respectively (Table 2). A majority of the non-poor farmers belonged to the Fulani ethnic group followed by the Mossi. None of the respondents from the indigenous group (Gourounsi) were considered non-poor.

The dominant farming system in Burkina Faso is the crop-livestock system. Livestock constitutes an invaluable asset for household wealth. Non-poor farmers owned more cattle (partly by definition, as number of cattle was a factor for categorizing households into wealth groups) and some farmers had over 20 head of cattle. Although cattle herding was done by all ethnic groups, it remains the main agricultural activity of the Fulani group that originates from the north. In the FGDs, participants indicated that poor households that owned few cattle are more likely to sell livestock during periods of poor harvest, or for anticipated expenditures such as school fees for children, etc. This finding differs slightly from another study in which smallholders and poor farmers were more likely to sell land for immediate benefits [18].

Table 2. Cross-tabulation results for categorical variables (N=200).

Households Characteristics		Wealth Categories of Households (%)			Total (%)	Chi-Square Tests	
		Non-Poor	Fairly Poor	Poorest		Value	Asymp. Sig. (2-Sided)
Ethnic group	Gourounsi	0.0	48.5	64.6	46.0	108.147	0.000
	Mossi	34.4	47.6	35.4	41.5		
	Fulani	65.6	3.9	0.0	12.5		
Level of education	No education	96.9	78.6	87.7	84.5	6.947	0.031
	Education	3.1	21.4	12.3	15.5		
Perceived tenure security	Insecure	100.0	51.5	35.4	54.0	36.595	0.000
	Secure	0.0	48.5	64.6	46.0		
Number of cattle	⩽9	0.0	85.4	100.0	76.5	128.713	0.000
	⩾10	100.0	14.6	0.0	23.5		

The level of education of the household head is likely to influence their livelihood. Across the wealth categories, 96.9 percent (non-poor), 78.6 percent (fairly poor) and 87.7 percent (poorest) could not read or write and were considered to be illiterate. The respondents who could read and write (were literate) included 3.1 percent (non-poor), 21.4 percent (fairly poor) and 12.3 percent (poorest). The overall level of illiteracy was 84.5 percent, and only 15.5 percent of the respondents were literate, with some education (Table 2).

3.2. *Indicators of Environmental Degradation: Differences among Wealth Categories*

This section presents the descriptive statistics as well as the mean comparisons (one-way ANOVA) among the wealth categories for the four numerical indicators used to quantitatively identify environmental degradation. The results indicate that all of the four numeric indicators of environmental degradation showed a statistical difference at the 5 percent significance level among all wealth categories (Table 3). This implies that each wealth category exhibited a different mean for each indicator of environmental degradation considered in the analysis.

Table 3. Continuous variables indicating environmentally degrading activities: Descriptive statistics and mean comparison (one-way ANOVA) (N=200).

Variables		Descriptive statistics			One-way ANOVA	
		N	Mean	Std. Deviation	F	Sig.
Mean annual deforestation (ha)	Non-poor	32	0.353	0.379	4.50	0.012
	Fairly poor	103	0.274	0.293		
	Poorest	65	0.175	0.228		
	Total	200	0.255	0.295		
Quantity of cotton produced (kg)	Non-poor	32	90.313	210.230	3.70	0.027
	Fairly poor	103	124.563	205.229		
	Poorest	65	47.692	102.313		
	Total	200	94.100	181.731		
Monthly fuel wood consumption (m^3)	Non-poor	32	2.281	1.039	20.60	0.000
	Fairly poor	103	1.849	0.776		
	Poorest	65	1.315	0.429		
	Total	200	1.745	0.805		
Overgrazing expressed by the number of cattle	Non-poor	32	41.500	34.586	104.90	0.000
	Fairly poor	103	4.282	3.197		
	Poorest	65	0.6462	0.694		
	Total	200	9.055	19.897		

In addition, categorical variable forms of these indicators were analyzed for association with each wealth category (Table 4). Other indicators expressed as categorical variables also were considered, including farmers' self-reported assessments of overgrazing, loss of soil fertility and tenure security (as an incentive for long-term investment in the land via FMNR for instance). Thus, the categorical format was used for all of the variables to measure their association to the wealth status. The indicator of overgrazing was assessed twice: first as a categorical variable stemming from prior classification (owning more than the mean number of cattle), and then according to farmers' self-reported assessments.

Table 4. Activities perceived as environmentally degrading and their relation to wealth categories (N=200).

Indicators of Environmental Degradation		Wealth Categories of Households (%)			Total (%)	Chi-Square Tests	
		Non-poor	Fairly poor	Poorest		Value	Asymp. Sig. (2-sided)
Derived categorical forms from our quantitative collected variables							
Deforestation 2003-2013	No	1.00	19.50	16.50	37.00	18.30	0.000
	Yes	15.00	32.00	16.00	63.00		
Cotton cultivation	No	13.00	30.50	25.50	69.00	9.57	0.008
	Yes	3.00	21.00	7.00	31.00		
Cutting and selling fuel wood	No	14.50	27.00	19.50	61.00	15.01	0.001
	Yes	1.50	24.50	13.00	39.00		
Overgrazing based on owning more than mean number of cattle	No	0.00	19.00	27.00	46.00	66.67	0.000
	Yes	16.00	32.50	5.50	54.00		
Variables collected in qualitative forms from field work							
Overgrazing (farmers' assessments of overgrazing on their own farms)	Low	0.0%	42.7%	84.6%	49.5%	110.259	0.000
	Moderate	6.3%	32.0%	15.4%	22.5%		
	High	93.8%	25.2%	0.0%	28.0%		
Soil fertility loss (farmers' assessments of soil fertility loss on their own farm)	Low	65.6%	49.5%	27.7%	45.0%	42.191	0.000
	Moderate	28.1%	23.3%	6.2%	18.5%		
	High	6.3%	27.2%	66.2%	36.5%		
Tenure insecurity (farmers' assessments of effect of tenure insecurity on farmer-managed natural regeneration)	No FMNR	100.0%	51.5%	35.4%	54.0%	36.595	0.000
	Yes FMNR	0.0%	48.5%	64.6%	46.0%		

3.2.1. Deforestation

Sixty-three percent of respondents expanded their fields into forests or cleared new farmland between the years 2003 and 2013, while 37 percent did not (Table 4). Furthermore, the non-poor farmers caused more deforestation than the poorest, with the difference in the changes in farm area during the 10-year period significant at the 5 percent level (Table A1). These findings are contrary to the poverty-induced environmental degradation that is considered to be driven primarily by the poor [1,2]. Some studies show that the poor are not so obsessed with their present and short-term nutritional needs that they completely ignore long-term livelihood security and resource productivity concerns [6,7,9,10]. Instead, poverty constrains farm households from adopting land management practices considered to be sustainable and thereby further intensifies land degradation.

3.2.2. Cotton Cultivation as an Environmentally Degrading Activity

Thirty-one percent of respondents cultivated cotton, while 69 percent did not (Table 4). Cotton cultivation is likely to follow the same pattern for fairly-poor and non-poor farmers, as there is no statistical difference between these two groups. Notwithstanding, significant differences were identified between the fairly-poor and poorest farmers (Table A2). This implies that the fairly poor and non-poor produced relatively higher quantities of cotton as a result of larger farm areas under cultivation and contributed more to environmental degradation than the poorest. Cotton cultivation in Burkina Faso requires animal traction and the use of pesticides and is capital intensive [15], which limits the participation of the poorest farmers. According to Ton [56], the expansion of cotton production in West Africa has been driven by households equipped with animal traction, and those with manual tools have been unable to participate effectively in this livelihood activity. Other studies in Burkina Faso determine that cotton cultivation is a driver of deforestation [14,15,57,58]. Here, the resultant pesticide contamination is additionally considered.

Agricultural poisoning was identified as a common problem in the region and attributed to the use of pesticides for the treatment of cotton. During one of our field visits in Vrassan village in January 2014, approximately eight cattle were found dead and many more affected (see Figure 3). A report from the district veterinary officer indicated that the cattle had been poisoned by eating fodder from a field that had been treated by herbicides used in cotton cultivation. Other studies in Burkina Faso identify similar cases in which cotton treatment resulted in agricultural poisoning that affected not only livestock but also human life [43,44].

Figure 3. Cattle suffering from agricultural poisoning emanating from cotton treatment in Vrassan village.

3.2.3. Cutting and Selling Fuel Wood

Sixty-one percent of households did not cut and sell fuel wood, compared to the 39 percent who engaged in this activity (Table 4). Comparisons among wealth categories indicated that the non-poor cut and sell larger quantities of fuel wood than the fairly-poor and poorest households. Likewise, fairly-poor farmers cut and sell more fuel wood than the poorest (Table A2). Thus, the relative contributions of non-poor and fairly-poor farmers to environmental degradation are higher than that of the poorest. The cutting of fuel wood in Burkina Faso is dominated by the use of rudimentary tools such as axe heads, which require the availability of some household work force, the potential to hire labor and the ability to provide better means of transportation.

Cutting and selling fuel wood in Burkina Faso is quickly degrading the environment and over 85 percent of urban and rural populations depend on wood fuel (fuel wood and charcoal) to supply household energy [13]. The increasing demand for fuel wood has led to illegal exploitation and unsustainable harvesting that does not allow sufficient time for regrowth and thereby affects wood volume. The poorest households cannot effectively benefit from this source of off-farm employment in the region. As such, poverty as a state of deprivation appears to reduce the chances that the poorest in the society engage in certain resource management strategies considered to be environmentally degrading [35]. The view corroborate our findings in which the poorest household equipped with relatively little resources (income to hire labor and pay for transportation) limits the extraction of wood from the forests.

3.2.4. Overgrazing

The assessment of overgrazing on their own farms (Table 4) showed that 84.6 percent of the poorest households perceive themselves as experiencing a low level of overgrazing. This self-perception is not different from our assessment based on the number of cattle owned because we found that the poorest were mostly associated with lower numbers of cattle. This indicator was significantly higher for the non-poor and fairly poor at a 5 percent level of significance (Table A2). Furthermore, 15.4 percent of households in the poorest category reported a moderate level of overgrazing, and none of them perceived themselves as experiencing high rates of overgrazing (Table 4). Finally, the fairly poor and the non-poor see themselves as experiencing high level of overgrazing (25.2 percent and 93.8 percent, respectively). These findings likely reflect the non-poor owning larger herds of cattle, which increases demand for fodder and likely causes higher rates of overgrazing. During the FGDs, the Fulani ethnic group was identified as owning large herds of cattle more often than the Mossi and Gourounsi ethnic groups. Increases in livestock number and a reduction in rangeland through land fragmentation have been identified as a threat to important fodder trees such as *Afzelia africana* and *Pterocarpus erinaceus* [19].

3.2.5. Perceptions of Soil Fertility Loss

A majority of the households reporting soil fertility loss were the poorest farmers, followed by the fairly poor. On the other hand, only 6.3 percent of the non-poor farmers perceived high rates of soil fertility loss (Table 4). Without inputs in the form of fertilizers and the application of land management techniques, soil fertility typically decreases over time. A study conducted in Ethiopia observes declines in soil fertility amongst farm households over time [36]. The lower rates of perceived soil fertility loss amongst the non-poor and fairly-poor farmers may be due to the greater availability of resources for improving the land, for the non-poor and fairly poor in comparison to the poor.

Loss of soil fertility continues to be a major problem that affects agricultural activities in different parts of the world, reducing productivity and in some cases causing crop failure [59,60]. Although considered to be a common problem, farmers' self-reported soil fertility loss is perceived by researchers to affect households differently. This is because knowledge and experience in soil conservation practices varies among different income and social groups within a community [61], a pattern that is consistent in our findings. Furthermore, impacts on farm households can depend on individual elements in the ecosystem that buffer the effects of anthropogenic and natural disturbances [62].

3.2.6. Farmers' Assessment of Tenure Insecurity on Assisted Natural Regeneration

Results show that many non-poor (100 percent), fairly poor (51.5 percent), and poorest (35.4 percent) households perceive their land tenure to be insecure (Table 2). Though these figures are much higher for the fairly-poor farmers (compared to the poorest ones), all the non-poor perceive their own tenure to be insecure (Table 2). Within the context of rural Burkina Faso, rights of limited use are often granted to migrants, while the indigenous groups are granted permanent use rights through inheritance. Land chiefs assign a portion of land to migrants without receiving monetary compensation, and the conditions that govern these land rights are uncertain. Ouédraogo [63] finds that land tenure types in Burkina Faso include: (i) rights of permanent use granted to the indigenous group; (ii) rights of permanent use acquired by claiming unclaimed forested land; (iii) rights of limited use extended to the indigenous group that might become permanent if held for more than one generation; (iv) rights of limited use granted to "strangers" (non-indigenous people) who are considered borrowers of the land. The first three tenure types are associated with indigenous groups, while migrant groups only have limited use rights. Farmers occupying borrowed lands are less likely to participate in FMNR because they lack a full bundle of rights to the land and trees including the right of transfer [63]. However, longer periods of management (approximately 30–40 years) give them some degree of ownership, especially if the land has been improved [39]. Despite these farmers' perceived sense of ownership, the indigenous population considers land occupied by migrants as borrowed. This is because land allocation is not linked to cash transfers, although borrowers do show gratitude through gifts.

Although rights of limited use are granted to migrants, without violation of local rules they can occupy such land as long as desired. One local rule is that they are not allowed to plant economic trees without permission from the land chiefs. Traditionally, loaned lands were not withdrawn unless there was a serious violation of rules of conduct. However, as land reserves are becoming exhausted in the southern region of the country, some cases of withdrawal without violation of local laws have been identified. This finding is consistent with an earlier study in Burkina Faso where land withdrawal occurred without violation of local laws [64]. As a result, perceptions of tenure security may be changing for borrowed lands, as supported by our findings here.

Based on farmers' self-reported assessments, the entire non-poor group reported that their own insecure tenure affected their practice of assisting the natural regeneration of trees on farms, as did 51.5 percent of the fairly poor and 35.4 percent of the poorest (Table 4). Farmers with insecure tenure are likely not to carry out long-term investments such as fallows, assisted natural regeneration of trees, etc. to improve the land. Thus, in-migrants (with insecure tenure due to their status as land borrowers) are hesitant to participate in long-term land management strategies.

However, FMNR is widely promoted in the Sahel because it has proven to foster tree ownership and land tenure security for farmers. Furthermore, it provides numerous environmental benefits such as restoration of tree cover, increased biodiversity that increases resilience to climate change, etc. Studies in Burkina Faso [65] and the Sahel [48,49] acknowledge the importance of FMNR in addressing desertification and land degradation while providing economic benefits to farmers. Farmers engaged in assisting natural regeneration contribute less to environmental degradation than they would in the absence of this activity. Thus, tenure security is an incentive for FMNR, a finding consistent with a previous meta-analysis conducted in tropical developing countries [66].

3.3. Land Management Practices Adopted by Respondents

Five common land management practices in the study area were investigated to assess whether poverty constrains the adoption of practices considered to protect the environment. These practices included fallowing, construction of planting pits, composting, use of stone bunds and use of live hedges (Figure 4).

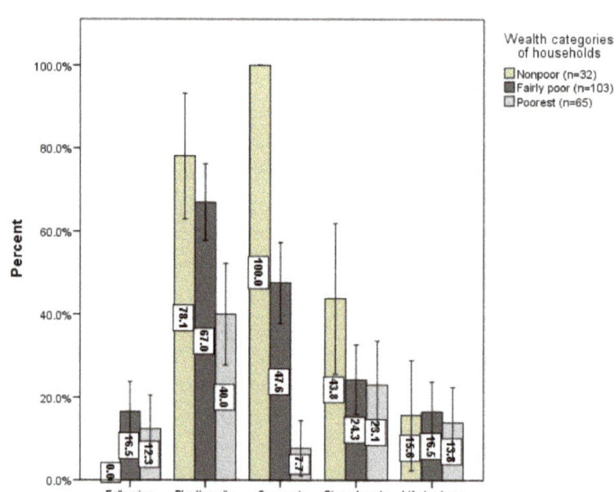

Figure 4. Land management practices adopted by respondents across poverty levels (N= 200).

3.3.1. Fallows

None of the non-poor farmers reported having fallow areas. On the other hand, 16.5 percent of the fairly-poor and 12.3 percent of the poorest farmers practiced fallowing. Improved fallow areas are considered to be a low-cost practice for improving soil fertility in the Sahel. This is made possible by a variety of tree species that regenerate naturally in the study area such as *Parkia biglobosa*, *Vitellaria paradoxa*, etc. Because migrant status is a proxy for insecure tenure to land and trees, the migrant-status of the non-poor may explain why no respondents in this group practiced fallowing. Aside from soil fertility improvement, fallows are used for other purposes as well, as sources of fuel wood, fodder, etc. Due to low inputs in agriculture in the study area, farmers without fallows are more likely to expand their fields into forest areas in order to increase production. Thus, fallows are important for both soil fertility improvement and livelihood value, a premise consistent with several past studies conducted in the West African Sahel [67,68].

3.3.2. Planting Pits

Also known as *zaï*, a planting pit is a soil water conservation technique commonly practiced in the Sahel. The percent of respondents who adopted this practice was lowest among the poorest (40 percent), compared to the non-poor (78.1 percent) and fairly poor (67 percent), for whom the use rates were relatively high. *Zaï* practice is labor intensive, and a household work force or the ability to hire labor is a prerequisite for its adoption. Two previous studies in Burkina Faso estimated that installing *zaï* on a hectare of land requires six to twelve weeks of daily work for one person, depending on soil condition [69,70].

Despite the labor demand, *zaï* is a critical practice in Burkina Faso due to the effects of climate variability and change on food security and livelihoods. Participants in the FGDs indicated that water harvested through this technique contributes to assisting natural regeneration in less than four years. An earlier study in Burkina Faso finds that *zaï* practice rehabilitates land within three to five years [71].

3.3.3. Composting

Results show that the respondents who apply compost on their farms include non-poor (100 percent), fairly-poor (47.6 percent) and poorest farmers (7.7 percent) (Figure 4). The FGDs revealed that although compost on fields is known to improve soil fertility, its adoption is dependent on the resources available to farmers. Studies in Burkina Faso [60,69] and the Sahel [62] acknowledge the importance of compost for soil fertility improvement. Compostable material such as crop residues have to be transported to a compost pit and then crushed with additives to produce compost (Figure 5). To construct and maintain such facilities requires money that the poorest farmers cannot afford. During the FGDs it was indicated that the poorest households are more likely than others to make use of community compost facilities. Due to the large time requirement for producing compost (2–3 months), those who use the community facilities have to work together to produce compost and then share it.

Most of the non-poor farmers belong to the Fulani ethnic group who own larger herds of cattle than other wealth categories. The dung from these cattle is vital in producing compost. As such, the household socioeconomic characteristics of farmers are important for adopting this technology and often constrain poor farmers with little resources at their disposal. A study in Burkina Faso finds that socioeconomic factors affect the adoption of compost technology in favor of households with greater available resources [69].

Figure 5. A compost facility owned by a household in Kou Village.

3.3.4. Stone Bunds

The respondents who adopted stone bunds across the wealth categories included non-poor (43.8 percent), fairly-poor (24.3 percent) and poorest farmers (23.1 percent). Among the fairly poor and poorest farmers, the adoption rates for stone bunds were similar, while for the non-poor farmers it was relatively high. Stone bunds are a soil conservation technique that involves the laying of stones on fields (Figure 6) to check runoff and to control soil erosion. This technique traps soil and allows it to accumulate over time. Another study in Burkina Faso indicates that although soil properties did not change significantly over a five-year period due to stone bunds, there was an increase in depth of the top layer of soil [70]. Farmers who have stones on their farms need to transport them to areas within the farm where the bunds are to be constructed. In the absence of available stones on farmland, households that want to engage in this practice must transport the stones from outside.

Figure 6. Use of stone bunds on farmland in Vrassan Village.

3.3.5. Live Hedges

The poorest farmers had the lowest percentage of adopters (13.8 percent) of live hedges on their farms. Among the fairly-poor farmers, 16.5 percent adopted this practice, and for the non-poor, the adoption rate was 15.6 percent. Live hedges are also known as living fences and are created by the planting of trees on part or all of the farm boundaries (see Figure 7). This practice provides a buffer against wind and water erosion. Two studies in Burkina Faso also report that dense tree cover buffer the effects of wind [72] and rainfall [73] and in turn reduce soil erosion. The relatively low adoption rates amongst the non-poor could be related to findings that all farmers in this wealth category perceived their own tenure to land and trees to be insecure. Aside from reducing the intensity of soil erosion, live hedges are important sources of fodder and fuel wood. Therefore, live hedges provide both livelihood benefits and environmental benefits to farmers.

Figure 7. Live hedge in Cassou Village with *Azadirachta indica*.

4. Limitations

This study faces several limitations. First, as part of the process of grouping households into different wealth status categories, one of the roles of the FGDs was to identify every household in each category. With a total of 900 households in the four villages, having detailed information for all households was challenging. To overcome this limitation, the twenty participants from the four villages agreed on a common list of local indicators to use in categorizing the households. The researchers then collected interview data from households based on these indicators and later classified households into different categories of wealth. Thus, the PPA approach could only be implemented in a limited way. However, the study maintained the key aspect of participant input into how poverty was defined.

In addition to limitations arising from the application of the PPA, the use of PPA itself introduced a key limitation into the study's analytical methods and findings. Both the number of cattle and the amount of land held by a household were used as independent variables in the categorization of households into the wealth groups. The use of these variables in wealth categorization is a direct reflection of local definitions of household wealth and a function of the PPA process. However, these variables also play a key role in the analysis of household engagement in environmentally degrading activities. The study treated number of cattle as a dependent variable to assess overgrazing. Furthermore, several of the environmental degradation dependent variables logically correlate with farm size: annual mean deforestation, as measured in hectares, likely reflects farm size, as does cotton production in kilograms. We nonetheless consider our findings to have merit, precisely because they highlight the role of wealth, as defined locally, as a driving force in environmentally degrading activities at this research site.

Next, self-reported assessments by farmers can be biased, especially if the issue under investigation is considered sensitive. Such data can only be gathered indirectly because a direct question might be too sensitive and can affect the engagement of respondents. Because parkland tree species are protected by law, farmers are likely not to provide information if such species are fast degrading on their farmlands. Also, unequal and small sample sizes might not effectively reveal differences between groups. Finally, applying a unified list of indicators across multiple villages requires careful consideration of potential differences in the socio-cultural settings. In our case, it was easier because these villages are close to each other and share ethnicity and culture.

5. Conclusions

This case study of the poverty–environment nexus not only reveals that the relationship between wealth or poverty status and environmentally degrading activities is site specific in general, as well as activity specific; it also provides insights into resource management practices across different wealth groups at this site. The current study reveals that non-poor and fairly poor farmers engaged more often in environmentally degrading activities compared to the poorest farmers. Approximately 93 percent of non-poor and 22 percent of fairly-poor farmers experienced high rates of overgrazing, while none of the poorest farmers experienced high rates of overgrazing. This is because the non-poor and fairly-poor farmers possess the available resources to own large herds of cattle. This study in fact categorized households into wealth status groups in part based upon number of cattle owned. Equipped with capital, the non-poor and fairly-poor farmers engaged more effectively in the cutting and selling of fuel wood and cotton cultivation than did the poorest farmers. Based on farmers' self-reported assessment of the role of their tenure security in relation to assisting natural regeneration of indigenous tree species, tenure insecurity constitutes a major constraint on FMNR. FMNR is known to reduce environmental degradation, and due to their perceived inability to participate in this practice, the non-poor contribute more to environmental degradation. Parklands in the Sahel have existed for centuries because of the regeneration of indigenous tree species that provide multiple livelihood and environmental benefits.

Furthermore, deforestation through field expansion proved to be significant for the fairly-poor and non-poor households. Cotton cultivation, which can motivate field expansion and lead to heavy pesticide use, is capital intensive. The lack of available resources among the poorest farmer limits their effective participation in this practice. Other activities considered environmentally degrading, such as the cutting and selling of fuel wood, are dominated by the fairly poor and non-poor farmers.

On the other hand, self-reported assessment of soil fertility loss was highest among the poorest farmers. In addition, the adoption rates of land management practices considered to improve the land (and help ameliorate soil fertility loss) were relatively low for the poorest households. Some of these land management practices, e.g., planting pits, use of compost and stone bunds, are both labor and capital intensive, which may explain the low rates of adoption among the poorest farmers.

Therefore, the results of our study indicate that the non-poor and fairly-poor farmers contribute toward environmentally degrading activities relatively more, while poverty constrains the adoption of sustainable land management practices for the poorest farmers. Further research is needed in the following areas: (i) a comparison of natural resource management strategies for in-migrant farmers versus indigenous farmers; (ii) the relationship among ethnicity, poverty, and environmental degradation; (iii) the role of tenure security, land quality, and land fragmentation on sustainable land management; and (iv) the factors influencing FMNR in Burkina Faso.

Acknowledgments: This work was supported by WP3 (National Policies and Capacity Strengthening) of the Building Biocarbon and Rural Development in West Africa Project (BIODEV) funded by the Ministry of Foreign Affairs of Finland. We also thank the International Tropical Timber Organization (ITTO) for providing funding for this research. We thank Oumarou Bognini for his assistance during the field work and the people of the local communities who developed the local indicators for poverty. We are thankful to Mr. Rabdo Abdoulay for producing a map for the study area.

Author Contributions: Daniel Etongo conceived the paper rationale and carried out data collection. Daniel Etongo and Nadia Djenontin analyzed the data. Daniel Etongo wrote the general draft and Nadia Djenontin and Markku Kanninen edited and contributed to the interpretation of results and discussion. All authors participated in the final draft.

Conflicts of Interest: The authors declare no conflict of interest.

Appendix

Table A1. Multiple comparison Scheffe Test for variables for which homogeneity of variance is assumed (N=200).

Dependent Variable			Mean Difference (I–J)	Std. Error	Sig.	95% Confidence Interval	
						Lower Bound	Upper Bound
Annual average deforestation	Non-poor	Fairly poor	0.08	0.06	0.40	−0.07	0.22
		Poorest	0.17 *	0.06	0.01	0.02	0.33
	Fairly poor	Non-poor	−0.08	0.06	0.40	−0.22	0.06
		Poorest	0.09	0.05	0.10	−0.01	0.21
	Poorest	Non-poor	−0.17 *	0.06	0.01	−0.33	−0.02
		Fairly poor	−0.09	0.05	0.10	−0.21	0.01

* The mean difference is significant at the 0.05 level.

Table A2. Multiple comparison Scheffe Test for variables for which homogeneity of variance is not assumed (N=200).

Dependent Variable				Mean Difference (I–J)	Std. Error	Sig.	95% Confidence Interval	
							Lower Bound	Upper Bound
Number of cattle	Scheffe	Non-poor	Fairly poor	37.22 *	2.82	0.000	30.27	44.17
			Poorest	40.85 *	3.01	0.000	33.44	48.27
		Fairly poor	Non-poor	−37.22 *	2.82	0.000	−44.17	−30.27
			Poorest	3.64	2.20	0.259	−1.80	9.07
		Poorest	Non-poor	−40.85 *	3.01	0.000	−48.27	−33.44
			Fairly poor	−3.64	2.20	0.259	−9.07	1.80
	Dunnett C	Non-poor	Fairly poor	37.22 *	6.12		22.15	52.28
			Poorest	40.85 *	6.11		25.80	55.90
		Fairly poor	Non-poor	−37.22 *	6.12		−52.28	−22.15
			Poorest	3.64 *	0.33		2.86	4.41
		Poorest	Non-poor	−40.85 *	6.11		−55.90	−25.80
			Fairly poor	−3.64 *	0.33		−4.41	−2.86
Monthly fuel wood consumption (m^3)	Scheffe	Non-poor	Fairly poor	0.43 *	0.15	0.016	0.06	0.79
			Poorest	0.97 *	0.16	0.000	0.57	1.36
		Fairly poor	Non-poor	−0.43 *	0.15	0.016	−0.79	−0.06
			Poorest	0.53 *	0.12	0.000	0.25	0.82
		Poorest	Non-poor	−0.97 *	0.16	0.000	−1.36	−0.57
			Fairly poor	−0.53 *	0.12	0.000	−0.82	−0.25
	Dunnett C	Non-poor	Fairly poor	0.43	0.19		−0.06	0.92
			Poorest	0.97 *	0.19		0.49	1.44
		Fairly poor	Non-poor	−0.43	0.19		−0.92	0.06
			Poorest	0.53 *	0.09		.31	0.76
		Poorest	Non-poor	−0.97 *	0.19		−1.44	−0.49
			Fairly poor	−0.53 *	0.09		−0.76	−0.31
Quantity of cotton produced (kg)	Scheffe	Non-poor	Fairly poor	−34.25	36.29	0.641	−123.77	55.27
			Poorest	42.62	38.73	0.547	−52.90	138.14
		Fairly poor	Non-poor	34.25	36.29	0.641	−55.27	123.77
			Poorest	76.87 *	28.40	0.027	6.79	146.94
		Poorest	Non-poor	−42.62	38.73	0.547	−138.14	52.90
			Fairly poor	−76.87 *	28.41	0.027	−146.94	−6.79
	Dunnett C	Non-poor	Fairly poor	−34.25	42.31		−137.58	69.08
			Poorest	42.62	39.27		−53.78	139.02
		Fairly poor	Non-poor	34.25	42.31		−69.08	137.58
			Poorest	76.87 *	23.87		19.95	133.79
		Poorest	Non-poor	−42.62	39.27		−139.02	53.78
			Fairly poor	−76.87 *	23.87		−133.79	−19.95

* The mean difference is significant at the 0.05 level.

References

1. Bojö, J.; Bucknall, J.; Hamilton, K.; Kishor, N.; Kraus, C.; Pillai, P. *Environment Chapter, Poverty Reduction Strategy Papers' Source Book*; World Bank: Washington, DC, USA, 2001.
2. Bosch, C.; Hommann, K.; Rubio, G.M.; Sadoff, C.; Travers, L. *Water, Sanitation and Poverty Chapter, Poverty Reduction Strategy Papers' Source Book*; World Bank: Washington, DC, USA, 2001.
3. Mabogunje, A.L. Poverty and environmental degradation: Challenges within the global economy. *Sci. Policy Sustain. Dev.* **2002**, *44*, 8–19. [CrossRef]
4. World Commission for Environment and Development. *Our Common Future*; Oxford University Press: Oxford, UK, 1987.
5. Moseley, W.G. Environmental degradation and the "poor" smallholders in the West African Sudano-Sahel: Global discourses and local realities. In *African Environment and Development: Rhetoric, Programmes, Realities*; Moseley, W.G., Logan, B.I., Eds.; Ashgate Publishing: Aldershot, UK, 2004; pp. 41–62.
6. Ravnborg, H.M. Poverty and environmental degradation in the Nicaraguan Hillsides. *World Dev.* **2003**, *31*, 1933–1946. [CrossRef]
7. Swinton, S.M.; Quiroz, R. Is poverty to blame for soil, pasture and forest degradation in Peru's Altiplano? *World Dev.* **2003**, *31*, 1903–1919. [CrossRef]
8. Gray, C.L.; Moseley, W.G. A geographical perspective on poverty-environmental interactions. *Geogr. J.* **2005**, *171*, 9–23. [CrossRef]
9. Duraiappah, A.K. Poverty and environmental degradation: A review and analysis of the nexus. *World Dev.* **1998**, *12*, 2169–2179. [CrossRef]
10. Angelsen, A. The poverty of the environment and the environment of poverty. In Poverty and the Environment, Proceedings of the CROP/ADIPA/UNCTAD workshop, Sabah, Malaysia, October 1995; Angelsen, A., Vainio, M., Eds.; Comparative Research Programme on Poverty (CROP) Publications: Bergen, Norway, 1998; pp. 2–18.
11. Boyce, J.K. Inequality as a cause of environmental degradation. *Ecol. Econ.* **1994**, *11*, 169–178. [CrossRef]
12. Leach, M.; Mearns, R.; Scoones, I. Environmental Entitlements: Dynamics and Institutions in Community-Based Natural Resource Management. *World Dev.* **1999**, *2*, 225–247. [CrossRef]
13. Forest Investment Program. *REDD Preparation Plan*; Ministry of Environment and Sustainable Development: Ouagadougou, Burkina Faso, 2012.
14. Kambire, H.W.; Djenontin, I.N.S.; Kabore, A.; Djoudi, H.; Balinga, M.P.B.; Zida, M.; Assembe-Mvondo, S. *La REDD+ et l'Adaptation aux Changements Climatiques au Burkina Faso: Causes, Agents et Institutions*; CIFOR: Bogor, Indonesia, 2015.
15. Kaminski, J.; Headey, D.; Bernard, T. The Burkinabe cotton story 1992–2007: Sustainable success or sub-saharan mirage? *World Dev.* **2011**, *39*, 1460–1475. [CrossRef]
16. Ouedraogo, I.; Savadogo, P.; Tigabu, M.; Cole, R.; Odén, P.C.; Ouadba, J.M. Is rural migration a threat to environmental sustainability in Southern Burkina Faso? *Land Degrad. Dev.* **2009**, *20*, 217–230. [CrossRef]
17. Johnson, D.L.; Ambrose, S.H.; Bassett, T.T.; Bowen, M.L.; Crummey, D.E.; Isaacson, J.S.; Johnson, D.N.; Lamb, P.; Saul, M.; Winter-Nelson, A.E. Meanings of environmental terms. *J. Environ. Qual.* **1997**, *26*, 581–589. [CrossRef]
18. Abu Hammad, A.; Tumeizi, A. Land degradation: Socioeconomic and environmental causes and consequences in the eastern Mediterranean. *Land Degrad. Dev.* **2012**, *23*, 216–226. [CrossRef]
19. Mortimore, M. *The Future of Family Farms in West Africa: What Can We Learn from Long-Term Data?*; Drylands Issue Paper No. 119; IIED: London, UK, 2003.
20. Belem, M.; Bayala, J.; Kalinganire, A. Defining the poor by the rural communities of Burkina Faso: Implications for the development of sustainable parkland management. *Agrofor. Syst.* **2011**, *83*, 287–302. [CrossRef]
21. United Nations Development Programme (UNDP). *Human Development Report 2006—Beyond Scarcity: Power, Poverty and the Global Water Crisis*; United Nations Development Programme (UNDP): New York, NY, USA, 2006.
22. Chomitz, K. *At loggerheads? Agricultural Expansion, Poverty Reduction, and Environment in the Tropical Forests*; World Bank Policy Research Report; World Bank: Washington, DC, USA, 2007.

23. Ouedraogo, I.; Tigabu, M.; Savadogo, P.; Compaoré, H.; Odén, P.C.; Ouadba, J.M. Land cover change and its relation with population dynamics in Burkina Faso, West Africa. *Land Degrad. Dev.* **2010**, *21*, 453–462. [CrossRef]
24. Paré, S.; Söderberg, U.; Sandewall, M.; Ouadba, J.M. Land use analysis from spatial and field data capture in Southern Burkina Faso, West Africa. *Agric. Ecosyst. Environ.* **2008**, *127*, 277–285. [CrossRef]
25. Ingram, K.T.; Roncoli, M.C.; Kirshen, P.H. Opportunities and constraints for farmers of West Africa to use seasonal precipitation forecasts with Burkina Faso as a case study'. *Agric. Syst.* **2002**, *74*, 331–349. [CrossRef]
26. Driessen, P.; Deckers, J.; Spaargaren, O. *Lectures Notes on the Major Soils of the World*; FAO World Soil Resources, Report-94; Food and Agriculture Organization of the United Nations: Rome, Italy, 2001.
27. Institut National des Statistiques et de la Démographie (INSD). *Résultats Préliminaires du Recensement Général de la Population et de l'Habitat de 2006*; Institut National des Statistiques et de la Démographie (INSD), Direction de la Démographie: Ouagadougou, Burkina Faso, 2007.
28. Fontes, J.; Guinko, S. *Carte de Végétation et de l'Occupation du sol du Burkina Faso*; Projet Campus; UPS: Toulouse, France, 1995.
29. Balen, J.; McManus, D.; Yue-Sheng, L.; Zheng-Yuan, Z.; Li-Ping, Y.; Utzinger, J.; Williams, G.M.; Li, Y.; Ren, M.Y.; Liu, Z.C.; et al. Comparison of two approaches for measuring household wealth via an asset-based index in rural and Peri-urban settings of Hunan province, China. *Emerg. Themes Epidemiol.* **2010**, *7*, 7. [CrossRef] [PubMed]
30. Laderchi, C.R.; Saith, R.; Stewart, F. Does it matter that we do not agree on the definition of poverty? A comparison of four approaches. *Oxf. Dev. Stud.* **2003**, *31*, 243–274. [CrossRef]
31. Weng, X. *The Rural Informal Economy: Understanding Drivers and Livelihood Impacts in Agriculture, Timber and Mining*; IIED Working Paper; IIED: London, UK, 2015; Available online: http://pubs.iied.org/16590IIED.html?r=p (accessed on 15 March 2016).
32. Booysen, F.; van der Berg, S.; Burger, R.; van Maltitz, M.; de Rand, G. Using an asset index to assess trends in poverty in seven sub-Saharan African countries. *World Dev.* **2008**, *6*, 1113–1130. [CrossRef]
33. Narayan, D.; Chambers, R.; Shah, M.K.; Petesch, P. *Voices of the Poor: Crying out for Change*; Oxford University Press USA for the World Bank: New York, NY, USA, 2000.
34. Ouedraogo, I.; Mbow, C.; Balinga, M.; Neufeldt, H. Transitions in land use architecture under multiple human driving forces in a semi-arid zone. *Land* **2015**, *4*, 560–577. [CrossRef]
35. Swinton, S.M.; Escobar, G.; Reardon, T. Poverty and environment in Latin America: Concepts, evidence and policy implications. *World Dev.* **2003**, *31*, 1865–1872. [CrossRef]
36. Teshome, A.; de Graaff, J.; Ritsema, C.; Kassie, M. Farmers' perceptions about the influence of land quality, land fragmentation and tenure systems on sustainable land management in the north western Ethiopian Highlands. *Land Degrad. Dev.* **2014**. [CrossRef]
37. Ravnborg, H.M.; Escolán, R.M.; Guerrero, M.P.; Méndez, M.A.; Méndoza, F.; de Páez, E.M.; Motta, F. *Developing Regional Poverty Profiles Based on Local Perceptions*; CIAT Publication No. 291; Centro Internacional de Agricultura Tropical (CIAT): Cali, Colombia, 1999.
38. Etongo, D.; Djenontin, I.N.S.; Kanninen, M.; Fobissie, K.; Korhonen-Kurki, K.; Djoudi, H. Land tenure, asset heterogeneity and deforestation in Southern Burkina Faso. *For. Policy Econ.* **2015**, *61*, 51–58. [CrossRef]
39. Gray, L.C. What kind of intensification? Agricultural practices, soil fertility and socioeconomic differentiation in rural Burkina Faso. *Geogr. J.* **2005**, *171*, 70–82. [CrossRef]
40. Reenberg, A.; Oksen, P.; Svendsen, J. Land use Changes vis-à-vis agricultural development in Southeastern Burkina Faso: The field expansion dilemma. *Geogr. Tidsskr. Dan. J. Geogr.* **2003**, *103*, 57–69. [CrossRef]
41. Jones, D.W.; Dale, V.H.; Beauchamp, J.J.; Pedlowski, M.A.; O'Neill, R.V. Farming in Rondomia. *Resour. Energy Econ.* **1995**, *17*, 155–188. [CrossRef]
42. Ouédraogo, M.; Tankoano, A.; Ouédraogo, Z.T.; Guissou, I.P. Etude des facteurs des risques d'intoxications chez les utilisateurs de pesticides dans la region cotonnière de Fada N'Gourma au Burkina Faso. *Environ Risque Santé* **2009**, *8*, 343–347.
43. Toe, A.M.; Kinane, M.L. Les pesticides au Burkina Faso/ Pesticide in Burkina Faso, Dakar-Fann. Available online: https://searchworks.stanford.edu/view/9154884 (accessed on 26 July 2015).
44. Toe, A.M.; Ouedraogo, M.; Ouedraogo, R.; Ilboudo, S.; Guissou, P.I. Pilot study on agricultural pesticides poisoning in Burkina Faso. *Interdiscip Toxicol* **2013**, *6*, 185–191. [CrossRef] [PubMed]

45. Brocard, D.; Lacaux, J.P.; Eva, H. Domestic biomass combustion and associated atmospheric emissions in West Africa. *Glob. Biogeochem. Cycle* **1998**, *12*, 127–139. [CrossRef]
46. Kituyi, E.; Marufu, L.; Huber, B.; Wandiga, S.O.; Jumba, I.O.; Andreae, M.O.; Helas, G. Biofuel consumption rates and patterns in Kenya. *Biomass Bioenergy* **2001**, *20*, 83–99. [CrossRef]
47. Niemeijer, D.; Mazzucato, V. Soil degradation in the West African Sahel: How serious is it? *Environ. Sci. Policy Sustain. Dev.* **2002**, *44*, 20–31. [CrossRef]
48. Larwanou, M.; Saadou, M. The role of human interventions in tree dynamics and environmental rehabilitation in the Sahel zone of Niger. *J. Arid Environ.* **2011**, *75*, 194–200. [CrossRef]
49. Place, F.; Binam, J.N. *Economic Impacts of Farmer Managed Natural Regeneration in the Sahel: End of Project Technical Report*; Free University, International Fund for Agricultural Development (IFAD): Amsterdam, NY, USA, 2013.
50. Weston, P.; Hong, R.; Kaboré, C.; Kull, C.A. Farmer-managed natural regeneration enhances rural livelihoods in dryland West Africa. *Environ. Manag.* **2015**, *55*, 1402–1417. [CrossRef] [PubMed]
51. Nabahungu, N.L.; Visser, S.M. Farmers' knowledge and perception of agricultural wetland management in Rwanda. *Land Degrad. Dev.* **2013**, *24*, 363–374. [CrossRef]
52. Karltun, E.; Lemenih, M.; Tolera, M. Comparing farmers' perception of soil fertility change with soil properties and crop performance in Beseku, Ethiopia. *Land Degrad. Dev.* **2013**, *24*, 228–235. [CrossRef]
53. Morgan, G.A.; Leech, N.L.; Gloeckner, G.W.; Barrett, K.C. *SPSS for Introductory Statistics: Use and Interpretation*, 2nd ed.; Lawrence Erlbaum Associates: Mahwah, NJ, USA, 2004.
54. Hilton, A.; Armstrong, R.A. Is one set of data more variable than another? *Microbiologist* **2006**, *7*, 34–36.
55. Field, A.P. *Discovering Statistics Using IBM SPSS Statistics: And Sex and Drugs and Rock'n'roll*, 4th ed.; Sage: London, UK, 2013.
56. Ton, P. *Cotton Production and Rural Livelihoods in West Africa*; Oxfam: Oxford, UK, 2001.
57. Bonnassieux, A. Filière coton, émergence des organisations de producteurs et transformations territoriales au Mali et au Burkina Faso. *Les Cahiers d'Outre-Mer* **2002**, *220*, 421–434. [CrossRef]
58. Sawadogo, L. *Adapter les Approches de l'Aménagement Durable Des Forêts Sèches Aux Aptitudes Sociales, Economiques et Technologiques en Afrique*; Center for International Forestry Research (CIFOR): Bogor, Indonesia, 2006.
59. Ndiaye, O.; Zoungrana, P. Techniques de maintien de la fertilité des sols: Une Revolution Silencieuse de l'Agriculture. Available online: http://www.inter-reseaux.org/auteur/philippe-zoungrana (accessed on 30 December 2014).
60. Ouédraogo, E.; Mando, A.; Zombré, N.P. Use of compost to improve soil properties and crop productivity under low input agricultural system in West Africa. *Agric. Ecosyst. Environ.* **2001**, *84*, 259–266. [CrossRef]
61. Reij, C.; Tappan, G.; Belemvire, A. Changing land management practices and vegetation on the Central Plateau of Burkina Faso (1968–2002). *J. Arid Environ.* **2005**, *63*, 642–659. [CrossRef]
62. McClintock, N.C.; Diop, A.M. Soil fertility management and compost use in Senegal's Peanut Basin. *Int. J. Agric. Sustain.* **2005**, *3*, 1473–5903. [CrossRef]
63. Ouédraogo, M. *Land Tenure and Rural Development in Burkina Faso: Issues and Strategies*; Issue Paper No. E112; International Institute for Environment and Development (IIED): London, UK, 2002; Available online: http://www.iied.org/ (accessed on 4 June 2014).
64. Bai, Z.G.; Dent, D.L.; Olsson, L.; Schaepman, M.E. Proxy global assessment of land degradation. *Soil Use Manag.* **2008**, *24*, 223–234. [CrossRef]
65. Lenhardt, A.; Glennie, J.; Intscher, N.; Ali, A.; Morin, G. *A Greener Burkina: Sustainable Farming Techniques, Land Reclamation and Improved Livelihoods*; Development Progress Case Study Report; Overseas Development Institute: London, UK, 2014; Available online: http://www.odi.org/sites/odi.org.uk/files/odi-assets/publications-opinion-files/9153.pdf (accessed on 5 January 2015).
66. Robinson, B.E.; Holland, M.B.; Naughton-Treves, L. Does secure land tenure save forest? A meta-analysis of the relationship between land tenure and tropical deforestation. *Glob. Environ. Chang.* **2013**, *29*, 281–293. [CrossRef]
67. Ajayi, O.C.; Franzel, S.; Kuntashala, E.; Kwesiga, F. Adoption of improved fallow technology for soil fertility management in Zambia: Empirical studies and emerging issues. *Agrofor. Syst.* **2003**, *59*, 317–326. [CrossRef]

68. Kenta, I.; Shinjo, H.; Tanaka, U.; Tobita, S.; Funakawa, S.; Kosaki, T. Fallow Band System, a land management practices for controlling desertification and improving crop production in the Sahel, West Africa, Effectiveness in desertification control and soil fertility improvement. *Soil Sci. Plant Nutr.* **2011**, *57*, 573–586.
69. Somda, J.; Nianogo, A.J.; Nassa, S.; Sanou, S. Soil fertility management and socio-economic factors in crop-livestock systems in Burkina Faso: A case study of composting technology. *Ecol. Econ.* **2002**, *43*, 175–183. [CrossRef]
70. Zougmore, R.; Gnankambary, Z.; Guillobez, S.; Stroosnijder, L. Effects of stone lines on soil chemical characteristics under continuous sorghum cropping in semiarid Burkina Faso. *Soil Tillage Res.* **2002**, *66*, 47–53. [CrossRef]
71. Kaboré, D.; Reij, C. *The Emergence and Spreading of an Improved Traditional Soil and Water Conservation Practice in Burkina Faso*; EPTD Discussion Paper No. 114; International Food Policy Research Institute: Washington, DC, USA, 2004; Available online: http://www.ifpri.org/publication/emergence-and-spreading-improved-traditional-soil-and-water-conservation-practice-burkin (accessed on 31 December 2014).
72. Leenders, J.K.; Sterk, G.; van Boxel, J.H. Wind erosion reduction by scattered woody vegetation in farmers' fields in northern Burkina Faso. *Land Degrad. Dev.* **2014**. [CrossRef]
73. Grimaldi, S.; Angeluccetti, I.; Coviello, V.; Vezza, P. Cost-effectiveness of soil and water conservation measures on the catchment sediment budget—The Laaba watershed case study, Burkina Faso. *Land Degrad. Dev.* **2013**. [CrossRef]

Section 2:

Dynamic Smallholder Livelihoods

land

MDPI

Article

Smallholders, Agrarian Reform, and Globalization in the Brazilian Amazon: Cattle versus the Environment

Ritaumaria Pereira [1,*], Cynthia S. Simmons [2] and Robert Walker [3]

1 Center for Sustainability and the Global Environment, University of Wisconsin-Madison, Madison, WI 53726, USA
2 Department of Geography, University of Florida, Gainesville, FL 32611, USA; cssimmons@ufl.edu
3 Center for Latin American Studies, University of Florida, Gainesville, FL 32611, USA; roberttwalker@ufl.edu
* Correspondence: rpereira@wisc.edu; Tel.: +55-91-988442415

Academic Editors: Claudia A. Radel and Jacqueline M. Vadjunec
Received: 13 October 2015; Accepted: 24 June 2016; Published: 7 July 2016

Abstract: Smallholder farming in the Brazilian Amazon has changed markedly over the last few decades, following a pervasive swing to cattle production observed across the basin. These changes have brought opportunities for accumulating a modicum of wealth that were not available in the early stages of colonization. At the same time, they have reconfigured livelihood systems away from diversified agriculture to a strong engagement with the cattle economy. They are also exposing smallholders to new forms of exploitation by transnational corporations, seeking to pass risk upstream to less powerful economic agents who provide inputs to production, such as calves. The case of Southeastern Pará provides a natural laboratory for investigating such phenomena, which the article considers through the presentation of data from field research conducted in the region over the past decade. Here, agrarian reform efforts have been particularly intense, and social movements have often espoused a green rhetoric in favor of diversified agriculture, even though smallholders show little interest in anything but cattle. Household level incentives promote Amazonia's emergent cattle economy, demonstrating how global production networks have reached into the basin, where production relations between smallholders provisioning calves to large ranching operations often resemble what has been referred to in the literature as *"contract farming" land grabs*, given the exploitive terms of trade.

Keywords: global cattle economy; production chains and networks; settlement projects; Amazonia; land grabs

1. Introduction

Since the 1960s, the Brazilian government has advanced a series of comprehensive plans to integrate the Amazon region with the economic and political core of the country, including construction of a highway network to make accessible this vast frontier, subsidies and fiscal incentives to attract capital investment, and colonization programs to promote orderly in-migration and occupation necessary to secure Brazil's territorial claim while in the same efforts provide land to the landless. Amazonia's development history reveals that government motivations and subsequent policies have vacillated greatly between a national desire for economic growth via the support of private interests and the populist demand for land and opportunity, and the outcomes from these efforts likewise present mixed results. On the one hand, the highway network opened the frontier to waves of immigration with population increases from four million in 1960 to more than 24 million by 2010 [1], and reduced transportation costs making the region attractive to agroindustry [2]. Indeed, Brazil's economy has mostly thrived since the turn of the millennium, with agroindustry in Amazonia an important part of this success story, particularly the expansion of cattle ranching throughout the basin [3]. On the other hand, the elaborate colonization programs intended to rectify Brazil's "rural problem" and ameliorate growing populist discontent were unable to completely satisfy the societal demand for land, and led to conflicts and rapid rates of deforestation [3].

Despite this checkered development history, there is little doubt that the livelihood systems of the smallholders who were originally targeted for colonization have evolved dramatically over the past few decades, from mostly subsistence farming based strictly on household resources, to small-scale ranching tied to the global economy via production networks. At the same time, a powerful agrarian reform movement has successfully pressured the government to provide land to the landless in thousands of new settlement projects (SPs) intended to support family farming. Amazonia's smallholders are no longer autarchic peasants living beyond the agricultural frontier. They are now fully engaged agents in Brazil's export economy, and, frequently, energetic participants in the region's agrarian politics. Although a potential development success in certain terms, this social transformation of Amazonia's rural population presents more dubious results for the environment, and for the participating households themselves. The livelihood systems are highly focused on animal husbandry, an agricultural activity held responsible for the lion's share of Amazonian deforestation. Further, the new economic status of the former peasant has also brought new potentials for exploitation on the part of large-scale ranchers and the emergent food-processing sector.

The goal of the present paper is to examine the productive activities of smallholders living in SPs, and how expanding engagement with the global cattle economy is affecting land-use decisions in agrarian reform settlements that were created as a mechanism to alleviate poverty and promote food security through diversified and sustainable crop production. It does so by first charting the path of cattle expansion in the Brazilian Amazon, and linking this to skewed patterns of land ownership that have arisen in its wake. The paper then explores how agrarian reform movements have arisen as a consequence, with a particular emphasis on the formation of SPs through political mobilization that has been referred to as *direct action land reform*, or DALR [4]. Next, research results are presented that document the circumstances that have encouraged smallholders in Southeastern Pará to adopt livelihood systems focused on cattle, which is in contradiction to the environmental requirements for SP formation. In addition, findings give evidence to the emergent cattle production chain, including the practice of contract "calving" arrangements that link smallholders in SPs to large ranching operations and the appearance in the region of heavily capitalized corporate interests engaged in export.

1.1. Cattle and Agrarian Reform

Amazonia's cattle economy has been key to Brazil's emergence as an economic power, but it came with a high cost to both the environment and society. In the Brazilian Amazon, it is mainly large-scale producers who are driving commercial agricultural development, while smallholders are finding it economically viable to replace forest and old croplands by planting pasture [5]. Walker et al. [6], Poccard-Chapuis et al. [7], and Vosti et al. [8] attributed this gradual transformation of crops and forestlands into planted pasture in small-scale farms to a widespread process of *pecuarização*, or expansion of the cattle ranching economy toward the Amazon's geographical limits. Participation of smallholders in the cattle economy has been analyzed in studies developed by Homma and Walker et al. [9,10]. Both studies compared cattle herds and pasture clearance in areas owned by small producers (land < 100 ha). Homma found that small producers living near the Transamazon highway in Altamira raised an average of 1.4 animals per 6.4 hectares of pasture in 1975 [9]. In the same region, in 2000, Walker et al. found that small producers had increased the average number of animals to 33, which required an average of 37 hectares of pastureland [10]. These findings show that, over time, more animals were being raised on each property. This in turn implies that more area was needed for pasture and consequently less was available for agriculture. In fact, Homma observed that the amount of land attributed to pasture was similar to the amount used for cropping in the properties visited in 1975 [9], while Walker et al. found that only one-tenth of the total land allocated to pasture was used for cropping [10].

Farm level dynamics such as these underlie a pervasive shift to cattle ranching in the Amazon at large. Despite early prognostication that the region was not viable for agriculture and ranching in particular offered little promise of success without massive government subsidies [11], animal husbandry has proven more successful and resilient [12]. Ranching has a long Amazonian history, but when the military regime began its push in the mid-1960s to open the north, the region accounted for only about 8 percent of Brazilian stock [13]. By 1990, the Amazonian herd reached about 20 million animals, and by 2005 it added another 50 million head, growing to over 70 million animals dispersed widely across the basin. As of 2013, the cattle herd in Amazonia was over 80 million animals, accounting for more than 25 percent of Brazilian export of beef destined to satisfy demands around the world, with markets in Latin America, the European Union, the Middle East, and Asia, including China [2]. Along with cattle expansion, pasture increased from 0.7 percent of the total region in 1970 to 10 percent by 2004 [14]. Implicated in this process is deforestation, since more than 80 percent of cleared areas in the Brazilian Amazon are dedicated to pasture formation [10,15–17].

Unfortunately, land-extensive cattle ranching has also resulted in the replication of land inequality observed in other parts of Brazil, as early smallholders were forced from their land due to their inability to compete in markets or through violent appropriation by largeholders, which pushed them further into forest frontiers or growing frontier boom towns. By 2007, almost 60 million hectares of land in Amazonia were owned by one percent of establishments that formed an elite group of wealthy ranchers, while a limited 4.4 million hectares were occupied by small farmers, who represented 45 percent of the establishments [18]. Displaced farmers forced to urban areas exacerbate already high rates of unemployment and poverty [19]. In turn, growing cadres of the poor in cities are easily mobilized to participate in *direct action land reform*, or DALR, which is characterized by the militant occupations, carried out by hundreds of landless farmers on large ranches deemed illegal or unproductive by movement leaders [4]. This has enflamed land conflict in the region, as the landless and their advocates engage in DALR to confront large landowners and pressure the government to follow through on agrarian reform promises.

The government has responded with new policy initiatives, especially in the aftermath of the Eldorado do Carajás massacre in 1996 when 19 landless farmers were killed by the military police during a protest [3]. Rather than return to the large-scale, state-led colonization programs of the 1970s, the government shifted focus with the creation of a *Novo Mundo Rural*, or new rural world, based on the promotion and support of small farmers in SPs. To date, nearly 9250 SPs have been created in Brazil, housing more than 968,000 families; most are located in the north and northeast, the poorest regions in the country [20]. Unlike traditional state-led colonization efforts, policy mandates that SPs be initiated at the grassroots level, an action that usually initiates with the involvement of a social movement organization, or SMO, such as the Landless Rural Workers' Movement (MST). SMOs often play active roles in the management of SPs acting as representatives to the National Institute for Colonization and Agrarian Reform (INCRA), negotiating budgets for transfer payments and distribution, and facilitating extension services [2,4,21].

INCRA's administrative requirement is that participants first establish an official SP association, and next create a Development Plan (PDA) that outlines the social, economic, and environmental strategies to be pursued in achieving sustainable development. The expectation is that smallholders will have a greater investment in the SP, and chances for success will be higher given their engagement and commitment, as opposed to the colonization programs of the 1970s [22]. After the association and PDA are established, the government provides resources for housing and infrastructure, and additional support through a myriad of credits aimed at expanding formal and technical education, improving income of residents, and promoting sustainable economic activities beyond traditional methods of agriculture. Policy explicitly prohibits monoculture systems, such as cattle ranching, and suggests that credit and extension must focus on diversified and sustainable agriculture practices [23,24].

However, contrary to INCRA policies and the views of the SMOs, which are decidedly "green," a significant number of the smallholders living in SPs have focused production on cattle [4].

For example, in a study of SPs in Southeastern Pará, Simmons and colleagues in 2006 found that 75 percent of the occupants viewed cattle as their primary agricultural activity [4]. Such findings mirror recent ethnographic studies across the basin, suggesting that smallholders are shifting from subsistence agriculture to a pasture-based farming system [10,25,26]. Even traditional groups that originally engaged in extractive activities, such as the rubber tappers in the Chico Mendes reserve, have instead opted for cattle ranching [17].

The extent to which land use in the SPs reflects this more general interest in cattle on the part of smallholders points to an apparent disconnect between policy and practice. This situation is potentially problematic given Brazil's recent adherence to the principles of the UN program to reduce emissions from forest degradation and deforestation (i.e., UN-REDD). REDD+ policies recognize the negative role played by ranching as a driver of deforestation, and the importance of involving the region's residents in sustainable development activities to mitigate environmental degradation (http://www.un-redd.org/ and www.imazon.org.br). If REDD is to be successful in Brazil, policymakers need to understand the agrarian sector, including the productive activities of the many smallholders presently residing in Amazonian SPs. Thus, it is essential to identify the factors motivating these farmers to participate in the expanding global cattle economy, as opposed to agroforestry or other green alternatives. This article directly addresses the gap in our knowledge on this front, and presents the results of field research examining the farming systems and decision-making of SP residents in the lower Amazon Basin. In doing so, it identifies the potential barriers to the implementation of more diversified and environmentally sustainable production systems that align with INCRA policy and new REDD+ programs. The research also provides evidence of the insertion of the SPs into the global cattle economy through contract farming, and subsequent implications for the social welfare of resident smallholders.

1.2. Conceptual Framework

Land Change Science (LCS) studies of Amazonian smallholders often rely on peasant economics and theories of the household life cycle that address behaviors of decentralized land managers who clear the forest in order to provide familial subsistence [26–28]. In these accounts, decisions are economically rational, if highly constrained by resource endowments. While acknowledging the utility of these foundational theories, the present paper heeds the call by Munroe et al. [29] to engage with approaches from New Economic Geography that redirect our attention from decentralized land managers to corporate actors, governance systems, the cultural embeddedness of economic activities, and globalized production and distribution networks [30]. In particular, we incorporate aspects of the production network perspective, and maintain that smallholders and ranchers, who in previous decades were adversaries in the struggle for land, today engage one another in input-output relations in the mutual interest of providing beef to an expanding global market. These new production relations are further motivated and maintained by Brazil's contradictory agrarian reform programs, and the unintended outcomes of environmental regulations. In the presentation and discussion of our results we describe the emerging global cattle economy in the region with specific attention to smallholder production systems and exchange mechanisms with ranchers. We provide evidence of exploitive production relations in the region, which may portend a new wave of contention brought on by what has been referred to in the literature as a land grab through informal contract farming [31–33]. Our theoretical contribution resides in building from theories of traditional peasant households and economics, and wedding the land grab concept to the emergence of new productive relations in the Amazon, specifically Southeastern Pará.

2. Methods

The present article attempts to shed light on the factors influencing the dominance of cattle-based systems in agrarian reform settlements, which contradict the agro-ecology and diversified production systems promoted by social movements. To this end, it presents the findings of several research activities undertaken in Southeastern Pará, a region that has long been targeted by development policies, and has therefore experienced significant growth associated, in part, with an expanding cattle sector. Today, this region is one of Brazil's largest beef exporting areas to international markets, as a result of concerted government investments to eradicate Foot and Mouth Disease (FMD), and its subsequent designation as a FMD-free zone in 2007, open to international export under requirements of the World Organization for Animal Health [12]. The region also has a long history of land conflict, pitting the landless against large ranchers, and triggering political mobilization for the purposes of agrarian reform that has resulted in the creation of many SPs [4].

The research presented here was conducted in 2010, 2011, and 2012, summing to a total of six months of fieldwork that entailed: (1) household surveys with 104 residents, selected from a subset of households visited in 2006, from SPs in Southeastern Pará (Figure 1); (2) interviews with 30 key informants that included community leaders, SMO representatives, and government officials, as well as key actors in the region's cattle production chain; and (3) the collection and assessment of regional, economic and environmental data from government sources, such as the Brazilian Institute of Geography and Statistics (IBGE), the National Institute for Space Research (INPE), National Institute of Colonization and Agrarian Reform (INCRA), Fundação Casa da Cultura de Marabá, and the Brazilian Agricultural Research Corporation (EMBRAPA).

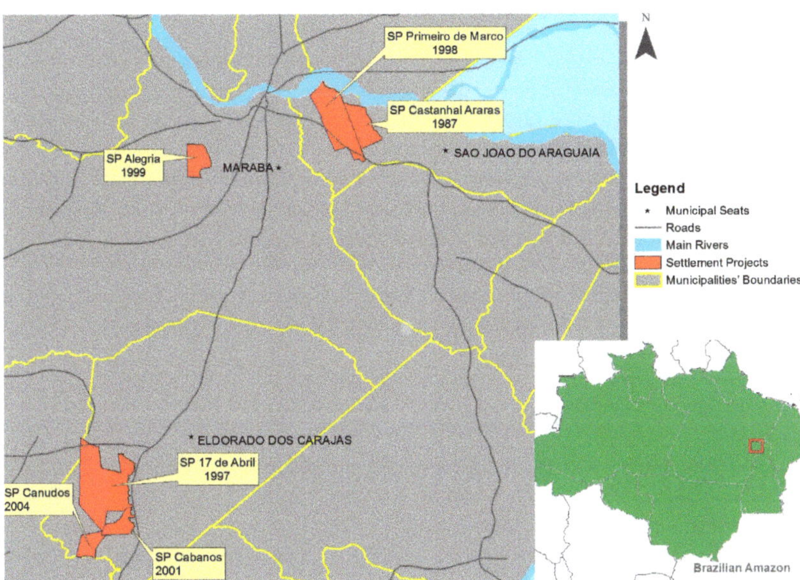

Figure 1. Study area: location and year of creation of settlements in Pará, Brazil. Source: INCRA, 2010 (personal contact).

The study sites include six SPs in three municipalities located in Southeastern Pará, which corresponded to a subset of sites visited during the Simmons-led, NSF-supported project in 2006 that resulted in the collection of extensive surveys on household characteristics and agronomic activities of residents [4]. The data identified those households engaged in cattle (i.e., head of cattle; area in pasture), but the survey instrument was limited in terms of explicating their production systems and the importance of institutional mechanisms for credit and agrarian support. Consequently, a subset of 121 households engaged in cattle from the 2006 fieldwork were selected for subsequent research and a new round of surveys was devised with the objective to elicit greater details on (1) cattle production (i.e., head of cattle for meat, for milk, quantity sold, to whom sold, prices and costs); (2) cropping system (i.e., crop type, yield, area planted, quantity sold); and (3) the types of credit and technological extension provided for each activity.[1] Additionally, the household surveys enabled the collection of detailed data on income (e.g., government transfer) and expenses (e.g., food, revolving credit payments) that permitted an assessment of the importance of various economic activities on the livelihoods of the SP residents. The surveys also included open-ended queries to probe the respondents directly on their economic decisions, incentives/motivations, and potential barriers to production.

The data collection involved both random and systematic sampling methods. For the initial 2006 data collection, the SP was divided into quadrants to ensure geospatial representation, and then every 3rd house along the road was selected; if the landowner was not present, the next house was visited, and so on until a 20 percent sample of households was reached. In 2010, a pilot study was conducted to test the survey instrument and ensure that the 2006 sampling strategy could be repeated, confirming our ability to reach the same small farmers for the follow-up surveys. Final data collection occurred in 2011, during which time 104 of the 121 households (85 percent) were located and surveys completed. The data presented here are from the 2011 fieldwork, which queried respondents about agricultural involvement and productivity measures from the previous year (2010). The analyses undertaken deploy multiple modes of assessment based on free-form interviews of key informants conducted in 2010, 2011, and 2012, and systematic surveys of settlement households in 2011. We use the interview data to provide our framing context, and also to identify biophysical, political, and historical circumstances of which individual households may be unaware. As for the survey data, we use descriptive statistics to portray household income and expenditures, as well as farming system attributes in the region. We also employ applications of inference to gauge the statistical significance of the key issues and arguments addressed in the present article. Thus, our analytical approach is a multi-method one that yields both context and detail.

Finally key informant interviews were conducted in 2011 and 2012 with public officials, SMO representatives, community leaders, and actors engaged in the cattle production chain. These interviews served multiple research objectives. First, they provided insight and context regarding the factors that impacted SP residents' decisions to invest in cattle, and the potential barriers to adoption of a more diversified agriculture system. Second, they provided necessary details that informed our understanding of the role of smallholders in the regional global cattle economy.

3. Results and Discussion

3.1. Household and Farming System Characteristics

Descriptive statistics for socio economic characteristics of households in the 2011 sample show that 71 percent of household heads were male, with a mean age of 53 years (see Table 1). On average, four people lived on each property, with a residence of 11 years on site. In terms of education, the average for the household head was three years of formal education, with 23 percent of those interviewed illiterate, and only one person in the entire sample who reached the maximum level of formal education (15 years with a college degree). The property size ranged from 8 to 50 hectares with a mean of 33 hectares, indicating that all properties fell within the regional limits (5 to 70 hectares) of the *Módulo Rural*, which is a classification for properties eligible for government agricultural assistance programs and deemed a size sufficient for family farming that ensures food security, as well as social and economic progress, which is the goal of the SPs (See Table 1).

Of the estimated 53 percent of households that reported involvement with cropping in 2010, most stated that the land dedicated to crops was small, and only 20 percent were able to report the area planted, which averaged 4 hectares. An estimated 7 percent indicated that they sold limited production

[1] The 2006 fieldwork resulted in 161 household surveys in the six SPs, and 121 respondents (75 percent) reported engagement in cattle activities.

to local markets. Corn, rice, and manioc were the three most important crops planted, primarily for family consumption or feed for animals, with the exception of two households that sold small quantities of manioc flour to neighbors (see Table 2). Beyond annuals production, a variety of fruit trees were planted in small numbers, with a handful of farmers selling banana and cacao. Despite the plans for diversified food production elaborated in the PDAs, smallholders indicated minimal involvement with cropping, and an estimated 47 percent of households stated that they planted no crops at all.

Table 1. General characteristics of households (N = 104).

Variable	Mean (Std. Dev.)	Min	Max
Age	52.73 (12.18)	19	78
Number of people	4.47 (2.57)	0	14
Length on property (years)	11.71 (4.70)	3	25
Lot size (hectares)	33.68 (10.17)	8	50
Education (years)	3.13 (3.19)	0	15

Table 2. Cropping system characteristics for primary crops in 2010 (N = 104).

		Area Planted (Hectares)				
Crop Type	**No. Smallholders** [1]	**Mean**	**Median**	**Min**	**Max**	**No. Who Sold Crops?**
Corn	16	0.37	0	0	5	0
Rice	13	0.08	0	0	1	0
Manioc	13	1	0	0	10	2
Banana	4	0	0	0	0	1
Cupuaçu	3	1.67	0	0	5	0
Sugarcane	1	5	5	5	5	0
Orange	2	0	0	0	0	0
Açai	1	0	0	0	0	0
Passion fruit	1	1	1	1	1	0
Cacao	1	5	5	5	5	1

[1] A limited 55 out of the 104 households reported that they planted crops, and only 11 could define the area planted.

In contrast to lack of involvement in cropping, 71 percent of respondents reported having cattle in 2010. This is not surprising given that the focus of this research was smallholder cattle production activities in SPs, thus, the main criterion for sample selection was household engagement with cattle in the 2006 fieldwork by Simmons and colleagues. Of the 29 percent reporting no cattle in 2010, all indicated that they had cattle previously, but needed to sell animals in order to pay accumulating debts, and their plans were to re-build their herds in the future. Other smallholders stated that they rented pasture, at about 9 percent of the sample. A comparison of cattle data from the 2006 and 2011 surveys reveals some notable changes. The mean number of animals (cows) per property increased from 19 to 21, which corresponds with pasture expansion by almost 19 percent, from an average 18 hectares to 22 hectares per property (see Table 3). Despite these increases, there was a decrease in animal density per lot across the five-year period. The decline in cattle density would appear to contradict the expected *pecuarização* of Amazonia, and the overall importance of cattle in the region [12]. Indeed, the stocking density is below smallholder averages in other parts of Amazonia (i.e., 1 animal/hectare), which is reflective of a highly unproductive land use that raises concerns about smallholder welfare and environmental sustainability. However, further analysis of the data shows that while mean changes in animals between 2006 and 2011 were not statistically significant, changes in area under pasture and animal density (cows/hectare) were. These findings suggest that decreases in density were due to increases in pasture, rather than reduction in animals.

Table 3. Cattle, pasture, and stocking density: Changes between 2005 and 2010 (N = 104).

Farm System Characteristics	2006 Mean (Std. Dev.)	2011 Mean (Std. Dev.)	Diff Mean (Std. Dev.)
Cattle (heads)	19 (19.47)	21 (27.08)	1.21 (26.75)
Pasture Area (hectares) [1]	18.22 (12.41)	21.66 (11.98)	3.44 (14.24)
Density (animals/hectare) [1]	1.30 (1.15)	0.95 (1.03)	−0.35 (1.39)

[1] Statistically significant at 5 percent.

In general, the livestock system of smallholders is distinct from that of large ranchers, who specialize in commercial production with animals genetically manipulated to produce high quality beef in a short period. For their part, smallholders have a mixed herd (dairy and beef) rather than pure-bred cattle, because such animals are less expensive and they better meet the dual needs of milk production and calving. Of the 74 households reporting cattle in 2010, 30 percent reported engagement in dairy, 9 percent in meat production, and 61 percent of smallholders reported mixed dairy-meat activities (see Table 4). When it comes to animals sold in 2010, regardless of reported activity, calves were by far the most important, with 85 percent of the sample reporting calf sales compared to only 31 percent who sold cows for meat (see Table 4). Furthermore, almost three times as many calves were sold than cows, and those reporting meat or mixed meat-dairy production actually sold more calves. From the interviews with smallholders it became apparent that calving operations were the primary focus of their activities, and the vast majority of sales were to middlemen working with large ranchers in the region. In general, farmers stated that they sold primarily male calves to middlemen and ranchers, and cows over five years of age and sickly to the local butcher. Finally, nearly all households with cattle received credit originally provided for dairy cattle, but most of them are engaged in mixed production (dairy and beef), and only 13 percent said they produced a limited quantity of milk for sale in local markets, the remaining produced milk for self consumption.

Table 4. Cattle production system characteristics in 2010 (N = 104).

	No. (% Total Smallholders with Cattle)	No. (%) Who Sold Animals?	Type	Quantity		Mean	Buyer	No. (%) with Credit
				Min	Max			
Dairy	22 (30%)	18 (82%)	Calves	1	15	5.4	66% middlemen/local ranchers	18 (82%)
		6 (27%)	Cows	1	8	2.3	local butcher	
Meat	7 (9%)	7 (100%)	Calves	1	15	5	71% to middlemen	7 (100%)
		2 (29%)	Cows	1	1	1	local butcher	
Mixed	45 (61%)	41 (91%)	Calves	1	30	8.1	96% middlemen/local ranchers	38 (84%)
		15 (33%)	Cows	1	5	2.4	local butcher	
TOTAL WITH CATTLE: 74 (100%)								

3.2. Income and Expenditures

Overall, non-cattle agricultural production ranked only fourth as a source of income, providing less than 10 percent of the annual average (See Table 5). Off-farm employment provided the highest income (28 percent), although only 30 percent of the sample reported involvement in such activities, and the average income was skewed as a result of a select few residents with formal sector jobs (e.g., teacher, ambulance driver). By and far the most important income source was (1) government transfers, at 28 percent average annual income, that were received by 68 percent of the sample, followed by (2) the sale of calves to large ranchers at 17 percent, involving 60 percent of those interviewed (Table 3). Government support included: (1) *bolsa-família* (the Family Allowance); (2) *rural retirement*; and (3) *disability retirement*. *Bolsa-família* is a governmental direct cash transfer program designed to reduce poverty, with payment tied to child vaccinations and school attendance. This monthly household stipend changes according to per person income, number of children and adolescents up to 17 years old, and number of pregnant and lactating women in the family. For the sample, the amount varied from US $16 to $151 per month. *Rural retirement* benefits can be claimed by all rural workers who prove they have worked and contributed to social security for at least 180 months, and have met the retirement age (60 years for males and 55 years for females). The retirement income is always equivalent to one minimum wage, which at the time data were collected was US $270[2] per month [30]. *Disability retirement*, also equivalent to one minimum wage, is a benefit granted to employees incapable of working due to illness or accident, in accordance with a certification by a medical group approved by the office of Social Security.

Table 5. Summary of annual income sources in 2010, in Brazilian reais (BRZ R$), at household level ($N$ = 104).

Income Source	Mean (Std. Dev.)	Min	Max	No. (%) Smallholders [3]
Off-farm Jobs [1]	4004 (7773)	0	37,200	37 (30%)
Government payments	3929 (4837)	0	21,924	71 (68%)
Calves	2377 (3477)	0	27,000	65 (60%)
Crop Sales	1291 (3279)	0	18,000	30 (29%)
Milk	776 (2544)	0	17,217	13 (13%)
Off-farm Day Labor [2]	772 (2112)	0	12,000	21 (20%)
Other Livestock	514 (1697)	0	11,800	35 (34%)
Cows	453 (1048)	0	5000	27 (26%)
Pasture Rental	231 (872)	0	6000	10 (10%)
Total Annual Income	14,405 (10,443)	1200	56,108	104

[1] Off-farm jobs refer to formal employment off the property, meaning that a family member has a steady job elsewhere; [2] Off-farm day labor refers to wages earned by family members for working other properties; [3] Respondents reported more than one income source.

In terms of monthly expenses, an estimated one-third of smallholder income is used to pay principal and interest on loans acquired during initial creation of the SPs (31 percent), and an additional 28 percent of monthly income is used to purchase food at local grocery stores. Indeed, despite claims extolling the virtue of and commitment to agro-ecology and food sovereignty advocated in SP policies, the vast majority of households (88 percent) reported that they bought most of their food at grocery stores. Corroborating this finding, one smallholder reported that, "It's cheaper to buy vegetables and meat in the grocery stores than produce them myself" (personal communication, April 2011). Another substantial expense stems from monthly installments to pay off credit for the purchase of durable goods, such as motorcycles, refrigerators, and TVs, amounting to about 9 percent of monthly income. Interestingly, the increase in debt from the purchase of durable goods coincides with the arrival of electricity in the SPs as part of the *Luz Para Todos* program (Light for All), initiated in 2003 to eradicate electricity exclusion in Brazil's rural areas [26]. By the time of the 2011 fieldwork, 71 percent of households stated they had access to electricity for the first time, with average monthly costs of US $21, or 3.5 percent of total income. Other monthly expenses, in order of importance, include medication, water, and rent for housing nearer the city center, which is necessary for sending children to school.

3.3. Smallholder Decision Making: Crops versus Cattle

In terms of institutional support for agriculture, only 33 people said they received credit for non-cattle (or alternative) activities. Of those who did receive credit for alternative production such as cropping (e.g., cassava, banana, coconut) or small animals (e.g., poultry), a limited 3 people are still involved in those activities in 2011. When queried as to the reasons for no longer engaging in these activities, respondents identified, in order of importance: the absence of technical support, accidental fires, and their lack of experience with intended crops. As one smallholder put it, "Crops did not develop since there was no technical support or evaluation of soils and water in the region. There was no contract establishing banana sales, and I would not know where to sell the product even if it had

[2] The minimum wage in Brazil changes every 1 January. In 2012 this value reached US $308.

worked out well" (personal communication, April 2011). In addition to criticism associated with lack of technical support and market strength, it was also mentioned with high frequency that soils were not sufficiently fertile, and that additional inputs would be needed, such as expensive fertilizers and tractors, which only large-scale farmers can afford.

All extension agents interviewed recognized the potential viability in the region for the production of cassava flour, banana, pineapple, passion fruit, *cupuaçu* and *açaí*. However, these options are limited by a number of problems that must be resolved before these crops can be productive. Foremost among them is the absence of a well-developed production chain for those products. Furthermore, roads are in poor condition during the extended rainy season, making it impossible to bring perishable products to the city in time to get a good return. For instance, *acerola* cherry and *açai,* two viable and profitable crops for the region, need to reach the market within 24 h after harvest, and *cupuaçu* within a window of four days (personal communication, extension agent, Marabá, April 2011). When it came time to recommend productive activities for the region, these logistical barriers were of greater concern than the physical and chemical characteristics of the soils. Another problem relates to the lack of continuity in technical support for diversified crop production after the contract with INCRA has ended. A final obstacle to crop production stems from the lack of laboratories in the region qualified to perform scientific analysis of soils, which is essential for any agricultural plans.

Of those 33 smallholders who had invested in alternative production, only nine (27 percent) considered this more profitable than investments in cattle. The most frequent explanation related to inadequate lot size for cattle to be economically viable and the long-term unsustainability of pasture in an area where soil quality is questionable. Indeed, smallholders understand that productivity and profitability are related to seasonality, investment in technology and soil quality in the region. The importance of seasonality became an issue for consideration during our data collection. For example, when the questionnaire was tested in July 2010, smallholders from SP Castanhal Araras reported that the region was not good for anything, and that after 23 years of settlement creation, there was not a single profitable plantation operating in the region. However, during the final data collection in 2011, which happened during *cupuaçu* harvest season, we learned that many smallholders were engaged in production of that fruit.

One respondent reported that it was possible to sell up to 60 kilograms of *cupuaçu* pulp per day, at a price up to US $2.50 per kg. However, price instability for crops in the region, such as cupuaçu, made the activity less appealing than investing in cattle, which have stable prices and demand. Furthermore, *cupuaçu* is harvested during the rainy season, which corresponds to a time when road conditions are the most problematic. Agricultural extension agents confirmed that all crop production during the rainy season, even that of farmers engaged in banana and cassava, has experienced serious problems due to the lack of options for sales. As a result of these failures, in the short-term many smallholders, even those who received credit for cropping, shifted their land use to pasture and began to invest in cattle.

Given an understanding of these economic gains coupled with the need to reduce deforestation in the region, i.e., by the inclusion of Marabá in the List of priority Municipalities[3], a number of strategies to include smallholders in diversified initiatives have been developed by agronomists. However, inadequate lot size was cited by extension agents and smallholders as being the main barrier to their involvement in such green economic activities. For instance, an economically viable project for carbon sequestration requires an area of at least 1000 hectares, which excludes the smallholders who participated in this research, as the average property size is 33 hectares. To be feasible, smallholders would need to form a cooperative. Although not impossible, cooperative formation has proven to be a challenge in the SPs located in this region and elsewhere. An additional constraint is the long-term planning time horizon to see a profit, since it would be several years before the standing forest generates

[3] Marabá was added to the list of priority municipalities in reduction of deforestation and degradation in 2009 through MMA DECREE No. 102 OF 24 MARCH 2009. This decree provides a list of municipalities located in the Amazon biome, where priority actions focus on prevention, monitoring and control of illegal deforestation.

income from carbon sequestration at the level required by the project. According to key informants, the opportunity costs for forest protection under programs, such as REDD, are too great, and smallholders see no motivation to keep their land in forest if they are unable to receive a profit.

Cattle, and in particular calving, is the only activity in the region that has a well-consolidated supply chain and does not depend on road conditions to reach the market in good condition. All told, credit availability for cattle production is the key motivation cited by smallholders in their decision to invest in cattle. Our 2011 interviews show that an estimated 79 smallholders reported that they received government credit exclusively for investment in cattle, and of those farmers, 63 (80 percent) still have animals on their property. According to our analysis, there was a statistically significant relationship between receiving credit for cattle and having cattle in 2010. Those no longer engaged in cattle activities, despite receiving credit, explained that they had no option but to sell their animals in order to pay debts, mostly related to medical expenses, and all expressed their intention to buy more cattle in the future.

Agricultural extension agents and smallholders alike have the perception that cattle provide better economic returns when compared to crops. This is especially so since herds move by themselves regardless of season, eliminating transportation costs, and they provide the added benefit of milk production for familial consumption. Therefore, once smallholders perceive that there is a well-established chain for calf sales, they open small plots of pasture without any credit and at the time of PDA development, they refuse to switch to more diversified production.

It is logical for smallholders to pursue an activity that requires less labor, receives easy credit, and has a stable market for their production. This fact perpetuates cattle production in the region. Even extension agents, who are tasked with explaining the benefits of agricultural diversification, argue that smallholders make more profit in cattle ranching. Given the small number of extension agents and high demand for their services, agents interviewed reported that they save time, and are, therefore, able to provide services to more SPs, if the recommended projects promote cattle as opposed to crop diversification, which requires feasibility studies and training for several crops, in accordance with the settler's history as PDA and policy require. According to a lawyer working with the landless movement (personal communication, April 2011), smallholders work by following the logic of immediate results, and the mechanics of bringing products to market in larger cities remain a challenge. Even new smallholders receiving a piece of land inside SPs know that crop diversification would be a better option for the environment; however, it is easier to get credit for cattle only, since rarely is credit made available for other production activities. In addition, a lawyer for the landless movement emphasized that some smallholders have previously worked on farms as *vaqueiros* (cowboys), which is an important factor when it is time to decide about investments in the lot: "They stay with the activity that they know when it comes to establishing their own land" (personal communication, April 2011).

3.4. Emergent Production Chains

In the SPs of Southeastern Pará, a multiplicity of factors potentially affect the switch from subsistence agriculture to cattle ranching, including the availability of credit and strong economic returns stemming from the global demand for beef [2,12,17]. Within the region, many smallholders started ranching in order to produce milk for home consumption and to sell to local micro-industries. It is important to recall that credit given for SP livestock was originally meant for milk production, although few households today are engaged in this activity. However, an important secondary product of dairy production are calves, which proved to be economically more profitable. Over time, many smallholders began selling off their male calves, which makes sense given calf prices rose from US $137 in 2005 to US $363 in 2011 [34,35]. The viability of such marketing has involved the development of commodity chain links between smallholders and large-scale ranchers, who buy calves to fatten on the way to the slaughterhouse. The precipitous rise in calf prices probably reflects the sudden insertion of Southeastern Pará into the global economy by way of a fast hook-up to transcontinental value chains in the provision of chilled beef. The expansion of meat processing capacity in the region has come quickly, from not even one modern slaughterhouse in 2000, to 14 modern facilities by 2014, capable of processing from 500 to 1200 animals per day (Table 6). This new industrial capacity also reflects global interests with significant transnational participation. For example, the Brazilian Company JBS, which is today the world's largest meat-processing corporation, possesses five of the region's slaughterhouses. Its record of corporate buyouts includes the acquisition of the formerly US-held businesses Swift & Company in 2007, Smithfield Foods in 2008, and Pilgrim's Pride in 2009.

Table 6. Slaughterhouses with federal inspection (SIF) in the south of Pará in 2005 and 2015.

SIF [1]	Location	Company Name 2005	Classification [2] 2005	Company Name 2015	Classification 2015
807 [3]	Redenção	Redenção Frigorífico do Pará Ltda	MB2	JBS S/A	MB2
457	Maraba	Bertin Ltda.	MB3	JBS S/A	MB3
1110	Santana do Araguaia	Redenção Frigorífico do Pará Ltda	MB3	JBS S/A	MB3
4398	Xinguara	Frigoxim Comercial Ltda.	MB3	Xinguara Ind. E Com. S/A	MB3
593	Eldorado do Carajás	Frigorífico Industrial Eldorado Ltda.	MB4	JBS S/A	MB4
2350	Tucumã	Redenção Frigorífico do Pará Ltda.	MB4	JBS S/A	MB4
2583	Água Azul do Norte	Frigol Pará Ltda.	MB4	Frigol Pará Ltda.	MB3
4413	Xinguara	Mafripar Matadouro frigorífico Paraense Ldta.	MB4	Mercúrio Alimentos S/A	MB4
112	Rio Maria	Frigorífico Rio Maria Ltda.	MB4	Frigorífico Rio Maria Ltda.	MB4
2437	Sao Geraldo do Araguaia	-	-	Masterboi Ltda.	MB1
4150	São Félix do Xingu	-	-	Frigol Para Ltda.	MB3
1497	Tucumã	-	-	MFB Marfrig Frigoríficos Brasil	MB4
2258	Redenção	-	-	Abatedouro de Bovinos Sampaio Ltda. - ME	MB4
1745	Santa Maria das Barreiras	-	-	NPL Frigorífico e Indústria Ltda. [4]	MB5

[1] SIF (Sistema de Inspeção Federal, or Federal Inspection Service) is a mandatory registry for slaughterhouses selling to other states and countries; [2] The classification here includes 5 levels, from the biggest, MB1, to smallest, MB5, and is based on the capacity to slaughter animals per hour and quantity of veterinarians needed. In the biggest classification, for example, MB1, the industry can slaughter more than 80 animals per hour, industrialize 20 or more tons of meat per day and must have at least three veterinarians following all steps of the production. A slaughterhouse with classification MB2 has the same capacity of slaughtering, can or cannot have an industrialization area and two veterinarians are employed in the production. The remaining classes only employ 1 veterinarian and can slaughter between 40–80 animals/h (MB3), 20–40 animals/h (MB4) or less than 20 animals/h MB5 (Monteiro, 2012); [3] Note that the SIF does not change between years, and so this number shows that companies have been bought by larger companies along the years; [4] This company has been leased by JBS S/A since 2014. It is not operating.

Because of smallholders' lack of access to technology, largeholders still provide a decisive commercial link between smallholders and final consumption, via slaughterhouses in the region. This is because modern slaughterhouses, highly capitalized and corporate, only buy in truck-lots of at least 18 animals. Since smallholders possess neither the land nor the resources to produce this many animals at one time, largeholders with fattening operations purchase aggregated production from multiple smallholders within SPs in their vicinity. Indeed, our fieldwork establishes that more than two-thirds of smallholders resident in SPs sell calves to local largeholders. Consequently, such purchase arrangements, often informal in nature, put considerable market power in the hands of largeholders to set the terms of trade, leaving smallholders at a disadvantage.

This study uncovered several questionable instances where individual largeholders bought the entire production of a SP, coercing smallholders to accept payments in kind instead of monetary compensation, including pasture improvements, fence repairs, etc. In most cases there is no formal contract specifying the rights and duties of each actor. For instance, one settler stated that he owed money to the rancher for the expenses to build a fence, which was necessary to make his property ready to receive the ranchers' animals. It was agreed that the amount owed would be deducted from the final pasture rental payment, which was to be paid not in currency, but in calves born on the smallholder's property. However, at time of reconciling accounts, the smallholder owed more than the quantity of animals he was supposed to keep as payment for the land rental. In the end, the smallholder had no calves in payment, and he had a debt still owed to the rancher, which was in turn tacked onto the subsequent pasture-for-calves rental agreement.

The newly emergent production chain linking large and smallholders has the potential to undermine Amazonian family farming, with implications for food security [5]. Starting in 1989, crops, such as beans (*Phaseolus vulgaris*), cassava (*Manihot esculenta Crantz*), and coffee (*Coffea arabica* L.), which historically were important in that region, began to yield space to more commercial products such as soybeans (*Glycine max*) and cattle ranching [5]. During this evolution, soybeans utilized more hectares in Northern Mato Grosso, while cattle expanded toward the so-called arc of deforestation [36]. In this process, food security concerns stem from exposure to the vagaries of the market, as food insecurity can result from a market crash. However, investing in diversified production provides access to nutritious food, and does not expose the producer to economic crashes, for example, in case a disease affects commercial products, not only reducing production, but also in closing markets.

4. Conclusions

For a variety of reasons, the smallholders we spoke with were not using their land in a manner consistent with long-term sustainability, at least with respect to the environment. Instead, they have articulated with global production chains for beef, now governed in part by large transnational enterprises, including Brazil's JBS, the world's largest beef products corporation. The nature of credit provision appears to be a key factor undergirding this result. In particular, the financial capital available to smallholders in the form of low interest loans has incentivized cattle production at the expense of diversified systems. This outcome stands at odds with the original agricultural goal for both agrarian reform movements and the Brazilian state. Our informants point to the ease of obtaining credit for cattle, while quite the opposite situation prevails for those desiring to implement diversified farming systems. Credit for cattle is meant to stimulate the dairy sector, but smallholders have discovered that calving operations aimed at beef production provide a greater return. Although they do not express it this way, the smallholders in our sample find the opportunity costs of both diversified farming and dairy production to be too high, given the good prices and sure market for calves that have in all likelihood resulted from Brazil's emergence on the world stage as a premier beef supplier. Complementing the land-use choices of the smallholders in our sample is low soil fertility throughout the region, a circumstance that has been observed by extension agents in our research network. Newly available transfer payments (e.g., *bolsa-família*) have also helped build the smallholder cattle economy by reducing the supply of rural labor, critical for the implementation and maintenance of diversified farm systems with intensified production. The fact that smallholders invest in cattle as their main economic activity deserves attention for a number of reasons, three of which we now consider.

First, it is contrary to agrarian reform policy, meant to secure both livelihood and environmental sustainability by supporting diversified family farming. Second, the substitution of cattle herds for diversified agriculture exposes smallholders to price fluctuation affecting any monocultural system. Third, smallholder interest in cattle facilitates absentee ownership given reduced labor requirements. In other words, credit availability meant to help smallholders might reduce the numbers of family farms, and potentially open the door to land concentration. The importance of cattle to smallholders in Southeastern Pará (and Amazonia more generally) has a basis in the failure of government to adequately vet regional market opportunities in the settlement formation process, which requires federal approval of the PDA. Barriers to sustainable development in the Brazilian Amazon are historically tied to public policies that have brought people to a region without providing adequate infrastructure. Mistakes of the past are amplified at the present time by the provision of credits favoring cattle acquisition, and stimulating more deforestation than might otherwise be the case, given the extensive nature of ranching. Opportunity costs associated with diversified agriculture in Southeastern Pará appear to stimulate Amazonia's emergent ranching economy. This economic consequence, although it has provided smallholders a modicum of accumulation, contradicts the principle of sustainability, as originally conceived by both the social movements and the Brazilian government. Furthermore, smallholders in the SPs appear to be subject to exploitive terms of trade by virtue of the informal calving contracts that often leave them in debt. Such arrangements make these SP residents "contract" ranchers, who bear a disproportionate share of risk in the supply chains to which they deliver calves. Thus, like contract farmers around the world, the smallholders in our study sample arguably maintain a tenuous link to the land they own, and, as such, are subject to a form of land grab now endemic to global production networks. Consequently, the cattle economy could add a new phase to this region's intransigent history of land conflict.

Acknowledgments: The research for this article was support by grants from: Inter American Foundation (IAF), # G14060059; Latin American and Caribbean Environmental Economics Program (Laceep), GSEBSA-130; and the National Science Foundation (Geography and Spatial Sciences) under projects # 0522062—Brazil's Direct Action Land Reform Movement: Environmental Impacts and Socio-Spatial Dynamics, and # 1541489—Contentious Land Change in Amazonia. Any opinions, findings, conclusions, or other statements are those of the authors and do not necessarily reflect those of the agencies that helped support this work. The authors would like to acknowledge the helpful suggestions of the anonymous reviewers and the special issue editors, Claudia A. Radel and Jacqueline M. Vadjunec, whose efforts have much improved this final manuscript.

Author Contributions: Ritaumaria Pereira oversaw the overall development of this project. All authors contributed with the research design. Ritaumaria Pereira conducted the field research in 2010, 2011, and 2012. Cynthia S. Simmons led the field research in 2006. All authors contributed with literature review, analysis, and writing.

Conflicts of Interest: The authors declare no conflict of interest.

References

1. IBGE. Censo Demográfico. Available online: www.sidra.ibge.gov.br (accessed on 22 July 2011).
2. Arima, E.; Barreto, P.; Brito, M. *Pecuária na Amazônia: Tendências e Implicações para a Conservação*; IMAZON: Belém, Brazil, 2005.
3. Simmons, C.S. The political economy of land conflict in the eastern Brazilian Amazon. *Ann. Assoc. Am. Geogr.* **2004**, *94*, 183–206. [CrossRef]
4. Simmons, C.; Walker, R.; Perz, S.; Aldrich, S.; Caldas, M.; Pereira, R.; Leite, F.; Fernandes, L.C.; Arima, E. Doing it for themselves: Direct action land reform in the Brazilian Amazon. *World Dev.* **2010**, *38*, 429–444. [CrossRef]
5. Simon, M.F.; Garagorry, F.L. The expansion of agriculture in the Brazilian Amazon. *Environ. Conserv.* **2005**, *32*, 203–212. [CrossRef]
6. Walker, R.; Homma, A.; Scatena, F.; Conto, A.; Rodriquez, C.; Ferreira, C.; Mourão, P.; Carvalho, R. *Amazonia: The Environment and Agricultural Development*; EMBRAPA: Brasilia, Brazil, 1998.
7. Poccard-Chapuis, R.; Turrand, J.F.; Pketty, M.G.; da Veiga, J. *Cadeia Produtiva de Gado de Corte e Pecuarização da Agricultura Familiar na Transamazônica*; EMBRAPA: Belem, Porgutal, 2001.
8. Vosti, S.A.; Baz, E.M.; Crpentier, C.L.; D'Oliveira, M.V.N.; Witcover, J. Small-scale managed forestry at the Brazilian agricultural frontier: Adoption, effects and policy issues. In Proceedings of the International Conference on Rural Livelihoods, Forests and Biodiversity, Bonn, Germany, 19–23 May 2003.

9. Homma, A. *Programacao das Atividades Agropecuarias, sob Condicoes de Risco, nos Lotes do Núcleo de Colonização de Altamira*; Universidade Federal de Viçosa: Minais Gerais, Brazil, 1976.
10. Walker, R.; Moran, E.; Anselin, L. Deforestation and cattle ranching in the Brazilian Amazon: External capital and household processes. *World Dev.* **2000**, *28*, 683–699. [CrossRef]
11. Hecht, S.B. Environment, development, and politics: Capital accumulation and the livestock sector in eastern Amazonia. *World Dev.* **1985**, *13*, 663–684. [CrossRef]
12. Walker, R.; Browder, J.; Arima, E.; Simmons, C.; Pereira, R.; Caldas, M.; Shirota, R.; De Zen, S. Ranching and the new global range: Amazônia in the 21st century. *Geoforum* **2009**, *40*, 732–745. [CrossRef]
13. Uhl, C.; Arima, E. Ranching in the Brazilian Amazon in a national context: Economics, policy, and practice. *Soc. Nat. Resour.* **1997**, *10*, 433–451.
14. IBGE. Pesquisa Pecuária Municipal. Available online: www.sidra.ibge.gov.br (accessed on 15 September 2015).
15. Veiga, J.B.; Tourrand, J.F.; Piketty, M.G.; Poccard-Chapuis, R.; Alves, A.M.; Thales, M.C. *Expansão e Trajetórias da Pecuária na Amazônia*; UnB: Brasilia, Brazil, 2004.
16. Faminow, M. Spatial economics of local demand for cattle products in amazon development. *Agric. Ecosyst. Environ.* **1997**, *62*, 1–11. [CrossRef]
17. Gomes, C.V.A.; Vadjunec, J.M.; Perz, S.G. Rubber tapper identities: Political-economic dynamics, livelihood shifts, and environmental implications in a changing Amazon. *Geoforum* **2012**, *43*, 260–271. [CrossRef]
18. Agropecuario, C. *Brasil, Grandes Regioes e Unidades da Federacao*; IBGE: Brasilia, Brazil, 2006.
19. Simmons, C.S.; Prz, S.; Pdlowski, M.A.; Silva, L.G.T. The changing dynamics of land conflict in the brazilian amazon: The rural-urban complex and its environmental implications. *Urban Ecosyst.* **2003**, *6*, 99–122. [CrossRef]
20. INCRA. Assentamentos. Available online: http://www.incra.gov.br (accessed on 15 May 2011).
21. Wolford, W. Environmental justice and agricultural development in the Brazilian cerrado. In *Environmental Justice in Latin America*; Carruthers, D.V., Ed.; The MIT Press: Cambridge, MA, USA, 2008; pp. 210–237.
22. Agrária, R. *Compromisso de Todos*; Presidência da República, Governo Fernando Henrique Cardoso: Brasília, Brazil, 1997.
23. INCRA. Crédito Instalação. Available online: http://www.incra.gov.br (accessed on 14 December 2011).
24. Gonçalves, E.C.; Fernandes, B.M. *Análise e Mapeamento dos Tipos de Assentamentos de Reforma Agrária e Reforma Agrária de Mercado no Brasil: Contribuição para Compreensão da Diversidade e a Atualidade da Reforma Agrária Brasileira*; NERA: Presidente Prudente, Brazil, 2009.
25. Walker, R. Mapping process to pattern in the landscape change of the Amazonian frontier. *Ann. Assoc. Am. Geogr.* **2003**, *93*, 376–398. [CrossRef]
26. Browder, J.O.; Pedlowski, M.A.; Walker, R.; Wynne, R.H.; Summers, P.M.; Abad, A.; Bocerra-Cordoba, N.; Mil-Homens, J. Patterns of development and land cover change in the Brazilian Amazon: A farm-level analysis of Rondonia'S settlement frontier, 1992–2002. *World Dev.* **2008**, *36*, 1469–1492. [CrossRef]
27. Mertens, B.; Poccard-Chapuis, R.; Piketty, M.G.; Lacques, A.E.; Venturieri, A. Crossing spatial analyses and livestock economics to understand deforestation processes in the Brazilian Amazon: The case of São Félix do Xingu in south Pará. *Agric. Econ.* **2002**, *27*, 269–294.
28. Ortiz, S. Peasant economics: Farm houseolds and agrarian development. *Am. Anthropol.* **1991**, *93*, 213–214. [CrossRef]
29. Munroe, D.K.; McSweeney, K.; Olson, J.L.; Mansfield, B. Using economic geography to reinvigorate land-change science. *Geoforum* **2014**, *52*, 12–21. [CrossRef]
30. Gereffi, G.; Korzeniewicz, M. *Commodity Chains and Global Capitalism*; Praeger Publishers: Westport, CT, USA, 1994.
31. Borras, S.M., Jr.; Franco, J.C.; Gómez, S.; Kay, C.; Spoor, M. Land grabbing in Latin America and the Caribbean. *J. Peasant Stud.* **2012**, *39*, 845–872. [CrossRef]
32. Lee Peluso, N.; Lund, C. New frontiers of land control: Introduction. *J. Peasant Stud.* **2011**, *38*, 667–681. [CrossRef]
33. Smalley, R.; Corbera, E. Large-scale land deals from inside out: Findings from Kenya's Tana Delta. *J. Peasant Stud.* **2012**, *39*, 1039–1075. [CrossRef]
34. Anualpec. *Anuário da Pecuária Brasileira*; Instituto FNP: São Paulo, Brazil, 2006.
35. Anualpec. *Anuário da Pecuária Brasileira*; Informa Economics-FNP: São Paulo, Brazil, 2011.
36. Margulis, S. *Causes of Deforestation of the Brazilian Amazon*; World Bank: Washington, DC, USA, 2004.

land

MDPI

Article

"Nothing Is Like It Was Before": The Dynamics between Land-Use and Land-Cover, and Livelihood Strategies in the Northern Vietnam Borderlands

Sarah Turner [1,*] and Thi-Thanh-Hiên Pham [2]

[1] Department of Geography, McGill University, 805 rue Sherbrooke Ouest, Montréal, QC H3A 0B9, Canada
[2] Département D'études Urbaines et Touristiques, Université du Québec à Montréal, 315 rue Sainte-Catherine Est, Montréal, QC H2X 3X2, Canada; pham.thi_thanh_hien@uqam.ca
* Correspondence: sarah.turner@mcgill.ca; Tel.: +1-514-398-4955

Academic Editors: Claudia A. Radel and Jacqueline M. Vadjunec
Received: 28 July 2015; Accepted: 9 November 2015; Published: 18 November 2015

Abstract: Land uses are changing rapidly in Vietnam's upland northern borderlands. Regional development platforms such as the Greater Mekong Subregion, state-propelled market integration and reforestation programs, and lowland entrepreneurs and migrants are all impacting this frontier landscape. Drawing on a mixed methods approach using remote sensing data from 2000 to 2009 and ethnographic fieldwork, we examine how land-use and land-cover change (LULCC) has occurred across three borderland provinces—Lai Châu, Lào Cai and Hà Giang—with high proportions of ethnic minority semi-subsistence farmers. After a broad examination of regional land-use changes, we select three case studies to further analyze the underlying relationships between specific LULCC and local livelihood diversification strategies. These include specific patterns of urban growth due to a range of political decisions in Lai Châu and Lào Cai Provinces; reforestation due to non-timber forest (NTFP) product cultivation in the west of Lào Cai Province; and a stable landscape that restricts government attempts at refashioning upland livelihoods in the east of Hà Giang. Our findings point to the difficulties of completing LULCC maps for this highly heterogeneous region and the complexity of LULCC and livelihood interactions and relationships examined on the ground.

Keywords: livelihood diversification; ethnic minorities; northern Vietnam; land-use and land-cover change; reforestation; cardamom; urbanization; mixed methods

---

1. Introduction

Two hundred million people, more than half of whom are ethnic minorities, reside in the Southeast Asian Massif, a broad expanse of uplands extending southeast from the Himalayan Plateau and shared today among 10 countries. Dwelling on the physical, economic, and often cultural fringes of their respective states and predominantly undertaking semi-subsistence rural agriculture, residents of these frontiers face numerous land-use and livelihood challenges. While studies of the causal mechanisms of land-use and land-cover change (LULCC) in the Southeast Asian Massif are growing [1–5], there remains an urgent need to refine our understandings of how and why potential factors—be they socio-economic, political, or ecological—underpin land-use changes in these uplands. Not only do we need to better comprehend how specific land policies and market-led changes, for instance, drive LULCC, but also how depleted soil nutrients, population growth, or household agricultural knowledge can trigger land-use change [6]. Furthermore, since rural household livelihoods often have direct relationships with land use, the interactions between land-use change and livelihoods merit greater attention [7–10].

The livelihoods of those residing in the Southeast Asian Massif are undergoing important transformations due to market integration, agrarian transformations, globalization, and the closing of land frontiers. Indeed, political scientist James C. Scott in his 2009 book, *The Art of Not Being Governed*, noted that since 1945, these uplands have become increasingly claimed and incorporated by modern states through processes labeled as "development, economic progress, literacy, and social integration" ([11]; p. 4). For most local residents, this has resulted in the replacement of communal property with private land-use rights, the establishment of cash cropping, and a drive to convert shifting cultivators into permanent farmers. One such regional development approach is the Greater Mekong Sub-Region (GMS), which aims to connect the uplands of Vietnam, Cambodia, Laos, Myanmar, Thailand, and Yunnan Province, China. Via the GMS project alone, the Asian Development Bank (ADB) and regional states have poured more than US$14 billion into infrastructure, including highways and railways, telecommunications, and energy and urban development [12,13]. With this region also increasingly targeted by state officials and private entrepreneurs for natural resource extraction and agrarian possibilities, livelihood approaches and land-use patterns are being impacted profoundly [14,15].

Within the Southeast Asian Massif, Vietnam's northern borderlands—with their complex political context on the Chinese border, ethnic diversity, and dynamic frontier location—provide an important case study with which to examine the mechanisms of LULCC through a livelihood lens. These borderlands are undergoing critical socio-economic transformations created by both trans-national and local forces. Two of the most notable implications for LULCC are increasing urbanization and changing forest dynamics. Urbanization, supported by the government's pro-urban policies since economic reforms were initiated in the mid-1980s (*Đổi Mới*) [16], is increasing market interactions and opening up possibilities of new livelihood diversification trajectories for local populations. Forest changes, including deforestation, afforestation (often for timber production), and reforestation (natural succession, as well as production policies) are in turn impacted by interwoven and oftentimes conflicting land-use and forest policies [14], creating different livelihood opportunities as well as restrictions.

In this paper we examine how livelihoods and LULCC are intertwined within three of Vietnam's northern borderland provinces, namely Lai Châu, Lào Cai, and Hà Giang. Each province is home to a high proportion of ethnic minority semi-subsistence farmers. We focus specifically on the 14 border districts of these three study provinces (Figure 1) and temporally, on changes that have occurred from 2000 to 2009.[1] After examining meso-level results across our study region, we select three case studies and examine the underlying relationships between specific LULCC and local livelihood diversification strategies by combining quantitative mapping with qualitative fieldwork in a mixed methods approach. Such an approach is increasingly recognized and promoted by scholars from different disciplinary backgrounds to provide insight into the causes of land-use change [17]. We find that combining ethnographic fieldwork with land-cover change mapping reveals hidden causes of LULCC in these three provinces. For instance, specific patterns of urban growth have been caused by pro-urban policies, cross-border trade, political boundary redrawing, and resettlement in Lai Châu and Lào Cai Provinces. In the west of Lào Cai Province, reforestation has occurred due to the increased cultivation of cardamom, a non-timber forest product (NTFP), while in the east of Hà Giang a particular landscape has restricted government attempts at refashioning upland livelihoods from those already built around local maize cultivation.

---

1 These years correspond with the last two national censuses completed for Vietnam (1999 and 2009), and the most unobstructed satellite images available. They also cover a dynamic time in the contemporary political and economic history of the uplands.

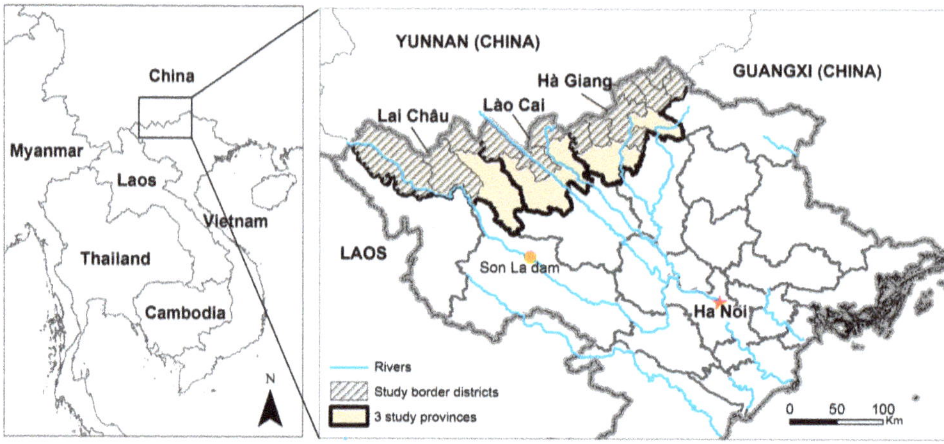

Figure 1. Location of the study border districts and provinces in northern Vietnam.

2. Conceptual Framework: Land Change Science, Frontiers and Livelihood Diversification

Conceptually, we draw from land change science (LCS), frontier studies, and livelihood diversification literature. LCS is an emerging interdisciplinary area of study that "focuses on material related to the nature of land use and land cover, their change over space and time, and processes that produce those patterns and changes"([9]; p. 3). We direct attention to three of the core objectives of LCS, namely the observation of land changes, understanding the causes of these changes, and assessing their impacts [9,18–20]. Of specific note here are causes of LULCC, which can be separated into predisposing environmental factors, underlying forces, and proximate causes [21]. Predisposing environmental factors include elevation, slope, soil conditions, and natural disasters. Underlying causes can include population, economic change, policy, ethno-cultural aspects, and technology [21]. Proximate causes are activities that directly alter land use. For example, one common proximate cause of deforestation is agricultural expansion, which could be the result of an underlying cause such as population growth or in-migration [22].

Frontier regions are both physical entities and social constructs [23,24]. Frontier studies are important to help analyze the degree to which the Sino-Vietnamese borderlands are experiencing natural resources extraction, agricultural intensification, in-migration, increasing cross-border exchanges, and state control [25,26]—all of which are altering land use and land cover in the region. Frontier regions offer numerous economic attractions such as access to natural resources, newfound business opportunities, and cross-border employment and trade prospects [13,27–32]. Agricultural expansion and natural resource exploitation are two particularly common vectors of frontier exploitation by both state and private entrepreneurs [33–36], often resulting in "spontaneous land opening, environmental degradation and the taming of ethnic minorities" ([37]; p. 14). Moreover, culturally speaking, frontier regions are often depicted as "backward" and in need of improvement, in contrast with the relative social and cultural "modernity" and dominance of the core [24,38], in this case the majority Vietnamese (Kinh or *Việt*) lowlands. Used to ease demographic pressure on core areas, state and private investment in frontier regions across the Southeast Asian Massif has also been argued to establish territorial domination by the state, further integrating ethnic minorities into the Nation [11,26,39].

For some, livelihoods in this region are changing substantially as one livelihood portfolio is replaced by another, such as when agricultural semi-subsistence is replaced by waged labor in a rural town. Yet more frequently, households undergo livelihood diversification, "the process by which rural families construct a diverse portfolio of activities and social support capabilities in their struggle for survival and in order to improve their standards of living" ([40]; p. 4). Livelihood diversification needs to be examined through a dynamic analysis incorporating the context-specific sociocultural (e.g., ethnic, gender, class), environmental, historical, and spatial dynamics of how people create and sustain viable and meaningful livings [41,42]. In turn, these responses are situated within broader spheres of institutional knowledge and power, with social actors supported or constrained by discrepancies in access to resources and information [43–45]. In rural locales in the Global South, livelihood diversification is often closely linked with changing land uses and, in turn, changes in land cover [10,46,47].

Together, these three bodies of literature inform each other and allow us to better interpret the LULCC that we find in Vietnam's northern uplands. This conceptual framing offers an integrated approach to examining the processes and results of changing land cover and land use as well as how they are interwoven with livelihood diversification decision making, power relations, access regimes, institutions, and state-society relations in this political and cultural frontier zone. Yet, combining these three conceptual approaches is challenging, as they are rooted in different paradigms, and, hence, use different research methodologies. Land change scientists usually use statistical models to identify environmental and socio-economic determinants of LULCC, but social scientists often criticize such methodological approaches as oversimplifying and generalizing social groups, and hence omitting intra-group livelihood diversity [17]. Combining such conceptual approaches thus needs to be done with caution, as explored below.

3. Context: Vietnam's Northern Uplands

The Socialist Republic of Vietnam recognizes 54 ethnic groups: the majority lowland Vietnamese and 53 "national minorities" (*các dân tộc thiểu số*). Though the latter comprise 14.8 percent of Vietnam's population [48], in the three upland provinces of Lai Châu, Lào Cai, and Hà Giang (Figure 1), ethnic minorities are in fact in the majority. They comprise 87 percent of the total population of 724,000 in Hà Giang Province, 85 percent in Lai Châu (total pop. 370,000), and 66 percent in Lào Cai (total pop. 614,000) [48]. Many of these ethnic minorities, such as the Hmong and Yao (Dao), belong to geographically-dispersed, politically-fragmented, lineage-based minority populations [49–51]. These populations frequently undertake diversified livelihoods through a composite agricultural system. A combination of maize and rice fields are commonly mixed with home gardens, small (officially banned) swidden plots, the gathering of firewood, honey, and herbs from forests, as well as small-scale commercial exchanges of cardamom, livestock, textiles, or homemade alcohol [15,52].

Land access and quality play central roles for these livelihoods, while it must be remembered that all land in Vietnam is officially owned by the state. Land-use rights have been assigned to households since the 1993 Land Law, following the dismantlement of socialist collectives in 1988, initiated by Resolution 10 [53,54]. Forests are also important for many upland livelihoods, providing fuel wood and non-timber forest products. The northern Democratic Republic of Vietnam (DRV) designated all hills and mountains with slopes over 25° as forestland in 1954 and founded a series of State Forest Enterprises (SFEs) [55,56]. These SFEs have had rather ambiguous functions, logging forests in some areas, while driving reforestation endeavors in others. Nonetheless, forest cover for the whole country dropped to 30 percent by 1985 due to the actions of these SFEs and individual farmers [57]. After the initiation of the *Đổi Mới* economic reforms of the mid-1980s that shifted Vietnam to a more open economy from a stricter socialist one [58], reforestation and forest protection became a priority in state policy (e.g., the 1991 Act on Forest Protection and Development and the 2004 Forest Protection and Development Law). A new management system was introduced with three classifications of forests: production forests (for timber extraction); protection forests (especially around watersheds); and special-use forests (focusing on strict preservation for biodiversity and cultural values) [59]. These classifications were integrated into the 1993 Land Law, which initiated a distribution of land tenure certificates providing production forest or protection forest allocations to upland households [60]. Through the late 1990s, a forest transition is thought to have occurred due not only to this classification of forests, but also due to a ban on opium and logging implemented in 1992/93, further reforestation programs including the "Greening the Barren Hills Program" (also named Program 327) from 1992, a similar program labeled Program 556, and the "Five Million Hectare Reforestation Program" from 1998. Nonetheless, the outcomes of these reforestation programs have been rather ambiguous, as case studies suggest that forest density and quality are diminishing [56,61–63].

The agricultural systems of upland ethnic minority groups have likewise undergone rapid transformations, including an expansion of wet rice farming due to migration of lowland Vietnamese from the Red River Delta since the 1960s and the *Đổi Mới* economic reforms. A move from swidden farming to fixed crops has been encouraged by the granting of long-term land-use rights to households as well as the introduction of high-yield hybrid rice and maize seeds since 1999 [26]. A transition from subsistence crops to cash crops in some regions has also been driven by the expansion of the market economy [2,52]. In addition, since 2000, new exogenous factors such as increasing weather extremes, and the 2008 food price crisis have further impacted livelihoods in these communities.

LULCC studies in upland Vietnam have started to analyze such transformations since the country's independence in 1954 [1,2,61,62,64–66]. However, all these studies have been conducted elsewhere in the country, either in the Central Highlands or in northern provinces at lower altitudes and more closely connected to the capital, Hanoi, than the borderlands. In addition, apart from our own prior research [14,67] including a systematic literature review, no study has addressed the relationships between LULCC and livelihoods along the northern borderlands with their ethnically diverse, quantitatively "poor" populations compared to the rest of the country [57] (but see [68] with one case study in a borderland district and [5] with a study neighboring, but not in, a borderland district). Consequently, LULCC dynamics in the northern border uplands—a political, resource, and cultural frontier—remain poorly understood.

4. Methods

Taking a mixed methods approach, this study draws on both quantitative and qualitative methods, namely land-cover change mapping and ethnographic interviews. More specifically, we take an "interactive design" to mixed methods, in that our approach emphasizes quantitative and qualitative methods equally. We do not, however, aim for data transformation (e.g., we do not take our qualitative data and transform them into quantifiable data for statistical analysis). Instead, we aim for "data importation", namely that data from one approach was reflected upon mid-stream in our analysis, and fed into the analysis of the other data set, and *vice versa* in an iterative manner [69]. We began by mapping land-cover change via remote sensing. We then chose cases representing the most important dynamics between LULCC and livelihoods, and drew on previous interviews completed since 1999 regarding livelihoods as well as additional interviews and observations focusing on the case sites during the summers of 2012, 2013, and 2014 to guide our interpretations. An important element in this interactive process was our long-term knowledge of the region, which allowed us to be confident about our choice of cases.

4.1. LULC and LULCC Mapping

Given the lack of accurate data regarding land use in rural Vietnam, especially historical data [70], we had to compromise somewhat methodologically, opting for an analysis of a mixed system of land-use and land-cover types. LULC maps were derived from Landsat thematic mapper (TM5) and enhanced thematic mapper (ETM+) images obtained from the United States Geological Survey (USGS). All images were taken during the winter season (Table 1) when sparse vegetable crops were growing, with no maize or wet rice. To identify land-cover types from our Landsat images, we drew on the Level I land-cover types proposed by Anderson in 1976 [71] and previous LULCC studies of Vietnam using Landsat images [52,61].

Table 1. Dates of images.

Province	Dates
Lai Châu	2009 November 03 2000 November 02
Lào Cai	2009 November 12 1999 December 27
Hà Giang	2009 November 05 2000 November 04

Initially, we identified five key land-*cover* types across the study region from satellite images (shrubs, bare soil, open canopy trees, closed canopy trees, and water). To provide further detail, we then added one land-*use* type, namely built-up areas, working from the five original land-*cover* types. It should be noted that there is rarely any grassland in this mountainous area; when small patches of grass are mixed up with shrubs, we identify them only as "*shrubs*". Cropped land is composed of subsistence crops (lowland and upland rice, corn, cassava) and cash crops (banana, pineapples, tobacco, among others). According to our interviews and local crop calendars, areas with subsistence crops are often covered by *bare soil* in the winter, when our images were acquired. This is also observed in other upland areas of Vietnam [4,52,61]. Pineapples are small and fairly dispersed plants, hence pineapple areas are spectrally similar to *bare soil* in the images. Areas having bananas remain vegetated and identified as *shrubs* in the Landsat images. However, bananas were not very common in the region in 2009 (interviews and observations), and were usually found next to houses or mixed with shrubs. Built-up areas are also identified as *bare soil* in the images, although dense urban areas have stronger spectral reflectance. To separate built-up from cropped land, we used information regarding road systems (see below). *Open canopy trees* in this region tend to be bamboo, plantations for timber production, or natural succession since the 1990s (interviews). *Closed canopy trees* are old and nearly intact forests that are often protected by the local government, such as special-use forests (often labeled national parks). This class has sometimes been called "mature and evergreen forest" in Southeast Asian forest studies [72]. Separating forests into two categories—open and closed canopy—is usual in Southeast Asian upland studies [4], although some authors tend to group them into one category [61,73]. Interpretation of classification results was done with these details in mind. Illustrations of these LULC types are presented in Figure 2, while we detail our classification method below.

The images were pre-processed to remove distortions caused by sensor errors, atmospheric interference, and surface irregularities. Cloud and cloud shadow masking was conducted on the 2000 image of Lai Châu and the 2009 image of Lào Cai (more specific technical details are provided in [14]). We used an object-based approach [74] to identify the five land-cover types listed above plus clouds and shadow. Although associated most often with very high-resolution images, this approach has proven accurate to produce rural land-cover types from middle-resolution images such as Landsat [75,76]. Cultivation systems in upland Southeast Asia are complex, mixing young forest, shrubs, and different types of crops [52,72]. In this context, an object-based approach is very helpful in creating segments (objects) and incorporating textural information of segments into classification. This approach has also allowed us to work at two spatial levels: segmentation of large-size patches (to separate large-scale forest and crops) and of small-size patches with relatively homogenous texture and signals (to identify smaller plots).

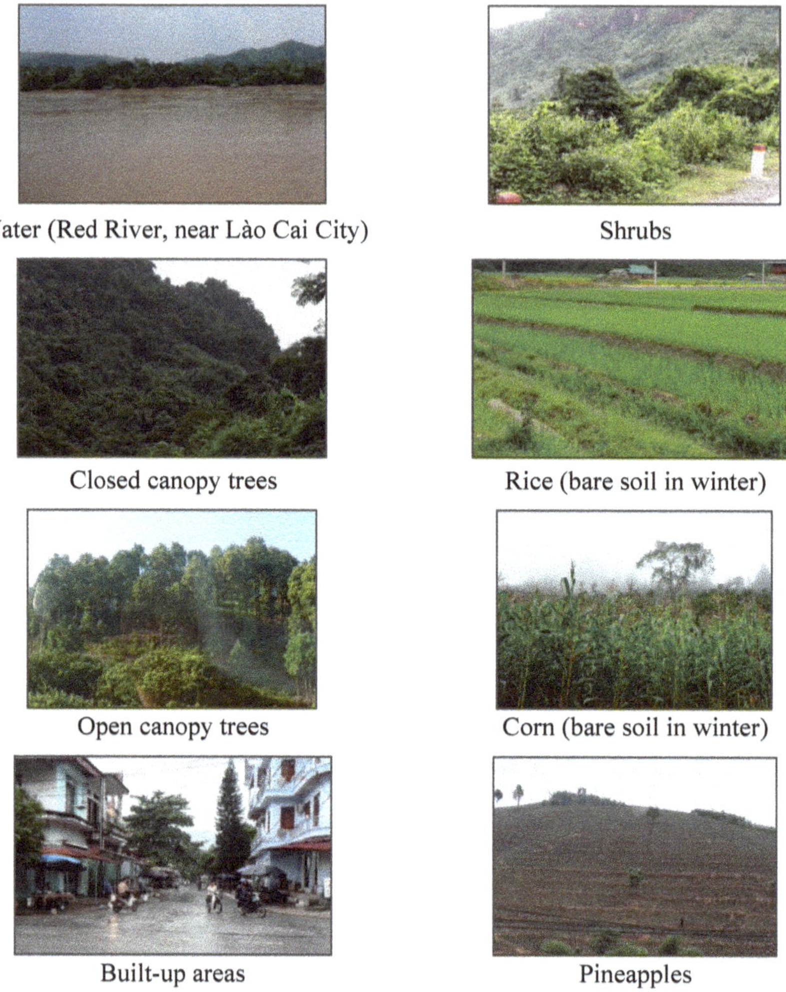

Figure 2. Illustration of the six LULC classes. (Photo credit: Lê Mạnh An, Thi-Thanh-Hiên Pham).

Segmentation parameters in eCognition (bands, scales, color/shape ratio, and compactness/smoothness ratio) were tested at different values. Segmentations and rule-based classifications were undertaken at different scales using the same band composition (bands 1–5, 7), color/shape ratio (0.2/0.8), and compactness/smoothness ratio (0.3/0.7). Segmentation values of 5, 10, 20, 50, 100, and 200 were tested, aiming at creating segments of different sizes. Segments were visually examined to determine the visibility of the main land-cover classes. We chose two values that produced the most homogenous segments in terms of spectral values and texture. The first segmentation was conducted at a scale of 50 to obtain large size objects. Rules were then used to classify those objects into *clouds and shadow*, *water*, and *bare soil* classes. The second segmentation was conducted at a scale of 10, then a second set of rules was used to classify segments into three vegetation classes of *shrubs*, *open canopy trees*, and *closed canopy trees*.

Bare soil was then separated into *bare soil* and *built-up*. Built-up pixels were assigned by evaluating road density. The road network was separated into solid (gravel or concrete) and non-solid (compacted soil susceptible to flooding during the rainy season) roads. A density map was developed by assigning solid roads an importance value of 3 and non-solid roads a value of 1. Pixels having high road density (3 m/km^2 in 1999 and 4m/km^2 in 2009) were visually compared to aerial images on Google Earth and those that coincided with more urban areas (built-up and having a street network) were renamed as *built-up*. The remaining *bare soil* pixels remained *bare soil*.

We did not use a set of training points *per se* to define rules of classification. Rather, we chose roughly 200 objects (segments) that corresponded to land-cover types that we knew on the ground from observations. Then we created rules composed of textural and spectral indicators for each class from the chosen objects. Refining and adjusting rules were based on interactive "trial and error". The classification process included initial field observations in summer 2013, and field verification in summer 2014 by the second author.

4.2. Assessing the LULC Mapping

A ground truth assessment was conducted for the 2009 classification by using ground control points (GCPs). There are no historical air photos for the region covering 1999/2000. There are a few photos on Google Earth from 2009/2010 but they are dispersed and cover only 15 percent of the study area. We hence opted for GCPs. We collected 365 GCPs: 142 points in Lào Cai Province (September 2012), 101 points in Lai Châu, and 122 points in Hà Giang (September 2013). Given the difficult access to many locations in the region, points were sampled along roads, focusing on typical land-cover types in the region. There were fewer points in Lai Châu given the lack of roads and access difficulties caused by landslides.

Each point was registered in GPS (precision of 5 m) and photographed in the four cardinal directions to capture potential mixtures of land cover. Descriptions of land use and land cover were included at each point and for the four directions. Land-cover types were assigned to each point based on photos and descriptions. Unfortunately, we do not have ground points for the 1999 images, but since we used the same image processing procedure for the 1999 and 2009 images, we believe that the accuracies of the 1999 images are similar to those of the 2009 images.

In order to evaluate the accuracy of the 2009 classification, we created a confusion matrix (Table 2) using the 365 GCPs. The overall accuracy is 71 percent, mostly due to confusions between *open canopy* and *closed canopy* classes. Accuracies varied from 46 percent (*open canopy*) to 96.88 percent (*closed canopy*). In this mountainous area undergoing complex forest transitions, *open canopy trees* are usually mixed with *shrubs*, making it difficult to separate them from Landsat images. The confusion between *open canopy* and *closed canopy* is most likely due to the fact that open forests in the three provinces are highly heterogeneous. What is defined on the ground as *open canopy* in areas with plantations (since the mid-1990s) could be similar spectrally to *closed canopy* in other areas. When grouping these two categories into "forest" (as done in several Southeast Asian LULCC studies [61,73]), we obtained a "user accuracy" of 79.30 percent and a "producer accuracy" of 72.22 percent, raising overall accuracy to 73.70 percent.

Other important confusions related to *bare soil*, our proxy for rice or corn fields. In these mountainous areas, dispersed houses and remote roads are often located close to rice or corn fields, creating a spectral mixture in the image. In Đồng Văn District, Hà Giang Province, this gets complicated further by limestone outcrops within fields and near houses (Figure 3a). Lastly, the confusion between *shrubs* and *bare soil* is explained by the fact that upland dry rice and maize are commonly planted in areas close to shrubs (Figure 3b).

Table 2. Confusion matrix and producer's and user's accuracy for accuracy assessment of the 2009 classification (based on ground control points).

		Ground Reference						Total	Producer Acc. (%)
		Water	**Closed Canopy**	**Open Canopy**	**Shrubs**	**Bare Soil**	**Built-Up**		
Classification	Water	**16**	0	0	0	1	2	19	84.21
	Closed canopy	3	**31**	10	7	3	0	54	57.41
	Open canopy	0	1	**23**	8	2	2	36	63.89
	Shrubs	0	0	13	**42**	7	2	64	65.63
	Bare soil	3	0	3	10	**109**	11	136	80.15
	Built-up	0	0	1	3	15	**37**	56	66.07
Total		22	32	50	70	137	54	365	
User Acc. (%)		72.73	96.88	46.00	60.00	79.56	68.52		

(a)

(b)

Figure 3. (**a**) Typical karst landscape and (**b**) mixture of maize and shrubs in Đồng Văn District, Hà Giang Province. (Photo credit: Sarah Turner and Lê Mạnh An).

4.3. Computing LULCC

To measure LULCC, we computed the percent change and the annual rate of change for each class, where *A1* is the cover of arable land at an initial time (t_1) and *A2* is the cover of arable land at a later time (t_2), and where $t = t_2 - t_1$ (Equations (1) and (2)) [77].

$$R = \frac{\mathrm{A2} - \mathrm{A1}}{\mathrm{A1}} \tag{1}$$

$$Rate = \ln(A2/A1)/t \tag{2}$$

We also computed the absolute values of changes for each LULCC type per district. In other words, we computed the magnitude of change (*R* in Equation (1)) without taking into consideration the direction of change (decreasing or increasing over time) because we wanted to capture the total amount of relative change. Finally, in order to adjust the changes by size of district, we computed the ratio of absolute value of change over the district area (without including areas covered by *clouds and shadow*). This allowed us to identify the most dynamic and the most stable districts in terms of LULCC. The higher the ratio, the more dynamic the district with regards to LULCC.

4.4. Qualitative Fieldwork

The qualitative fieldwork that supports this study was completed by the first author during repeated research visits to Lào Cai Province since 1999, Lai Châu Province since 2004, and Hà Giang Province since 2009, and by the second author in Lào Cai Province since 2012. This includes over 100 in-depth unstructured (conversational) interviews with ethnic minority farmers (Tày, Hmong, Yao, Nùng) in Lào Cai Province, and 50 each in Lai Châu and Hà Giang Provinces. Approximately 75 percent of farmer interviewees were women, as they were more likely to be in the house during the day, yet had a wide knowledge of land uses and changes. Farmer ages ranged from 25 to 80 years old. Interviews ranged in duration from 20 minutes to over an hour, focusing on local livelihood diversification, agricultural practices, land-use changes, and state-society relations. Twenty-eight semi-structured interviews were also completed with provincial officials (mostly Kinh) working in government departments linked to agriculture, planning, labor, and natural resources. The core themes of these interviews were livelihoods, market integration, land-cover change, and the impacts of state policies on each of these. Interviews were completed with the aid of local ethnic minority interpreters for farmers of the same ethnicity, or with Kinh interpreters or alone for Kinh farmers, urban dwellers, and state officials. All interviews were transcribed and coded using a mix of constant comparative, axial, and thematic qualitative coding approaches. Concurrently, observations of LULCC have been completed and noted annually.

5. Results

Our land-cover mapping shows that the most important changes in terms of area (Table 3) include an increase in *closed canopy forest* (2437 km^2), followed by a decrease in *open canopy forest* (roughly 1694 km^2), a decrease in *bare soil* (537 km^2), and a decrease in *shrubs* (324 km^2). Examining the 2000 and 2009 maps (Figure 4), spatial patterns for these changes are visible. The increase in *closed canopy forest* is most notable in all the border districts of Lào Cai Province (except Si Ma Cai and Lào Cai City), in Mường Tè District (Lai Châu Province), and in the southwestern parts of Hà Giang Province (also see Table 4). In turn, a decrease in *open canopy forest* has occurred in almost all these same locales where *closed canopy forest* increased.

Table 3. Land-cover change in the region, 2000–2009.

Land Cover	Area in 2000 (km^2)	Area in 2009 (km^2)	2009–2000 Area Change (km^2)	2009–2000 Percent Change (%)	Annual Rate of Change (%/year)
Closed canopy	1723.89	4161.03	2437.14	141.37	8.81
Open canopy	3281.12	1587.14	−1693.98	−51.63	−7.26
Shrubs	3357.08	3033.28	−323.80	−9.65	−1.01
Bare soil	3668.91	3141.54	−527.37	−14.37	−1.55
Built−up	50.58	108.29	57.70	114.08	7.61
Cloud shadow	1068.04	918.51	−149.53	−14.00	−1.51
Water	40.23	59.29	19.06	47.37	3.88
Total	13 189.86	13 009.07*			

* The Landsat scenes for the two years were slightly different, hence different total land-cover area.

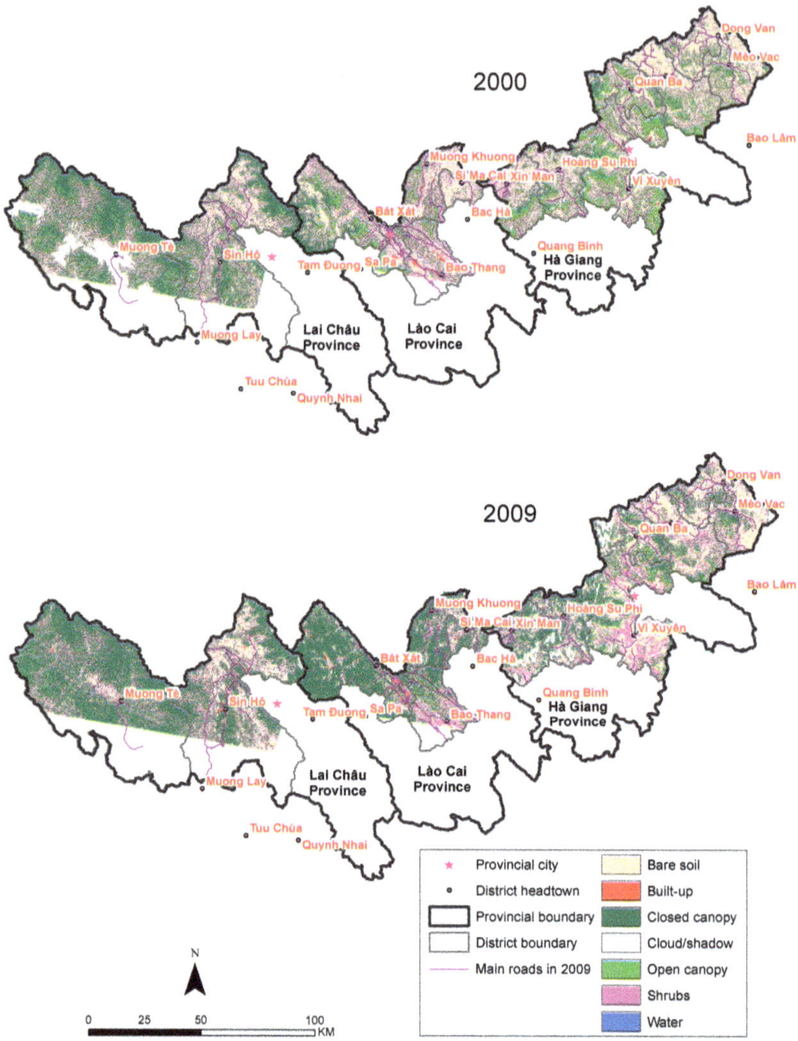

Figure 4. Land cover maps for the region, 2000 and 2009.

Table 4. Land-cover change in the region by district (km^2), 2000–2009.

Province/District	Closed Canopy	Open Canopy	Shrubs	Bare Soil	Built-up	Cloud Shadow	Water	Absolute Change (km^2)*	Absolute Change/Total Area**
Lai Châu Province									
Mường Tè	*503.14*	**−231.83**	−30.54	−75.52	3.90	−243.95	7.22	852.15	0.35
Sìn Hồ	93.47	**−175.11**	14.07	32.50	*17.05*	−71.93	−2.71	334.90	0.27
Phong Thổ	69.50	−77.95	−43.30	28.69	*11.45*	8.59	2.64	233.53	0.28
Lào Cai Province									
Bát Xát	*571.35*	**−218.24**	**−234.72**	**−149.47**	*13.27*	13.07	4.61	1191.66	1.13
Lào Cai City	71.92	−18.50***	−27.74	−47.02	17.76	0.65	2.94	185.89	0.87
Bảo Thắng	*105.41*	22.83	−32.25	−88.88	−7.94	0.85	−0.31	257.63	0.49
Mường Khương	*313.35*	−89.74	**−121.43**	**−112.09**	6.69	2.91	0.87	644.18	1.16
Si Ma Cai	91.42	−6.96	−18.21	−73.69	5.74	0.26	1.25	197.27	0.85
Hà Giang Province									
Xín Mần	*161.50*	**−120.03**	−51.27	−41.19	0.00	48.63	2.20	376.20	0.66
Hoàng Su Phì	*188.44*	**−116.39**	−62.59	−52.84	−0.99	43.73	1.12	422.37	0.67
Vị Xuyên	*131.31*	**−420.03**	*117.64*	*170.33*	−7.97	10.74	−1.66	848.94	0.58
Quản Bạ	74.81	**−119.75**	−3.97	11.60	−0.35	36.86	1.28	211.76	0.39
Yên Minh	22.00	−68.72	94.37	−47.10	−0.69	0.00	0.11	232.98	0.30
Đồng Văn	28.76	−11.20	37.74	−61.65	−0.22	−0.01	−0.48	140.05	0.31
Mèo Vạc	10.75	−42.36	38.41	−21.04	0.00	0.05	−0.03	112.59	0.20

Note: Blue italics and red bold values: important increases and decreases, respectively. * Absolute change is the sum of absolute values of each LULCC, excluding areas covered by *cloud/shadow*. ** Total areas do not include *cloud/shadows*. *** The change in *open canopy* class in Lào Cai city is numerically small, but was 37.83 percent of the total *open canopy* area in 2000.

Coverage by *shrubs* declined across our study region, with the exception of the eastern districts of Hà Giang (Vị Xuyên, Yên Minh, Đồng Văn, and Mèo Vạc). The most noticeable loss was in Bát Xát District, Lào Cai Province, and a close look at the transition map (Figure 5) shows this area changed mostly to closed canopy forest. Two main transitions to *shrubs* occurred; first, when *open canopy forest* converted to *shrubs* (in Vi Xuyên District, Hà Giang Province) and second, and more pronounced, when *bare soils* changed to *shrubs* (mostly in Yên Minh, Đồng Văn, and Mèo Vạc). It should be noted that this so-called transition from *bare soils* to *shrubs* probably reflects the confusion noted above in Section 4.2, with observations and interviews (2010) supporting the contention that very little change in areas of crop cover have occurred here. *Bare soils* were reduced in most districts. They changed to *closed canopy forest* in Mường Tè (Lai Châu Province), in west Bát Xát and Mường Khương (Lào Cai Province), and in Xin Mần and Hoàng Su Phì (Hà Giang Province). However there were some gains of *bare soils* in southern Hoàng Su Phì, adjacent to Vị Xuyên. Another noticeable change, although not as large in square kilometers as the aforementioned changes, was the expansion of *built-up areas* corresponding to urban growth in the region. This expansion occurred mostly in Sìn Hồ, Phong Thổ (Lai Châu Province), and in Bảo Thắng and Lào Cai City (Lào Cai Province) (Table 4). The most important transitions in the region are shown in Table 5 and illustrated in Figure 5.

Examining the magnitude of change *versus* stability in each district (Table 4, final two columns) we note some important tendencies. Districts with the highest amounts of change were Mường Tè, Bát Xát, Mường Khương, and Vị Xuyên, hence high rates of change are spread across the study region. When taking into account district size, Bát Xát, Lào Cai City, and Si Ma Cai were the most dynamic, all in Lào Cai Province. Inversely, the most stable districts were Sìn Hồ and Phong Thổ in Lai Châu Province, and Yên Minh, Mèo Vạc, and Đồng Văn in Hà Giang Province. Further analyses of the links between these LULCC patterns and livelihoods are explored next.

Table 5. Transitions of land-cover types in the region, 2000 to 2009 (km^2).

		In 1999							
		Cloud Shadow	**Water**	**Closed Canopy**	**Open Canopy**	**Shrubs**	**Bare Soil**	**Built-up**	**Total**
In 2009	**Cloud shadow**	506.81	1.52	110.40	142.63	93.13	61.92	0.87	917.28
	Water	8.51	15.45	1.68	3.86	8.41	19.70	1.82	59.45
	Closed canopy	139.74	0.97	1 285.01	**1 344.63**	**871.10**	**515.50**	1.66	4 158.61
	Open canopy	98.04	1.73	115.34	650.42	472.50	245.52	3.21	1 586.76
	Shrubs	103.44	4.30	87.95	**662.15**	1 155.14	**1 005.88**	13.39	3 032.25
	Bare soil	66.78	14.72	114.72	**456.69**	**723.57**	1 744.98	18.33	3 139.79
	Built-up	2.58	1.61	0.98	6.08	**24.66**	**61.06**	11.29	108.27
	Total	925.90	40.31	1 716.09	3 266.45	3 348.51	3 654.56	50.58	13 002.40

Note: Grey cells indicate amount of land-cover type that has not changed, bold numbers indicate important changes.

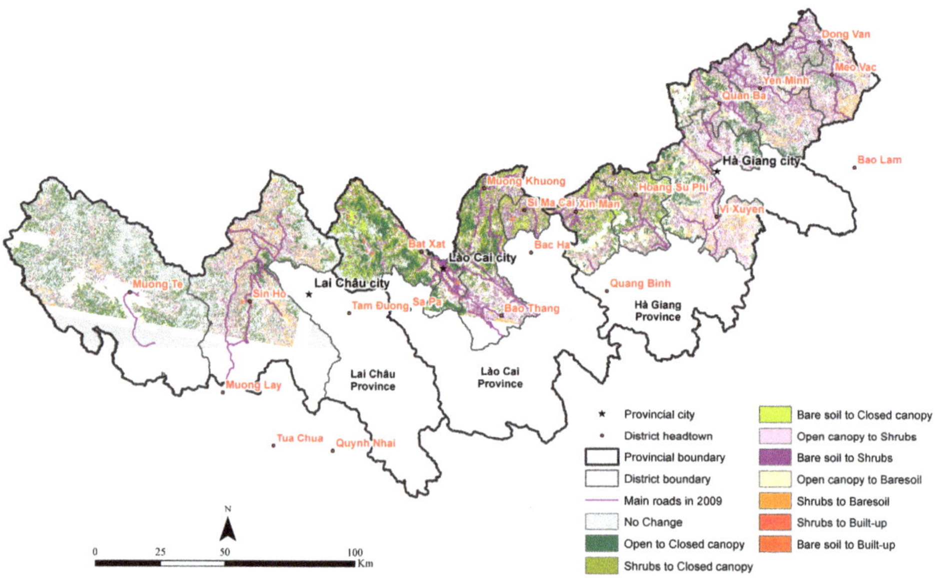

Figure 5. Transitions of land-cover types in the region between 2000 and 2009.

5.1. Examining the Underlying Relationships between LULCC and Livelihood Diversification

Drawing on our quantitative findings above, in this section we focus on three case studies, one from each of our study provinces. Specific case studies cannot represent the entire spectrum of change across these uplands, and our cases were chosen instead to highlight the breadth of LULCC types underway, the diversity of causes, some important impacts on local livelihoods, livelihood diversification responses, and some surprise findings.

5.1.1. Urbanization in the Uplands and the Case of Lai Châu's Changing Capitals

Across our study site there was a 114 percent increase in *built-up areas*, notably in Sìn Hồ and Phong Thổ (Lai Châu Province), as well as in Bát Xát, Lào Cai City, and Bảo Thắng (Lào Cai Province). Over half of this expansion transpired in the Red River valley, within the city municipality of Lào Cai, and to its immediate west in Bát Xát District. This increase is concurrent with national strategies to encourage cross-border trade and to stimulate the Greater Mekong Subregion North-South Economic corridor, begun in 1998. The Vietnamese government is investing heavily in Lào Cai City and its environs, supporting trade and improving road infrastructure. This has included the creation or upgrading of border crossing points for heavy vehicles carrying merchandise from both Bát Xát District and Lào Cai City to and from Yunnan, China. At the provincial level, the *2020 Lào Cai Economic Development Plan* continues to prioritize construction, renovation, and improvement of facilities at the Lào Cai–Hekou (Yunnan) border crossing, as well as the construction of new, smaller border crossings [78].

Livelihood opportunities have undergone notable diversification in Lào Cai City. In 1999 this was a sleepy, dusty frontier town with most activity focused on the border crossing with Hekou Town, Yunnan. By 2009, the metropolises on both sides of the border had expanded considerably, with Lào Cai City boasting a new six-story border gate administrative center and a brand new multi-story shopping center. The city's main marketplace was also renovated and expanded, selling numerous electronics and plastic goods imported from China. As one Kinh trader noted, sweeping her hands over her stall's plastic toys: "All of this, I get it all from China" (interview, August 2013). While these expanding trades and services have provided new livelihood opportunities, it is important to note that overwhelmingly, it is ethnic Kinh (Vietnamese lowlanders) who have taken up employment opportunities here, rather than ethnic minorities from the surrounding countryside. Without the formal education skills, social capital, and financial reserves necessary to enter urban livelihoods, upland urbanization has remained strongly ethnically divided. When asked why they do not move to Lào Cai City for work opportunities, a group of Hmong ethnic minority young women responded, "Why would we go there? It's scary—close to the border we might get kidnapped, and we don't want to work in the city anyhow. Our home is here" (interviews, January 2009; July 2014).[2] As such, young ethnic minority men and women tend to remain based in rural villages—either that of their parents or, for women after marriage, that of their husband—and maintain agricultural livelihoods, plus small-scale trade for some. While there is migration of ethnic minority individuals over the border for agricultural or mining labor, this remains strongly gender-divided for men only and is limited in size. This urban-rural ethnic divide is also apparent elsewhere in our study region, in Lai Châu and Hà Giang cities.

Growth in *built-up areas* in these borderlands is also due to the establishment of new administrative districts, and thus the need for new head towns. This transformation occurred in August 2000 for Si Ma Cai Town, which became the head town of a new district with the same name in the east of Lào Cai Province [79]. This resulted in a new suite of roads, public services, and administrative buildings [80], plus all the officials who go with such a bureaucracy. Locals whom we talked to in Si Ma Cai, who had lost their land to these new infrastructure developments, were none too pleased with the limited compensation (interviews, March 2009).

The cause of *built-up areas* in Lai Châu Province, the farthest west of our study provinces, is perhaps even more controversial. Here, urban "upgrading" has also occurred, but this time largely due to the creation of a large reservoir for a dam project further south in Sơn La Province (see Figure 1 for the dam's location). Lai Châu Province consists of Lai Châu City and seven districts (with the most recent, Nậm Nhùn, created in 2012) with a combined total of 98 communes [81]. The province's total land area is 9,112.3 square kilometers [82]. The province is considerably smaller than it used to be, however, as a new province called Điện Biên was created to the west of Lai Châu province in 2004, carving off Lai Châu Province's more industrialized and financially prosperous southern and western parts [48,83]. To the south of Lai Châu Province, construction began on the Sơn La dam in 2005; when it was completed in 2012, it was the largest hydropower plant in Southeast Asia, with a 224 km^2

[2] An increasing number of young ethnic minority women have been trafficked to China to become wives for rural farmers there. A common kidnapping approach is for a young man—often from the same ethnicity and posing as a friend of a friend—to lure a young woman to Lào Cai City on his motorbike 'to have fun'; she is then abducted and taken over the border (interviews with Hmong and Yao youth, 2009, 2014).

reservoir [84]. The project required the displacement of over 91,000 ethnic minority individuals across three provinces, including Lai Châu Province, beginning in 2005 [85]. In Lai Châu, the resettlement of certain villages was delayed because of conflicts over land prices and an absence of infrastructure at resettlement sites [84]. These sites are mostly in rural areas and with resettlement being fairly recent, their impacts were not clear on our LULCC maps, although resettlement villages are now observable in the high-resolution imagery used by Google Earth.

Just south of our study area, it has been intriguing to observe the increase in urban land use around the newly relocated provincial capital, also called Lai Châu, inaugurated with its new name in 2004 [86]. Formerly a sleepy town called Phong Thổ, the "new Lai Châu" was given a substantial makeover to become the new provincial capital because the former Lai Châu Town was lost to the dam's reservoir.[3] Extensive new infrastructure including large government offices and wide boulevards were already complete by 2006—when we observed the roads being used for motorbike racing and not much else—and "new Lai Châu" was formally recognized as having gained "city status" in December 2013 [87]. By 2014, the total urban area of this new city was more than 7,000 ha with 52,500 residents and seven wards [87].

Back within our study site's border districts in Lai Châu Province, the most important growth of *built-up areas*—an increase of 17 km^2—has occurred in a linear fashion along Road 12, which crosses the district of Sìn Hồ from north to south (Figure 6). This increase is on par with the urban growth around Lào Cai City, in Lào Cai Province. While most interviewees suggested this growth was due to local household livelihoods diversifying into new, small-scale trade possibilities along the road as traffic (and tourism) increases with improved roads, some of this urban growth might also be due to resettlement from the dam. Only two interviewees of the 10 with whom we talked along Road 12 had moved to this area due to dam resettlement, but it would not surprise us if others had also done so.

These changes require further investigation as uncertainty also surrounds a new hydroelectric project, the Lai Châu dam, this time in the province itself (Nậm Hang Commune, Mường Tè District), which will be the third largest in the country after the Hoà Bình and Sơn La projects, also on the Đà River. The Lai Châu dam was initiated in 2011, was 80 percent completed in 2014, and is expected to be operational by 2017. This project alone requires the relocation of approximately 800 rural households [88] and may bring important LULCC to Mường Tè District and, in turn, changes in possible livelihoods.

Figure 6. Sìn Hồ Town on Road 12. (Photo credit: Lê Mạnh An)

[3] In 2004–2005, when former Lai Châu Province was split into Lai Châu and Điện Biên, former Lai Châu Town was renamed Mường Lay Town, now part of the new province of Điện Biên. Most of this town is now under the reservoir for the Sơn La dam.

5.1.2. The Impacts of Increased Cash Needs on Closed Canopy Forest: The Case of Bát Xát District

In terms of LULCC in rural areas of our study region, there was a total increase of 531 km^2 in *closed canopy forest* (over 300 percent). This is in line with a national study that also concluded that the extent of the forest network increased during this time period (though it did not note density, diversity, or quality) [70]. Our findings show that most forest change happened in Bát Xát and Mường Khương Districts (Lào Cai Province), Mường Tè (Lai Châu Province), and Hoàng Su Phì and Xín Mần (Hà Giang Province).

Lào Cai Province lies in the center of our three study provinces. It is divided into eight administrative districts with a combined total of 164 communes, plus Lào Cai City. The province's total land area is 6,383.9 square kilometers [82]. In the west of Lào Cai Province, Bát Xát District is characterized by high elevations and steep slopes, a sparse road network, and relatively few marketplaces. With a population of approximately 70,000, the dominant ethnic group in the district is Hmong (approx. 26 percent) [48,89]. In 1999, this region was principally classified as upland fields, identified as bare soil in our maps, with dry rice and maize production. From our interviews with Hmong and Yao uplanders here, we found that a transition towards state-sponsored high-yield hybrid rice had taken place alongside a decrease in upland fields/*bare soils*; the total decrease in agricultural land being nearly 150 km^2, or 54 percent of the district. Interviewees explained that many farmers had chosen to begin using intensive rice farming techniques as the state discouraged swidden agriculture with dry rice and maize. As one Hmong interviewee in Mường Hum, a market town within Bát Xát District, put it, "The corn and the dry rice, we had to stop" (August 2013). While reducing the land used for agriculture, this switch has had important repercussions for livelihood strategies, as farmers are now obliged to purchase (infertile) hybrid rice and maize seeds annually as well as chemical pesticides and fertilizers. Another Hmong farmer explained, "We need to buy the seeds every year now, rather than saving some, because the Chinese [hybrid] seeds are not so good to save" (March 2007). As a result, semi-subsistence households are far more dependent on the cash economy than in the past (interviews 2004, 2007, 2013, see also [26]).

Interviews with minority farmers in the district suggest that the increase in forest cover in Bát Xát District (one of the most dynamic districts across our fieldwork sites; see Table 4) could be the result of a declining agricultural footprint, reforestation policies, and an increase in cardamom cultivation to meet the increasing cash needs of local households (interviews 2013). Highly sought after in China, cardamom grows in the shade of mature trees and requires relatively little labor to cultivate [90] (Figure 7). In Bát Xát District, ethnic minority farmers are increasing the amount of cardamom they cultivate to earn cash income, with some farmers recently reporting an annual harvest of 500 kg from their cardamom plots if the weather is good and noting, "There's far, far more [cardamom] than ten years ago" (interview, August 2013). From Bát Xát, the cardamom is transported across the border at nearby local level crossings by the farmers themselves or by Kinh intermediaries who purchase it at the "farm gate" and transport it to Lào Cai City, where most often it is collected by Chinese wholesalers. The cardamom then makes its way to wholesale markets and processing plants in Kunming, the capital of Yunnan Province, or in neighboring Guangxi Province to satisfy the sizeable demand across China (interviews 2013, 2015).

As the Vietnamese state attempts to homogenize agriculture in the uplands using hybrid varieties, and as demand for cash income among local farmers consequently rises, ethnic minority households are reacting by supporting reforestation through the protection and restoration of old growth forests for cardamom cultivation. Surprisingly, therefore, market integration and agrarian change have led to forest conservation and expansion in a rather roundabout way. Nonetheless, it is not clear how long this situation will continue. Despite the cash income that cardamom can bring, the increasing frequency of extreme weather events—especially extreme cold and hail storms—is resulting in unpredictable cardamom yields and unstable incomes for farmers, with one noting fluctuations of 60 to 400 kilograms a year (interviews, August 2013, June 2014). Additionally, prices fluctuate substantially from one year to the next, in part due to varying demand in China and in part due to volatile relationships between individual intermediary traders and cultivators, who tend to be of different ethnicities (interviews 2013) [15].

(**a**) (**b**)

Figure 7. (**a**) Forest containing cardamom fields above a Hmong hamlet and (**b**) cardamom growing in the shade of mature trees Bát Xát District, Lao Cai Province. (Photo credit: Sarah Turner).

5.1.3. When Geomorphology Tests Livelihood Limits: Đồng Văn District, Hà Giang Province

The three districts with the least LULCC are the karst hill and mountain-peaked districts of Sìn Hồ and Phong Thổ (Lai Châu Province), and Đồng Văn and Mèo Vạc (Hà Giang Province). Here, the ratio of change is the lowest, between 0.20 and 0.30 (Table 4). To explain this lack of LULCC, we focus on the case of Đồng Văn and Mèo Vạc in Hà Giang Province. Hà Giang, the farthest east of our study provinces, is divided into 10 administrative districts with a combined total of 195 communes, plus Hà Giang Town; its total land area is 7945.8 square kilometers [82]. The province spans three officially recognized distinct agro-ecological zones. The northern borderland districts of Mèo Vạc, Đồng Văn, and Quản Bạ, within our case study, are in Zone 1. This is a reasonably homogenous high plateau agro-ecological zone. About 90 percent of the surface area is limestone, reflecting the region's karst geography. This zone supports the cultivation of maize, often on steep slopes, and the raising of livestock, mostly cattle, horses, goats, and poultry. Zone 1 also includes Yên Minh District, which is somewhat different, having large areas of open land both with and without forest cover. Zone 2 covers the mountainous western borderland districts of Hoàng Su Phì and Xín Mần, also in our case study. This zone has an average elevation of 1,600 meters, many steep slopes, and poor soil. Agriculture includes rice and maize, cash crops such as tea, and livestock rearing. Zone 3 includes Hà Giang Town as well as the districts of Bắc Mê, Bắc Quang and Vị Xuyên, which fall outside this study. These are less mountainous (500 to 1,000 meters), with old forests and valleys alternating with rivers and large streams (interviews and observations 2009, 2010) [91].

On our LULCC maps, the northernmost districts of Đồng Văn and Mèo Vạc recorded important areas of *bare soil* (234.5 km^2 and 285 km^2 in 2009, or roughly 50 percent of the total area in both districts). Particularly interesting in these two districts are the ways by which local ethnic minority farmers maintain livelihoods, as these *bare soils* on the maps represent a specific karst landscape with extremely small pockets of usable soil just a few meters square between large numbers of rocky outcrops. In these small pockets, minority farmers have diligently added soil and traditional fertilizers (a mix of household fire ash and cattle dung) to be able to grow traditional local maize. Running up the stalks of the maize are beans, and other vegetables such as pumpkins are sometimes interspersed among these (Figure 8).

As noted earlier, across much of the upland northern provinces—and indeed elsewhere in Vietnam—farmers are being strongly encouraged by the government to switch to hybrid varieties of rice and maize seeds to increase outputs. A Yao farmer in Đồng Văn market explained, "We have to spend 3 million VND on fertilizer for the hybrid corn seeds. I get it in advance from the government and pay them back with some of my crop. Each year I have to buy new seeds" (June 2010). Our research and that of our former graduate students has shown that these crops can actually increase food insecurity in these uplands rather than improve it, due to the agro-ecological limits of the new seeds, difficulties with the appropriate quantity of seeds being supplied, and (un)timely distribution [26,92].

Notably, hybrid maize is *not* grown in the small pockets of land in Mèo Vạc and Đồng Văn's karst landscape because farmers insist it requires a more level growing surface. This points to the importance of landraces and traditional ecological knowledge for maintaining livelihoods here. Local maize diversity is maintained by traditional seed-saving within households, as well as exchange among households [92]. Hmong farmer interviewees noted that they far prefer traditional maize to hybrids due to its suitability in the rugged topography and the means by which farmers can intercrop it—this is more difficult or impossible with hybrid maize, which is planted closer together. Traditional maize also has superior long-term storage properties, being less susceptible to mold, and was declared by numerous interviewees to taste far better. As noted earlier, it should be remembered that in Table 4, the increase and decrease in *bare soil* and *shrubs* in these districts might counter-balance each other to some degree, given the results of the confusion matrix and interviews confirming very little change in land uses. Overall, when considering *bare soil* and *shrubs* together, these districts have remained extremely stable in land-cover and land-use types compared to other areas of our case study. We interpret this as being due to a harsh terrain that is not conducive to the diversification of local livelihoods away from traditional, carefully adapted land uses.

(a) (b)

Figure 8. (**a**) Typical small fields with numerous karst outcrops, and (**b**) local maize with runner beans, Đồng Văn District, Hà Giang Province. (Photo credit: Sarah Turner).

6. Discussion: Market Integration, State Policies, and Land-Use Change

In general, land uses in our frontier study region are undergoing substantial and often conflicting transformations. In one interview, a forest ranger noted, "They [the state] are razing the mountains," in relation to new hydropower dam projects and recently built urban areas in Lào Cai Province, both reflecting the modernization drive of state and private actors in this frontier zone. Yet elsewhere in the same province, another ranger said, "Look at the hills there. I planted and protected them, those hills and forest there. I worked six to seven years there. People understand, they dare not burn forest to make new fields," reflecting an awareness (corroborated by local farmers) of the importance of forests for local watershed protection. Not only our interviews but also our maps demonstrated a complex LULCC picture. In some areas, forests were expanding, while in others they were declining; in yet others they remained remarkably stable due to the agro-ecological limits of the local geology, but what are the underlying causes of LULCC in this region? How are livelihood strategies related to these highly variable dynamics of land use across this upland border region?

During the last 25 years, upland northern Vietnam has been considered a frontier region by the state, a region to be modernized and civilized as rapidly as possible [93]. The 2013 Vietnam Constitution promotes "cultural development" (*phát triển văn hóa*), and "social progressiveness" (*tiến bộ*) while focusing on the "modernization" (*hiện đại hóa*) of the country [94]. The northern uplands are the focus of numerous policies that have drawn on these principles and concurrently have substantively challenged or changed the livelihoods and livelihood diversification strategies of ethnic minorities [2,3,68,90,95]. Infrastructure to encourage upland integration increased considerably between 1999 and 2009, marked by an expanded road network that reduced distances to marketplaces newly renovated by the state and connected upland and lowland towns and cities. Such infrastructure has created additional livelihood opportunities for local households to sell agricultural surpluses, gain access to market knowledge or resources, and acquire non-agricultural income [96]. Yet other state-driven interventions and market-oriented economic reforms, including the introduction of and strong state support for hybrid rice and maize seeds, have rarely favored ethnic minorities' indigenous knowledge and historical land uses, but instead have frequently challenged their land-use strategies, often with negative impacts on food security [3,26,45,97].

Ongoing state-supported "development" projects in these borderlands trigger multidimensional responses from local populations that face challenges accessing livelihood resources or who are confronted with the state's vision of what their livelihoods should be (resettled, based on hybrid crops, cash-based, and so on). In our study locale, many rural upland residents have maintained composite agricultural systems as their primary livelihoods, rather than transitioning to plantation crops such as coffee or rubber that are often (but not always) grown on a larger scale [98]. Instead, livelihood diversification has occurred in these borderlands for a range of other complex reasons. The state's drive for farmers to use hybrid rice and maize seeds, for instance, means farmers must have access to more cash for inputs than ever before. In turn, some individuals with the human capital—appropriate skills and know how—and access, have chosen to cultivate cardamom under the shade of closed canopy trees. Providing cash income in this way, farmers have let go of some of their former hillside swidden agricultural lands to focus on hybrid seed crops; in turn, forest cover is increasing. While this livelihood diversification approach and chain of events is a surprising avenue for forest cover regeneration, resulting in a specific LULCC, less surprising is the urban development occurring along the Red River, as again the state intervenes in frontier "development". This time, pushing for greater market integration via inclusion in the Greater Mekong Subregion, the state has heavily promoted urbanization and industrialization adjacent to the Lào Cai–Hekou border crossing, with imports and exports rising steadily (trade data shows the value of goods traded across this border crossing rose from US$92 million to US$993 million from 2000 to 2011 [99]). Urban livelihood opportunities here have diversified dramatically for Kinh residents and traders as this frontier city expands, and an ever-increasing range of goods and services has become available. The complexity of this region is again highlighted with the contrasting tale of eastern Hà Giang Province, where ethnic minority farmers have worked within fairly constraining agro-ecological limits to maintain livelihoods that are diverse within themselves, but yet have remained extremely stable over time. This stability has been reflected in the near lack of LULCC in the most north-eastern districts.

7. Conclusions

State policies for this frontier region, market opportunities (both state-supported and private), and agro-ecological conditions have resulted in highly complex and heterogeneous land uses and land covers. Yet can LULCC mapping really help us to unravel these changes? While completing this project we hit many roadblocks. Mapping LULCC in the mountainous regions of Vietnam is challenging, not only due to the political and physical complexities of accessing the field, but also due to a lack of fine-scale spatial data. Landsat images are the most available data, but the resolution is not fine enough to distinguish bare soils from built-up areas, shrubs from planted crops, and various crops from each other. We found many interesting tales on the ground of livelihood diversification at a scale the maps could not register. For instance, Hmong farmers in Mường Khương District, Lào Cai Province have been experimenting with pineapple plantations, bringing inputs (pineapple plants) and crop-cultivation skills with them from former employment in plantations just over the border in China. Will such crop diversification change land cover in important ways in this area? Others in Mèo Vạc District, Hà Giang Province have been experimenting with honey production and sales to Hanoi distributors, with beehives kept in forests near specific trees to produce highly desired flavored honey. Could honey be the new cardamom and start to protect (the limited) forests there? Only time will tell. As such, the interpretation of LULCC maps must be done with caution and be combined with in-depth, on-the-ground knowledge.

Nonetheless, at the same time, the LULCC maps we developed allowed us to take a step back from our ethnographic results and place them within the bigger picture. We knew that cardamom was an increasingly important crop for many ethnic minority livelihoods, but we needed to see these maps to begin to realize that there were dramatic changes occurring to forest cover that seemed to be connected. This motivated us to return to local villages and ask the relevant questions to confirm these causal links. Likewise, we knew of livelihood constraints in Hà Giang Province, but not the degree to which land-use types are static at the district level over time, proving that the state really does reach limits in trying to push specific agricultural policies. The numerous urban growth patterns across the region are noticeable as one drives through these built-up areas, but the causes of their growth became easier to deconstruct while examining the bigger picture. Although not part of this paper, our findings regarding urban change have stimulated us to start a new project examining the growth of small cities and towns in these uplands with a focus on how LULCC is occurring in peri-urban zones, the quality of life that local residents enjoy, and rural-urban migration processes.

Our mixed methods approach, drawing on an interactive design, provided scope for raw data to be examined and re-examined in an iterative process, moving back and forth between quantitative and qualitative results so as to socialize the pixel as well as pixelize the social [100]. Moreover, a conceptual framework drawing on land change science, frontier studies, and livelihood diversification literature allowed for the integration and recognition of a number of important variables across multiple scales. These ranged from frontier transformation and resource exploitation projects (hydropower dams, hybrid seed programs, urban infrastructure), to culturally-rooted livelihood diversification decisions (specific knowledges of agro-ecological limits, watershed needs, non-timber forest product responses to extreme weather events, and so on). We also found that a key benefit of a mixed methods approach emerged from the justifications and explanations we had to provide to each other regarding our contributions. The first author was forced to find answers beyond the household and village levels regarding processes she had previously most often examined at the micro-scale, and then be able to defend her arguments. Likewise, the second author found that she needed to justify and explain meso-scale results to the first author, who required convincing of the percentage changes that had been calculated. We constantly went back to check our fieldwork notes, examine the data again, and tighten our arguments. While all researchers should do this, of course, our different methodological perspectives and assumptions had to be explained and "typical findings" defended. We believe that this cyclical verification added further strength to our results. Our study thus contributes to the emerging trend of using grounded, in-depth fieldwork to help explain regional land change [17,101], while also shedding light on the benefits meso-scale studies can bring to micro-scale ethnographies.

Our findings point to the importance of policy makers having access to complementary methods and an integrated conceptual framework for implementing appropriate, sustainable land-use and livelihood policies in the region. Yet, at the same time, many upland residents may not be that receptive to state officials asking them probing questions regarding their land uses and livelihoods: farmers are sometimes cultivating cardamom within national parks where doing so is illegal; those being strongly encouraged to plant hybrid rice and maize often prefer to covertly maintain a more diverse livelihood approach that they know will work within local agro-ecological limits; and urban growth in this frontier locale includes illegal (smuggling, prostitution, trafficking, *etc.*) as well as legal livelihood opportunities. Perhaps this points to an important bridging role for non-state researchers here [102]. Nonetheless, if officials are sympathetic to local resident concerns over livelihood opportunities, greater knowledge of LULCC and livelihoods in this region could also forge important opportunities. For instance, officials recognizing an increase in closed canopy forest on a LULCC map could encourage the regulated cultivation of cardamom with strict firewood rules (for drying the pods) and set harvest dates to help protect forests, as has been implemented with success just over the border in Yunnan (interviews 2015). Likewise, greater awareness of agro-ecological limits—noting the lack of change in certain areas in LULCC tables and maps—could mean that unsustainable agricultural policy options are rejected. Instead, the traditional ecological knowledge of upland farmers could be acknowledged and taken on board, as is beginning to occur in lowland regions regarding the limits of hybrid seeds [103].

The question then turns to how to incorporate such findings into decision-making processes. With an extremely hierarchical state apparatus, this is not a particularly easy task in Vietnam [102,104]. We recommend a diverse approach, making the most of opportunities to collaborate with local academics such as members of the North-West Research Program at Vietnam National University (Hanoi), discussing findings at public conferences in the country and regionally, and contributing to meetings such as those held by the Ethnic Minority Working Group supported by the Non-Governmental Organization (NGO) Resource Centre in Hanoi. Through numerous discussions and collaborations it is hoped that greater awareness among policy makers and NGOs of these layers of LULCC complexity can effectively advance relevant research and culturally aware policies for appropriate household strategies and livelihood diversification across the uplands.

Acknowledgments: We would like to thank Lê Mạnh An for fieldwork assistance, Melody Lynch, Thomas Kettig, and Kate Trincsi for research assistance, and special issue editors, Claudia Radel and Jackie Vadjunec, for feedback on an earlier draft. We acknowledge funding from the Social Sciences and Humanities Research Council, Canada.

Author Contributions: Pham completed the remote sensing data analysis and initial interpretations; Turner completed the ethnographic data analysis and interpretations. Turner drafted the context and conceptual framework, with input from Pham. Using a mixed methods approach, explained in the paper, both authors worked together to explore emerging interpretations and themes from the combined data. Both authors contributed equally to the final paper in an iterative process.

Conflicts of Interest: The authors declare no conflict of interest.

References

1. Fox, J.; Rindfuss, R.R.; Walsh, S.J.; Mishra, V. *People and the Environment: Approaches for Linking Household and Community Surveys to Remote Sensing and GIS*; Kluwer Academic: Boston, MA, USA, 2003.
2. Castella, J.-C.; Manh, P.H.; Kam, S.P.; Villano, L.; Tronche, N.R. Analysis of village accessibility and its impact on land use dynamics in a mountainous province of northern Vietnam. *Appl. Geogr.* **2005**, *25*, 308–326.
3. Jakobsen, J.; Rasmussen, K.; Leisz, S.; Folving, R.; Quang, N.V. The effects of land tenure policy on rural livelihoods and food sufficiency in the upland village of Que, North Central Vietnam. *Agric. Syst.* **2007**, *94*, 309–319. [CrossRef]
4. Chi, V.K.; van Rompaey, A.; Govers, G.; Vanacker, V.; Schmook, B.; Hieu, N. Land transitions in northwest Vietnam: An integrated analysis of biophysical and socio-cultural factors. *Hum. Ecol.* **2013**, *41*, 37–50. [CrossRef]
5. Jadin, I.; Vanacker, V.; Hoang, H.T.T. Drivers of forest cover dynamics in smallholder farming systems: The case of northwestern Vietnam. *Ambio* **2013**, *42*, 344–356. [CrossRef] [PubMed]
6. Lambin, E.F.; Meyfroidt, P. Land use transitions: Socio-ecological feedback *versus* socio-economic change. *Land Use Policy* **2010**, *27*, 108–118. [CrossRef]

7. Chhabra, A.; Geist, H.; Houghton, R.A.; Haberl, H.; Braimoh, A.K.; Vlek, P.L.G.; Patz, J.; Xu, J.; Ramankutty, N.; Coomes, O.; Lambin, E.F. Multiple impacts of Land use/cover change. In *Land Use and Land Cover Change: Local Processes and Global Impacts*; Geist, H., Lambin, E., Eds.; Springer: Berlin, Germany, 2006; pp. 71–116.
8. Xu, J.; Fox, J.; Melick, D.; Fujita, Y.; Jintrawet, A.; Jie, Q.; Thomas, D.; Weyerhaeuser, H. Land use transition, livelihoods, and environmental services in Montane Mainland Southeast Asia. *Mt. Res. Dev.* **2006**, *26*, 278–284.
9. Aspinall, R.J.; Hill, M.J. *Land Use Change: Science, Policy and Management*; CRC Press: Boca Raton, FL, USA, 2008.
10. Makalle, A.M.P.; Obando, J.; Bamutaze, Y. Effects of land use practices on livelihoods in the transboundary sub-catchments of the Lake Victoria Basin. *Afr. J. Environ. Sci. Technol.* **2008**, *2*, 309–317.
11. Scott, J. *The Art of Not Being Governed: An Anarchist History of Upland Southeast Asia*; Yale University Press: New Haven, CT, USA, 2009.
12. Glassman, J. *Bounding the Mekong: The Asian Development Bank, China, and Thailand*; University of Hawai'i Press: Honolulu, HI, USA, 2010.
13. Turner, S. Under the state's gaze: Upland trading-scapes on the Sino-Vietnamese border. *Singap. J. Trop. Geogr.* **2013**, *34*, 9–24. [CrossRef]
14. Trincsi, K.; Hien, P.T.T.; Turner, S. Between state-led policies and mountains: Land use land cover change of ethnic minorities in Vietnam's borderlands. *Land Use Policy* **2014**, *41*, 484–497. [CrossRef]
15. Turner, S.; Bonnin, C.; Michaud, J. *Frontier Livelihoods. Hmong in the Sino-Vietnamese Borderlands*; University of Washington Press: Seattle, WA, USA, 2015.
16. Douglass, M.; DiGregorio, M.; Bunjamin, W.; Pichaya, V.; Taniguchi, K. Poverty: From basic needs to entitlements in Vietnam during urban transition. In *New Challenges for Sustainable Development in Millennia*; Liu, C.H., Yeh, J.-R., Huang, C.-H., Eds.; CIER Press: Taipei, Taiwan, 2003; pp. 127–174.
17. Brannstrom, C.; Vadjunec, J.M. Notes for avoiding a missed opportunity in sustainability science: Integrating land change science and political ecology. In *Land Change Science, Political Ecology, and Sustainability*; Brannstrom, C., Vadjunec, J.M., Eds.; Routledge: London, UK, 2013; pp. 1–23.
18. Gutman, G.; Janetos, A.C.; Justice, C.O.; Moran, E.F.; Mustard, J.F.; Rindfuss, R.R.; Skole, D.; Turner, B.L., II; Cochrane, M.A. *Land Change Science Observing, Monitoring and Understanding Trajectories of Change on the Earth's Surface*; Kluwer Academic: Dordrecht, The Netherlands, 2004.
19. Rindfuss, R.R.; Walsh, S.J.; Turner, B.L., II; Fox, J.; Mishra, V. Developing a science of land change: Challenges and methodological issues. *Proc. Natl. Acad. Sci. USA* **2004**, *101*, 13976–13981.
20. Turner, B.L., II; Lambin, E.F.; Reenberg, A. The emergence of land change science for global environmental change and sustainability. *Proc. Natl. Acad. Sci. USA* **2007**, *104*, 20666–20671.
21. Geist, H.J.; Lambin, E.F. Proximate causes and underlying driving forces of tropical deforestation: Tropical forests are disappearing as the result of many pressures, both local and regional, acting in various combinations in different geographical locations. *BioScience* **2002**, *52*, 143–150. [CrossRef]
22. López-Carr, D.; Davis, J.; Jankowska, M.M.; Grant, L.; López-Carr, A.C.; Clark, M. Space *versus* place in complex human–natural systems: Spatial and multi-level models of tropical land use and cover change (LUCC) in Guatemala. *Ecol. Model.* **2012**, *229*, 64–75. [CrossRef] [PubMed]
23. Alvarez, R.R., Jr. The Mexican–US border: The making of an anthropology of borderlands. *Annu. Rev. Anthropol.* **1995**, *24*, 447–470. [CrossRef]
24. Giersch, C.P. *Asian Borderlands: The Transformation of Qing China's Yunnan Frontier*; Harvard University Press: Cambridge, MA, USA, 2006.
25. Baker, J. *Small Town Africa: Studies in Rural-Urban Interaction*; Seminar Proceedings No. 23; Nordic Africa Institute: Uppsala, Sweden, 1990.
26. Bonnin, C.; Turner, S. At what price rice? Food security, livelihood vulnerability, and state interventions in upland northern Vietnam. *Geoforum* **2012**, *43*, 95–105.
27. Walker, R.; Homma, A.K.O. Land use and land cover dynamics in the Brazilian Amazon: An overview. *Ecol. Econ.* **1996**, *18*, 67–80. [CrossRef]
28. Carvalho, G.O.; Nepstad, D.; McGrath, D.; de Carmen Vera Diaz, M.; Santilli, M.; Barros, A.C. Frontier expansion in the Amazon: Balancing development and sustainability. *Environ. Sci. Policy Sustain. Dev.* **2002**, *44*, 34–44. [CrossRef]

29. Moller-Jensen, L.; Knudsen, M.H. Patterns of population change in Ghana (1984–2000): Urbanization and frontier development. *GeoJournal* **2008**, *73*, 307–320. [CrossRef]
30. Dillman, C.D. Urban growth along Mexico's northern border and the Mexican national border program. *J. Dev. Areas* **1970**, *4*, 487–508.
31. Kusakabe, K. The politics of "opening up": Female traders on borderlands of Cambodia, Laos, Thailand and Burma (Myanmar). In *On the Border of State Power: Frontier in the Greater Mekong Sub-Region;* Gainsborough, M., Ed.; Routledge: Abingdon, UK, 2008; pp. 60–74.
32. Anderson, J.B.; Gerber, J. *Fifty Years of Change on the US-Mexico Border: Growth, Development, and Quality of Life;* University of Texas Press: Austin, TX, USA, 2009.
33. Coxhead, I.; Shively, G.; Shuai, X. Development policies, resource constraints, and agricultural expansion on the Philippine land frontier. *Environ. Dev. Econ.* **2002**, *7*, 341–363. [CrossRef]
34. Jepson, W. Producing a modern agricultural frontier: Firms and cooperatives in eastern Mato Grosso, Brazil. *Econ. Geogr.* **2006**, *82*, 289–316. [CrossRef]
35. Agergaard, J.; Fold, N.; Gough, K.V. Global–local interactions: Socioeconomic and spatial dynamics in Vietnam's coffee frontier. *Geogr. J.* **2009**, *175*, 133–145. [CrossRef]
36. Eilenberg, M. Frontier constellations: Agrarian expansion and sovereignty on the Indonesian-Malaysian border. *J. Peasant Stud.* **2014**, *41*, 157–182. [CrossRef]
37. De Koninck, R. The theory and practice of frontier development: Vietnam's contribution. *Asia Pac. Viewp.* **2000**, *41*, 7–21. [CrossRef]
38. Lary, D. Introduction. In *The Chinese State at the Borders;* Lary, D., Ed.; UBC Press: Vancouver, BC, Canada, 2007; pp. 1–10.
39. Tan, S.B.-H. The struggle to control land grabbing: State formation on the Central Highlands frontier under the First Republic of Vietnam (1954–63). In *On the Borders of State Power: Frontiers in the Greater Mekong Sub-Region;* Gainsborough, M., Ed.; Routledge: Abingdon, UK, 2008; pp. 50–66.
40. Ellis, F. *Rural Livelihoods and Diversity in Developing Countries;* Oxford University Press: Oxford, UK, 2000.
41. Scoones, I. Livelihoods perspectives and rural development. *J. Peasant Stud.* **2009**, *36*, 171–196. [CrossRef]
42. Bebbington, A. Capitals and capabilities: A framework for analysing peasant viability, rural livelihoods, and poverty. *World Dev.* **1999**, *27*, 2020–2044. [CrossRef]
43. Long, N. *Development Sociology: Actor Perspectives;* Routledge: London, UK, 2001.
44. Sen, A. *The Standard of Living;* Cambridge University Press: Cambridge, UK, 1987.
45. Turner, S. Making a living the Hmong way: An actor-oriented livelihoods approach to everyday politics and resistance in upland Vietnam. *Ann. Assoc. Am. Geogr.* **2012**, *102*, 403–422. [CrossRef]
46. McCusker, B.; Carr, E. The co-production of livelihoods and land use change: Case studies from South Africa and Ghana. *Geoforum* **2006**, *37*, 790–804. [CrossRef]
47. Rigg, J. Land, farming, livelihoods, and poverty: Rethinking the links in the rural South. *World Dev.* **2006**, *34*, 180–202. [CrossRef]
48. General Statistics Office of Vietnam. *The 2009 Vietnam Population and Housing Census: Completed Results;* Central Population and Housing Census Steering Committee: Hanoi, Vietnam, 2010.
49. Dien, K. *Population and Ethno-demography in Vietnam;* Silkworm: Chiang Mai, Thailand, 2002.
50. MacKerras, C. Ethnic minorities in China. In *Ethnicity in Asia;* MacKerras, C., Ed.; RoutledgeCurzon: London, UK, 2003.
51. Michaud, J. *Historical Dictionary of the Peoples of the Southeast Asian Massif;* Scarecrow Press: Lanham, MD, USA, 2006.
52. Leisz, S.J.; Ha, N.T.T.; Yen, N.T.B.; Lam, N.T.; Vien, T.D. Developing a methodology for identifying, mapping and potentially monitoring the distribution of general farming system types in Vietnam's northern mountain region. *Agric. Syst.* **2005**, *85*, 340–363. [CrossRef]
53. Corlin, C. Hmong and the land question in Vietnam: National policy and local concepts of the environment. In *Hmong/Miao in Asia;* Tapp, N., Michaud, J., Culas, C., Lee, G.Y., Eds.; Silkworm: Chiang Mai, Thailand, 2004; pp. 295–320.
54. World Bank. *Vietnam Development Report 2008: Social Protection;* World Bank: Washington, DC, USA, 2008.
55. Sikor, T. Forest policy reform in Vietnam: From state to household forestry. In *Stewards of Vietnam's Upland Forests;* Poffenberger, M., Ed.; Asian Forestry Network: Berkley, CA, USA; Hanoi, Vietnam, 1998; pp. 18–38.

56. McElwee, P. Reforesting "bare hills" in Vietnam: Social and environmental consequences of the 5 Million Hectare Reforestation Program. *Ambio* **2009**, *38*, 325–333. [CrossRef] [PubMed]
57. McElwee, P. Becoming socialist or becoming Kinh? Government policies for ethnic minorities in the Socialist Republic of Vietnam. In *Civilizing the Margins: Southeast Asian Government Policies for the Development of Minorities*; Duncan, C., Ed.; Cornell University Press: Ithaca, NY, USA, 2004; pp. 182–213.
58. Hardy, A. *Red Hills: Migrants and the State in the Highlands of Vietnam*; Nordic Institute of Asian Studies Monographs: Copenhagen, Denmark, 2005.
59. Clement, F.; Amezaga, J.M. Linking reforestation policies with land use change in northern Vietnam: Why local factors matter. *Geoforum* **2008**, *39*, 265–277. [CrossRef]
60. Castella, J.C.; Boissau, S.; Thanh, N.H.; Novosad, P. Impact of forestland allocation on land use in a mountainous province of Vietnam. *Land Use Policy* **2006**, *23*, 147–160. [CrossRef]
61. Clement, F.; Orange, D.; Williams, M.; Mulley, C.; Epprecht, M. Drivers of afforestation in Northern Vietnam: Assessing local variations using geographically weighted regression. *Appl. Geogr.* **2009**, *29*, 561–576. [CrossRef]
62. Meyfroidt, P.; Lambin, E.F. The causes of the reforestation in Vietnam. *Land Use Policy* **2008**, *25*, 182–197. [CrossRef]
63. Sikor, T. The allocation of forestry land in Vietnam: Did it cause the expansion of forests in the northwest? *For. Policy Econ.* **2001**, *2*, 1–11. [CrossRef]
64. Müller, D.; Munroe, D.K. Tradeoffs between rural development policies and forest protection: Spatially explicit modeling in the Central Highlands of Vietnam. *Land Econ.* **2005**, *81*, 412–425. [CrossRef]
65. Leisz, S.J. Dynamics of land cover and land use changes in the upper Ca river basin of Nghe An, Vietnam. *Southeast Asian Stud.* **2009**, *47*, 287–308.
66. Dao, M.T.; Kono, Y.; Yanagisawa, M.; Leisz, S.J.; Kobayashi, S. Linkage of forest policies and programs with land cover and land use changes in the Northern mountain region of Vietnam: A village-level case study. *Southeast Asian Stud.* **2009**, *47*, 244–262.
67. Pham, T.T.H.; Turner, S.; Trincsi, K. Applying a systematic review to land use land cover change in northern upland Vietnam: The missing case of the borderlands. *Geogr. Res.* **2015**. [CrossRef]
68. Cuc, L.T.; Rambo, T. *Bright Peaks, Dark Valleys. Bright Peaks, Dark Valleys: A Comparative Analysis of Environmental and Social Conditions and Development Trends in Five Communities in Vietnam's Northern Mountain Region*; The National Political Publishing House: Hanoi, Vietnam, 2001.
69. Greene, J.C. *Mixed Methods in Social Inquiry*; Jossey-Bass: San Francisco, CA, USA, 2007.
70. Meyfroidt, P.; Lambin, E.F. Forest transition in Vietnam and its environmental impacts. *Glob. Change Biol.* **2008**, *14*, 1319–1336. [CrossRef]
71. Anderson, J.R. *A Land Use and Land Cover Classification System for Use with Remote Sensor Data*; US Government Printing Office: Arlington, VA, USA, 1976.
72. Hurni, K.; Hett, C.; Epprecht, M.; Messerli, P.; Heinimann, A.A. Texture-based land cover classification for the delineation of a shifting cultivation landscape in the Lao PDR using landscape metrics. *Remote Sens.* **2013**, *5*, 3377–3396. [CrossRef]
73. Cassidy, L.; Binford, M.; Southworth, J.; Barnes, G. Social and ecological factors and land-use land-cover diversity in two provinces in Southeast Asia. *J. Land Use Sci.* **2010**, *5*, 277–306. [CrossRef]
74. Baatz, M.; Hoffman, C.; Willhauck, G. Progressing from object-based to object-oriented image analysis. In *Object-Based Image Analysis: Spatial Concepts for Knowledge-Driven Remote Sensing Applications*; Blaschke, T., Lang, S., Hay, G.J., Eds.; Springer: Berlin, Germany, 2008; pp. 29–42.
75. Dorren, L.K.A.; Maier, B.; Seijmonsbergen, A.C. Improved Landsat-based forest mapping in steep mountainous terrain using object-based classification. *For. Ecol. Manag.* **2003**, *183*, 31–46. [CrossRef]
76. Vieira, M.A.; Formaggio, A.R.; Rennó, C.D.; Atzberger, C.; Aguiar, D.A.; Mello, M.P. Object based image analysis and data mining applied to a remotely sensed Landsat time-series to map sugarcane over large areas. *Remote Sens. Environ.* **2012**, *123*, 553–562. [CrossRef]
77. Puyravaud, J.P. Standardizing the calculation of the annual rate of deforestation. *For. Ecol. Manag.* **2003**, *177*, 593–596. [CrossRef]

78. People's Committee of Lào Cai. *Quy hoạch tổng thể phát triển kinh tế - xã hội tỉnh Lào Cai đến năm 2020* [The 2020 Lào Cai economic development plan], 2008. Available online: http://laocai.gov.vn/thongtinktxh/dinhhuongphattrien/quyhoachphattrienktxh/Trang/634046199107564190.aspx (accessed on 27 September 2015).
79. Socialist Republic of Vietnam. *Của Chính Phủ Số 36/2000/Nđ-Cp Ngày 18 Tháng 8 Năm 2000 Về Việc Điều Chỉnh Địa Giới Hành Chính Huyện Bắc Hà Để Tái Lập Huyện Si Ma Cai, Tỉnh Lào Cai* [Decree No. 36/2000 / ND-CP of August 18, 2000 of the Government on the adjustment of administrative boundaries of Bắc Hà district to reestablish Si Ma Cai District, Lào Cai Province], 2000. Available online: http://thuvienphapluat.vn/archive/Nghi-dinh/Nghi-dinh-36-2000-ND-CP-dieu-chinh-dia-gioi-hanh-chinh-huyen-Bac-Ha-de-tai-lap-huyen-Si-Ma-Cai-tinh-Lao-Cai-vb46707t11.aspx (accessed on 27 September 2015).
80. Socialist Republic of Vietnam. *Nghị định số 42/2009/NĐ-CP. Về việc phân loại đô thị* [Decree 42/2009/ND-CP 7th May, 2009 on the grading of urban centers], 2009. Available online: http://www.moj.gov.vn/vbpq/en/_layouts/printeng.aspx?id=10670 (accessed on 27 September 2015).
81. Socialist Republic of Vietnam. *Về Việc Điều Chỉnh Địa Giới Hành Chính Để Thành Lập Đơn Vị Hành Chính Cấp Xã, Cấp Huyện Thuộc Tỉnh Lai Châu* [Adjustments of administrative boundaries to establish Administrative Units at the commune and district Levels, Lai Châu Province], 2012. Available online: http://thuvienphapluat.vn/archive/Nghi-quyet-71-NQ-CP-nam-2012-dieu-chinh-dia-gioi-hanh-chinh-de-thanh-lap-don-vi-vb150870.aspx (accessed on 27 September 2015).
82. General Statistics Office of Vietnam. *Socio-economical Statistical Data of 63 Provinces and Cities*; Statistical Publishing House: Hanoi, Vietnam, 2009.
83. General Statistics Office of Vietnam. *The 1999 Vietnam Population and Housing Census: Completed Results*; Central Population and Housing Census Steering Committee: Hanoi, Vietnam, 2000.
84. Bui, T.M.H.; Schreinemachers, P.; Berger, T. Hydropower development in Vietnam: Involuntary resettlement and factors enabling rehabilitation. *Land Use Policy* **2013**, *31*, 536–544. [CrossRef]
85. International Rivers Network. Planned Dams in Vietnam, 2001. Available online: http://www.internationalrivers.org/de/resources/planned-dams-in-vietnam-4079 (accessed on 27 September 2015).
86. Socialist Republic of Vietnam. *Về Việc Thành Lập Thị Xã Lai Châu Và Thành Lập Thị Trấn Thuộc Các Huyện Tam Đường, Phong Thổ, Tỉnh Lai Châu* [On the establishment of the township of Lai Châu and district headtowns of Tam Đường, Phong Thổ, Lai Châu province], 2004. Available online: http://thuvienphapluat.vn/archive/Nghi-dinh/Nghi-dinh-176-2004-ND-CP-thanh-lap-thi-xa-Lai-Chau-va-thanh-lap-thi-tran-thuoc-cac-huyen-Tam-Duong-Phong-Tho-tinh-Lai-Chau-vb5530t11.aspx (accessed on 27 September 2015).
87. VietnamNews. PM Okays Establishment of Lai Châu City. 30 December 2013. Available online: http://vietnamnews.vn/domestic-press-highlights/249576/pm-okays-establishment-of-lai-chau-city.html (accessed on 27 September 2015).
88. Resettlement Problems Put Hydro-Plant behind Schedule. *VietnamNews*. 21 August 2014. Available online: http://vietnamnews.vn/society/259092/resettlement-problems-put-hydro-plant-behind-schedule.html (accessed on 27 September 2015).
89. General Statistics Office of Vietnam. *Results of the 2006 Rural, Agricultural and Fishery Census*; Statistical Publishing House: Hanoi, Vietnam, 2007.
90. Turner, S. "Forever Hmong": Ethnic minority livelihoods and agrarian transition in upland northern Vietnam. *Prof. Geogr.* **2012**, *64*, 540–553. [CrossRef]
91. Novellino, D. Indigenous highlands in transition: The case of Hà Giang Province in northern Vietnam. *Land Reform: Land Settl. Coop. Bull.* **2000**, *2*, 94–107.
92. Kyeyune, V. Yielding to High Yields: Hybrid Maize and Food Security in Hà Giang Province, Northern Vietnam. Master's Thesis, McGill University, Montreal, Canada, 2014.
93. Duncan, C. *Civilizing the Margins. Southeast Asian Government Policies for the Development of Minorities*; Cornell University Press: Ithaca, NY, USA, 2004.
94. Socialist Republic of Vietnam. *Hiến Pháp Nước Cộng Hòa Xã Hội Chủ Nghĩa Việt Nam (Sửa Đổi) [The Constitution of the Socialist Republic of Vietnam (Amended)]*; Socialist Republic of Vietnam: Hanoi, Vietnam, 2013.
95. Turner, S.; Michaud, J. Imaginative and adaptive economic strategies for Hmong livelihoods in Lào Cai Province, northern Vietnam. *J. Vietnam. Stud.* **2008**, *2*, 154–186. [CrossRef]

96. Alther, C.; Castella, J.C.; Novosad, P.; Rousseau, E.; Hieu, T.T. Impact of accessibility on the range of livelihood options available to farm households in mountainous areas of northern Viet Nam. In *Doi Moi in the Mountains: Land Use Changes and Farmers' Livelihood Strategies in Bac Kan Province, Vietnam*; Castella, J.C., Quang, D.D., Eds.; The Agricultural Publishing House: Hanoi, Vietnam, 2001; pp. 121–146.
97. Evans, G.; Hutton, C.; Eng, K.K. *Where China Meets Southeast Asia: Social and Cultural Change in the Border Regions*; St Martin's Press: New York, NY, USA, 2000.
98. Meyfroidt, P.; Vu, T.P.; Hoang, V.A. Trajectories of deforestation, coffee expansion and displacement of shifting cultivation in the Central Highlands of Vietnam. *Glob. Environ. Change* **2013**, *23*, 1187–1198.
99. Zhang, B.Q. 云南省边境口岸物流发展状况分析: 以河口口岸为例 [Analyzing logistics development at border crossing points in Yunnan: A case study on Hekou Border Crossing Point]. *J. Manag.* **2014**, *27*, 48–52.
100. Geoghegan, J.; Pritchard, L., Jr.; Ogneva-Himmelberger, Y.; Roy Chowdhury, R.; Sanderson, S.; Turner, B.L., II. "Socializing the pixel" and "pixelizing the social" in land-use and land-cover change. In *People and Pixels. Linking Remote Sensing and Social Science*; Liverman, D., Moran, E.F., Rindfuss, R.R., Stern, P.C., Eds.; National Academy Press: Washington, DC, USA, 1998; pp. 51–69.
101. Kull, C.A. Politicizing land use change in highland Madagascar: Struggles with air photo analyses and conservation agendas. In *Land Change Science, Political Ecology, and Sustainability*; Brannstrom, C., Vadjunec, J.M., Eds.; Routledge: London, UK, 2013; pp. 66–83.
102. Turner, S. *Red Stamps and Gold Stars. Fieldwork Dilemmas in Upland Socialist Asia*; UBC Press: Vancouver, BC, Canada, 2013.
103. Oxfam. *Growing a Better Future: Expanding Rights and Choices for Small-scale Farmers in Viet Nam, Brochure*; Oxfam: Hanoi, Vietnam, 2012.
104. Mackenzie, C.A.; Christensen, J.; Turner, S. Advocating beyond the academy: Dilemmas of communicating relevant research results. *Qual. Res.* **2015**, *15*, 105–121. [CrossRef]

land

MDPI

Article

Forest Transition in Madagascar's Highlands: Initial Evidence and Implications

William J. McConnell [1,*], Andrés Viña [2,3], Christian Kull [4] and Clayton Batko [5]

1 Center for Global Change and Earth Observations, Michigan State University, East Lansing, MI 48823, USA
2 Center for Systems Integration and Sustainability, Michigan State University, East Lansing, MI 48823, USA; vina@msu.edu
3 Department of Geography, University of North Carolina, Chapel Hill, NC 27599, USA
4 Institut de Géographie et Durabilité, Université de Lausanne, 1015 Lausanne, Switzerland; Christian.Kull@unil.ch
5 Department of Anthropology, Michigan State University, East Lansing, MI 48823, USA; clayton.batko@gmail.com
* Correspondence: mcconn64@msu.edu; Tel.: +1-517-884-1237; Fax: +1-517-353-2932

Academic Editors: Claudia A. Radel and Jacqueline M. Vadjunec
Received: 2 August 2015; Accepted: 9 November 2015; Published: 25 November 2015

Abstract: Madagascar is renowned for the loss of the forested habitat of lemurs and other species endemic to the island. Less well known is that in the highlands, a region often described as an environmental "basket-case" of fire-degraded, eroded grasslands, woody cover has been increasing for decades. Using information derived from publically available high- and medium-resolution satellites, this study characterizes tree cover dynamics in the highlands of Madagascar over the past two decades. Our results reveal heterogeneous patterns of increased tree cover on smallholder farms and village lands, spurred by a mix of endogenous and exogenous forces. The new trees play important roles in rural livelihoods, providing renewable supplies of firewood, charcoal, timber and other products and services, as well as defensible claims to land tenure in the context of a decline in the use of hillside commons for grazing. This study documents this nascent forest transition through Land Change Science techniques, and provides a prologue to political ecological analysis by setting these changes in their social and environmental context and interrogating the costs and benefits of the shift in rural livelihood strategies.

Keywords: afforestation; forest transition; Landsat; rural livelihoods

1. Introduction

Madagascar is still losing its natural forest cover, despite decades of conservation efforts [1]. Many temperate countries and a few tropical countries have undergone a "forest transition," whereby net deforestation is reversed, but Madagascar as a whole does not yet fit this pattern [2]. However, at the regional scale, the island nation exhibits large pockets where tree cover is expanding. In the highlands, for instance, which a century ago were largely grasslands devoid of tree cover, introduced eucalypts, pines, acacias, and diverse fruit trees have become ubiquitous.

While widespread and seemingly driven by rural people together with state foresters, this phenomenon presents several challenges for researchers and environmental managers. For one, methodologically, it has been difficult to map and quantify these tree cover dynamics. Traditional remote sensing techniques struggle with these sparse land covers, which are also very patchy and distributed across a landscape mosaic (unlike large blocks of closed canopy forest). Second, such tree cover—largely dominated by exotic species—is often ignored by conservationists for its lesser biological value, yet it turns out to be central to livelihoods and rural development, as well as being relevant to soil and water conservation, carbon budgets, invasion biology, and other aspects of environmental management.

In this paper, we seek to determine the degree to which the highlands of Madagascar are experiencing a sub-national forest transition, and to identify the causes and livelihood consequences of changes in tree cover in a region widely regarded as the scene of ecological and economic devastation. The specific objectives of the study were to document the dynamics of low-density woody cover in the highlands region, and to explain these dynamics and their livelihood implications. We use a simple, two-step mixed Land Change Science – Political Ecology approach [3], in which we first characterize the nature of tree cover change using remote sensing, and then subsequently paint a picture of the diverse factors—livelihood strategies, forestry and land tenure policies, struggles for resource control, ideas about proper environmental management, and plant invasiveness—that combine to cause these changes. More specifically, we applied novel remote sensing techniques involving "continuous" measures of tree cover at the pixel level, rather than delimiting strict forest cover categories. This allows us to characterize the extent and dynamics of low-density woody vegetation in the highlands over the past two decades. We then explain these changes, situating them in the context of land use and livelihoods in the highlands, setting the stage for further political ecological investigation. This study seeks to contribute to the rapidly maturing theoretical literature on the forest transition, which is briefly reviewed below, before an exposition of the novel remote sensing methods we bring to bear on the debate over forest change in Madagascar.

1.1. *Theoretical Motivations*

The realization that large parts of the world had undergone net increases in forest cover during the 20th Century has sparked one of the most important lines of inquiry in human-environment research over the past ten years, owing in part to growing interest in questions of carbon sequestration and other ecosystem services. The process of slowing deforestation and a subsequent shift towards net forestation in Western Europe and eastern North America as those economies shifted from agricultural to industrial bases has been coined the "forest transition" [4,5]. Detection of similar phenomena in East and South Asian countries [6,7] has fueled interest in the possibility that this phenomenon might be reproduced elsewhere—particularly in the tropics where deforestation continues to be a dominant and still growing trend [8]. This has become one of the most promising avenues for theoretical advancement in the emerging science of land change [9–12].

It was initially theorized that forest transitions mark the end of the process of clearing forest land to make way for agriculture. According to this explanation, industrialization leads to the concentration of population in urban centers, and consequent rural depopulation, spurring a spatial contraction of increasingly intensive agricultural production. It also leads to levels of affluence at which forests are increasingly valued for aesthetic purposes [4,5]. Cross-national analyses have provided some support for this explanation, but have also suggested a second—so-called "forest-scarcity"—path to the forest transition in developing countries, *i.e.*, the planting of trees in response to perceived shortage of forest products [13].

The development of forest plantations around town centers has been a central tenet of land-use theory since its inception; von Thünen [14] theorized the development of a zone of tree production in proximity to towns, just outside a more proximate zone dedicated to the production of perishable products. Initial attempts have been made to link the von Thünen model with forest transition theory under idealized conditions [15], but a great need exists for empirical examination of the actual patterns and processes of the suggested "forest scarcity" pathway [13].

As it begins to mature, forest transition theory has attracted critical review on several counts. In terms of establishing the facts of forest cover change, there is concern that the "dependent variable"—spatio-temporal forest cover patterns—requires more systematic treatment. Generally, the procedures used to assess forest dynamics from remote sensing at global and continental scales suffer from lack of standardization, sometimes failing to differentiate natural forest from exotic monoculture plantations [16]. This issue is particularly acute in sub-humid landscapes, where trees are very important elements of agroecosystems even at relatively low densities [17].

In terms of attributing changes in forest cover to particular causes, forest transition theory is cited for lack of attention to scalar dynamics—the recognition that different processes tend to be more strongly associated with land change at different levels of observation [18–21]. Finally, critics caution proponents of forest transition theory against a "universalist approach and a structuralist-functionalist account," calling instead for research emphasizing "historical contingencies, variable causes among cases, and other contextual specificities" ([16]; pp. 109, 111).

Place-based research on tropical forest transitions has so far focused largely on Latin America [5,22–25] and South and Southeast Asia [26–28] (see also [29,30]). Meanwhile, studies of the role of trees in African landscapes have not yet, to our knowledge, been framed within the forest transition debate. However, some of the foundational contributions to the fields of Political Ecology [31,32] and Environmental History [33] set out to challenge the dominant received wisdom of peasant-driven environmental degradation in Africa [34–37] (see also [38]). In one rather famous case, the landscape of Machakos, Kenya, was observed to have undergone significant intensification under conditions of rapid population growth, upending the conventional population-poverty spiral view [39].

African smallholders have traditionally retained beneficial trees within their fields, and are now experimenting with the incorporation of other woody species into their production systems, often in collaboration with government and NGO efforts [40–42]. These outside interventions, along with differential access to land and other resources constitute a crucial set of institutional factors shaping the distribution and consequences of trees in African landscapes [43,44]. Understanding such institutional factors is a key challenge for the development of forest transition theory [45], and the governance of forests and other resources more generally [46,47].

1.2. Forest Change Studies in Madagascar and Advances in Remote Sensing

A vast literature exists documenting the loss of forest in Madagascar, almost exclusively focusing on what is considered "natural" or pristine formations, under the assumption that greater height, greater canopy closure and higher species diversity are indicators of habitat quality for endangered wildlife, such as lemurs. These assumptions may generally hold for the purposes of conserving biological diversity. However, from the standpoint of local livelihoods, woody species play important roles at much lower density and diversity.

Detecting the presence of woody cover at low densities using traditional remote sensing techniques presents methodological challenges and this, along with the conservation motives of most land-cover change studies in places like Madagascar, means that changes involving such land covers are typically not captured. In particular, traditional satellite image processing techniques reflect cartographic technology favoring choropleth maps, in which the landscape is divided into units of internally homogenous land-cover types, with the number of categories limited to ease the readability of the map. In practice, however, it is quite rare for studies to actually specify what they consider to constitute forest cover. For example, among nearly two dozen studies of land change in Madagascar (See Table S01 in Supplementary Materials), we were able to identify only one that provided an explicit definition of forest. Harper and colleagues, in their review of half a century of deforestation and forest fragmentation in Madagascar, defined forest as "areas of primary vegetation dominated by tree cover at least seven meters in height, with neighboring tree crowns touching or overlapping when in full leaf" ([1]; p. 2). The issue of conflicting definitions of forest is treated explicitly by Dufils [48], and McConnell and Kull [49].

The past decade has witnessed rapidly growing interest in the development of techniques for the classification of remotely sensed imagery to obtain continuous depictions of the landscape, along gradients such as "percent tree cover" [50]. Such techniques allow the detection of trees and other woody vegetation at a range of densities, including those with important livelihood implications that are overlooked in most studies. At the same time, the rapid increase in the availability of recent high resolution commercial satellite imagery, through platforms such as Google Earth®, provides an unprecedented ability to collect the kind of information needed to "train" an automated classification of multispectral imagery. Of course, understanding the causes of the changes revealed through these remote sensing and image classification techniques still requires "boots on the ground." The present study brings together decades of such concerted fieldwork with novel remote sensing techniques to document and explain changes in the open woodlands, woodlots and orchards of highland Madagascar.

2. Study Area

Madagascar is classified as a low-income country by the World Bank, with 88 percent of the population subsisting on $1.25 per day, and four-fifths of the population consists of families operating farms averaging roughly 1 ha with manual labor and animal traction. Large, mechanized farms are the exception. The smallholder farms face high transport costs due to poor infrastructure, usually combining subsistence and market-oriented production. Most market production is destined for local or regional (within the island) consumers, though certain regions have found specialized export niches, such as green beans in the central highlands around Antananarivo [51–54].

Like elsewhere in Africa, farmers seek to diversify their livelihoods as a risk-reduction strategy. If they can, households combine crop production, livestock rearing, unskilled wage labor, seasonal or periodic migration (as labor on farms or in mining), small-scale artisanal or trading activities, urban employment, opportunistic resource exploitation, and other activities [53,55]. In this way, households respond to opportunities and constraints, such as the expansion of economic opportunities linked to preferential trade agreements [53], or the contraction of the livestock sector due to increased dangers of cattle rustling [56]. Natural resource extraction helps as a safety net in the case of crises, whether drought, cyclone, market shifts, or illness or death of household labor [57,58]. Woody tree species—whether native, planted, or invasive—often serve such a role, offering rural people either regular income or a kind of backup option to earn supplemental cash in times of difficulty [59,60].

The central highlands (defined by the current administrative regions of Bongolava, Analamanga, Itasy, Vakinankaratra, Amoron'i Mania and Haute Matsiatra; Figure 1) cover an area of approximately 96,000 km^2, contain three of the island's largest cities (including the capital), and support nearly half (*ca.* 7.5 million) of the island's human population. Total annual rainfall decreases westward, with the eastern portion of the highlands hosting tropical rainforests, while the western portion is mostly composed of low productivity rangeland. The human imprint on Madagascar's highlands is characterized by a mix of irrigation infrastructure and extensive cattle grazing, illustrating the blending of African and Asian cultural heritages [61,62]. Farmers cultivate irrigated rice and a number of upland crops, such as manioc, sweet potatoes, beans, and potatoes. All farms keep at least some animals, and the region's grasslands are typically burned nearly every year to provoke late dry season forage for cattle, and to prevent bush encroachment [63]. These fires, combined with the region's ubiquitous erosion gullies, known as *lavaka*, have given the highlands a reputation as a site of environmental degradation [64–66].

3. Methods

The application of forest transition theory in landscapes dominated by smallholder plots requires a multi-level approach that enables the consistent detection of woody cover dynamics across a broad area, combined with highly contextualized explanations of processes occurring in specific landscapes. We therefore begin with a broad scale assessment of land-cover change using innovative techniques for the classification of moderate resolution (*i.e.*, 30 m horizontal resolution) satellite imagery. The procedure yields change maps that highlight particularly dynamic areas, from which we select some for closer examination using field observations in combination with ancillary information.

3.1. Remotely Sensed Data, Digital Image Processing and Change Detection

Woodland dynamics in the highlands were evaluated using three multi-spectral scenes (path 159, rows 73 to 75 of the World Reference System-2 (WRS-2)) acquired by the Thematic Mapper (TM) sensor onboard the Landsat-5 satellite system and by the Operational Land Imager (OLI) sensor onboard the Landsat-8 satellite system on September 25, 1994 and September 16, 2014, respectively. The use of anniversary data from the dry season (*i.e.*, September) was preferred because it enabled the use of mostly cloud-free scenes. These scenes account for *ca.* 80 percent of the central highlands (Figure 1).

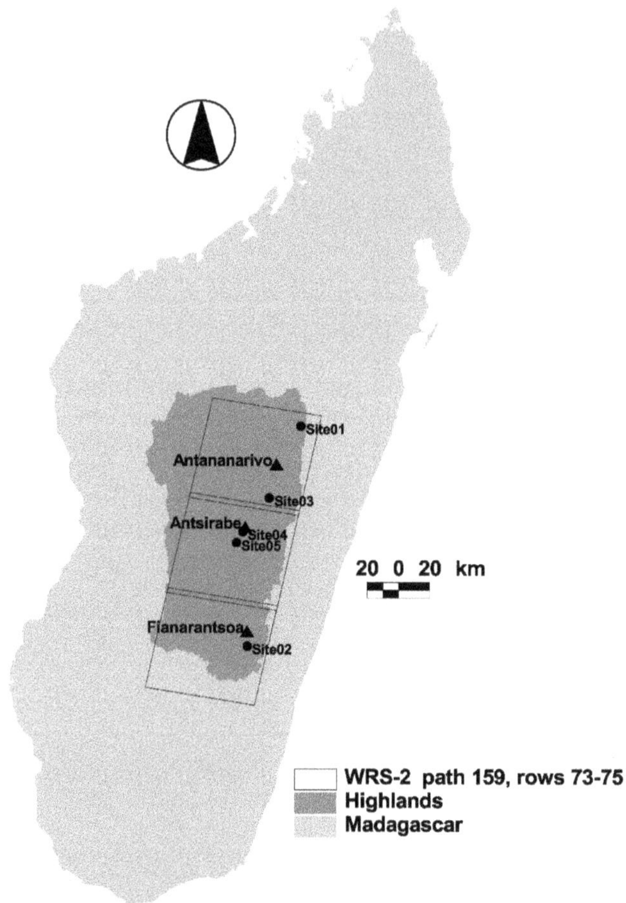

Figure 1. The study region was defined by the area within three Landsat scenes (path 159, rows 73 to 75 of the World Reference System-2 (WRS-2)), comprising *ca.* 80 percent of Madagascar's Highlands. Also shown are the locations of three main cities and five field sites visited between 2003 and 2006 whose land-cover dynamics are explained in more detail in the text.

The three scenes per date were mosaicked, and a relative radiometric normalization was applied to render the radiometric structure of the scenes from the two anniversary dates (*i.e.*, 25 September 1994 and 16 September 2014) comparable. The normalization consisted of the development of linear regression models for each of six optical bands (*i.e.*, bands 1–5 and 7 of the Landsat TM, and bands 2–7 of the Landsat OLI), to predict the brightness values of the 2014 imagery that would have been obtained had they been acquired under the same atmospheric conditions as the 1994 imagery. A total of 40 pseudo-invariant targets (*i.e.*, wet and dry targets present in both scenes assumed to be constant reflectors through time so that any changes in their brightness values could be attributed to sensor calibration, astronomic, atmospheric and phase angle differences) were used to develop the linear regression models. Table 1 shows the coefficients of these models.

Table 1. Coefficients of linear regression models (one for each spectral band) developed using pseudo-invariant targets. These linear models were constructed in order to match the radiometric structure of the six optical bands of the 16 September 2014 Landsat Operational Land Imager (OLI) imagery to that of the September 25, 1994 Thematic Mapper (TM) imagery.

Band	Slope	Intercept	R^2
Blue	0.013	−51.128	0.928
Green	0.007	−28.163	0.910
Red	0.008	−33.207	0.949
Near Infrared	0.006	−20.427	0.971
Shortwave Infrared 1	0.010	−44.685	0.986
Shortwave Infrared 2	0.007	−32.121	0.978

We estimated the probability of an area to exhibit woody cover using a fuzzy classification algorithm based on the principle of maximum entropy [67] using MaxENT, a general purpose machine-learning method for making predictions from incomplete information [68]. To calibrate and validate the maximum entropy classification algorithm, we developed a dataset of 280 "ground truth" polygons of the same area as a Landsat pixel (*ca.* 900 m^2) semi-randomly distributed throughout the study area, in which the interpretation of an initial set of 250 randomly distributed polygons was used to oversample in areas exhibiting some woody cover. Within each of these polygons we randomly distributed 10 points. Using high spatial resolution imagery available in Google Earth®, we visually ascertained the number of points per polygon coinciding with a tree canopy. To assess the reliability of the interpretation of imagery in Google Earth®, two image interpreters independently performed the point counts. The average point count between the two interpreters was obtained, and only those polygons that exhibited a point count difference of 10 percent or less between the two interpreters were used.

Following the classification of forested areas established by the United Nations Food and Agriculture Organization [69], we employed a threshold of 10 percent tree cover as a definition of woodland (encompassing closed canopy forest, plantations, treed savanna, riparian growth and sometimes dense shrubland), and considered a polygon to be at least 10 percent covered by woody vegetation if one or more points per polygon coincided with such cover. Out of the 280 ground truth polygons, 97 (*i.e.*, 34.6 percent) had 10 percent or greater woody cover. Using these data together with the MaxEnt algorithm, we created a woody cover probability (*i.e.*, the probability of having tree cover of 10 percent or higher) map from the multi-spectral bands (*i.e.*, bands 2–7) of the 2014 Landsat imagery. The output probability map was validated using a cross-validation procedure in which 2/3 of the ground truth polygons were used for calibration and the remaining 1/3 used for validation. To reduce dependence on a single random partition into calibration and validation, we generated 10 different random partitions to be used in 10 different classifications, which were then averaged. The 10 output probability maps were validated by means of a receiver operating characteristic (ROC) curve [70]. The ROC curve is a plot of the sensitivity values (*i.e.*, true positive fraction) *vs.* their equivalent 1-specificity values (*i.e.*, false positive fraction) for all possible probability thresholds. The area under the ROC curve (AUC) is a measure of model accuracy, with AUC values ranging from 0 to 1, where a score of 1 indicates perfect classification, a score of 0.5 implies a classification that is not better than random, and values lower than 0.5 imply a worse than random classification. The average AUC value was 0.872, which was significantly ($p < 0.0001$) different from 0.5 and denotes high classification accuracy. The average coefficients of the validated classification of the multi-spectral bands of the 2014 Landsat OLI imagery were then applied to classify the multi-spectral bands (bands 1–5 and 7) of the 1994 Landsat TM imagery.

A simple image differencing technique (in essence, a post-classification comparison [71]), in which the probability of woody cover in 1994 was subtracted from the 2014 probability, was performed on a per-pixel basis yielding a continuous measure of change over 20 years from the 1994 probability, which was used as the baseline. This procedure yields an output map that depicts spatial patterns of more subtle dynamics than is captured by more traditional image "hard" classification approaches in which each pixel is assigned to one of a fixed set of categories. The dynamics revealed by the multi-date Landsat imagery include: no significant change in probability, which we interpret as stability; an increase in probability, which we interpret as gain in woody cover; and a decrease in probability, which we interpret as loss in woody cover. These dynamics were then further explored using the high resolution imagery from Google Earth®, in conjunction with aerial photographs from the early/mid 1990s, along with ground level photos and field visits conducted midway through the study period (*i.e.*, 2003 and 2006).

3.2. *Explaining the Observed Woodland Dynamics*

Interpretation of the documented land change patterns was based on field observations and previous studies over the past two and a half decades, in conjunction with the published literature. In particular, we relied on two dozen case study sites (selected using a spatially stratified random sample) previously investigated using 1950s and 1990s air photos as well as field visits [72]. Then, in this two-step Land Change Science–Political Ecology study, we move from the more descriptive remote sensing and case study site discussion of tree cover changes to an investigation of the various factors lying behind these changes. Following a political ecological framework, attention was accorded to not only market incentives, demographic pressure, and official policies, but also to access to resources and to the livelihood strategies of rural residents including land tenure arrangements, de-facto rules around trees, and environmental ideologies, as well as the invasive characteristics of some of the trees themselves [55,73,74].

4. Results

4.1. *Remote Sensing and Case Study Analysis*

The results of the probability map differencing procedure are shown in Figure 2. Increases and decreases in probability between 1994 and 2014 (expressed in the map as gains and losses, respectively) represent increases and decreases in woody cover. Increases in woody cover are indicated in shades of green, and decreases in shades of red, while the initial cover (*i.e.*, baseline probability value in 1994) is given by shades of blue. The combinations of these colors yield gradations among six trends depicting woody cover dynamics between 1994 and 2014 over the study area. What is notable is the wide extent of increasing tree cover, particularly in interior highland areas often treated in broad scale assessments as essentially devoid of trees. Below, we describe in detail the three main categories of change shown in Figure 2 (landscape stability, decreases in woody cover, and increases in woody cover), linking the general patterns to specific dynamics revealed in drill-down analyses of the specific case study sites.

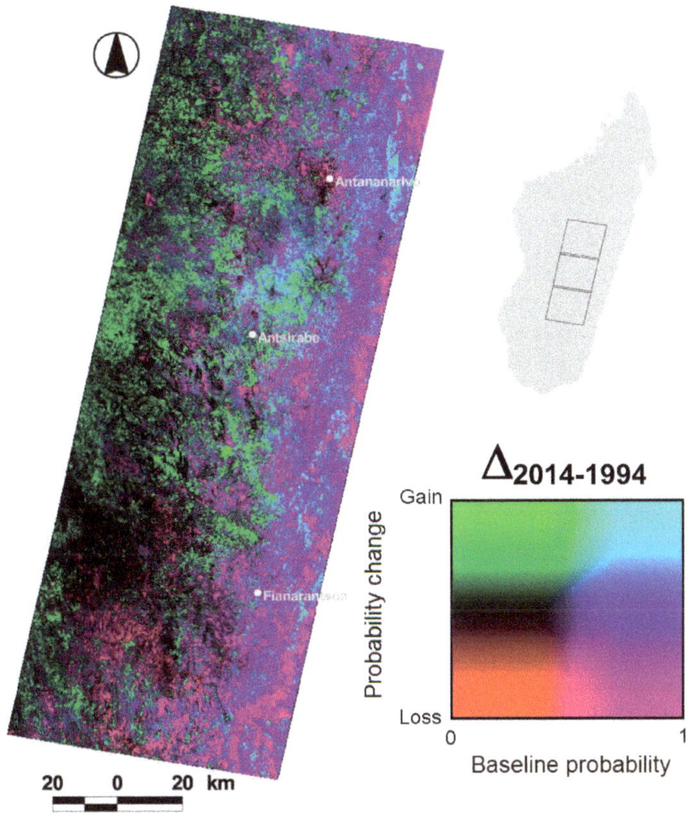

Figure 2. Spatial distribution of the temporal dynamics (*i.e.*, gain, loss, no change) of woody cover in the study area over a 20-year period (*i.e.*, 1994–2014), as a function of the baseline probability of woody cover (*i.e.*, probability of woody cover in 1994). Also shown are the locations of the three major cities (including the capital) in the Highlands of Madagascar.

First, stable landscapes are represented in Figure 2 with black and blue tones, representing herbaceous and forest landscapes, respectively. In the case of more dense (higher probability) woody cover in 1994, this stasis is rendered in dark blue, which one would expect from undisturbed forest. This dynamic is seen in the eastern portion of the study area, which contains the region's remaining dense forests, along the eastern escarpment of the highlands, as well as in some of the larger and longer-term zones of woodlot forests (e.g., east of Antananarivo).

The darkest tones in the map represent areas that had a low probability of woody cover at either date, and these are concentrated in the western, and especially southern, portion of the study area, which are also drier and less populated (Figure 2). These areas are often characterized by lateritic soils with minimal organic layers. Typically, herbaceous vegetation covers the uplands, with woody vegetation restricted to slope hollows and the edges of streams and seasonal drainages.

A ***second*** category of change is decrease in woody cover. A decrease in woody cover is unlikely when the initial probability was low, explaining the paucity of deep red in Figure 2. This dynamic could result, though, from the burning of grasslands containing shrubs and occasional trees, but occurs quite infrequently in our results. Much more common are decreases from a more dense initial cover, shown in magenta. This dynamic is particularly visible along the far eastern edge of the study area, and corresponds to significant degradation of woody cover, as would be expected from clearing of forest for agriculture (Figure 2). In this region, it is likely that some—or even most—of these forests had been subject to some degree of disturbance prior to 1994, but our study period only comprises the dynamics after 1994, which prevents us from distinguishing such prior disturbance. Elsewhere, however, our fieldwork and the literature allow us to identify the conversion of dense forest to other land-cover types. We recognize two distinct processes, one involving cyclical loss of canopy, known as a coppice. This entails removing the stems of tree species capable of re-sprouting from their stump, a property that has made several species of eucalyptus very popular in Madagascar. The large eucalyptus plantations around the capital (Antananarivo), including the Manjakandriana Massif, have been well-described [75]. Such coppicing appears in the change map as large magenta blocks around the black mass of the city in the northeast portion of the study area (Figure 2), interspersed with increases in woody cover (described below).

Smaller, more dispersed stands of eucalyptus have heretofore received less attention in the peer-reviewed literature, but are revealed in our results. A good example is our field site at Ambongabe, north of Anjozorobe (Site 01; Figure 3), where dense woody cover visible in the 1992 aerial photograph was much more open in a 2014 Digital Globe image, resulting in bright magenta tones in the change map. Zooming in on the high resolution imagery reveals this to be a well-organized hilltop plantation with large crowns in 1992, and much smaller crowns in 2014 (Ellipse 1 in Figure 3). Our fieldwork confirms this to be a eucalyptus plantation, which was probably coppiced twice during the intervening 20 years.

Another process explaining decreased woody cover, particularly in plantations, involves more complete harvesting. The study area includes several large government and private plantations subject to regular clearcutting and replanting, much as in other parts of the world. Some of our field sites, however, are characterized by a spectrally similar, yet institutionally different, process. The Landsat change map is dominated by bright magenta tones (representing decreased woody cover) at our Ambozontany field site (Site 02, Figure 4) where portions of the government's Haute Matsiatra pine plantations—shown in the 1991 aerial photos to be closed canopy plantations—had been cleared for upland cultivation when Digital Globe's sensor recorded an image in October of 2014. This process was well under way during a field visit in 2006, which revealed that this conversion represented the unsanctioned actions of communities who had previously been dispossessed of the land for the creation of the state plantations during the first decade following Independence. The village lands in the eastern third of the image have their usual eucalyptus woodlots, but the pine stands were rather threadbare (see Figure S1 in Supplementary Materials).

The ***third*** and final set of dynamics, representing increases in woody cover, is concentrated in very different parts of the landscape. In the Landsat change map cyan tones represent an increase from an already dense initial cover. These are predominantly found in the eastern area, where secondary growth often occupies land previously cleared for agriculture (Figure 2). They are also particularly visible north of Antsirabe, in a region with a long history of woodlot plantations that continues to intensify. These frequently co-occur with magenta tones, in landscapes where fallow periods are long enough for woody vegetation to reestablish itself following cultivation. This dynamic was also observed, however, farther west, where forests were rare or absent in 1994. For example, in the same field site at Ambongabe described above (Ellipse 2 in Figure 3) as showing evidence of eucalyptus coppices, we see evidence of hilltops covered in brush in 1992 that later showed tree canopies.

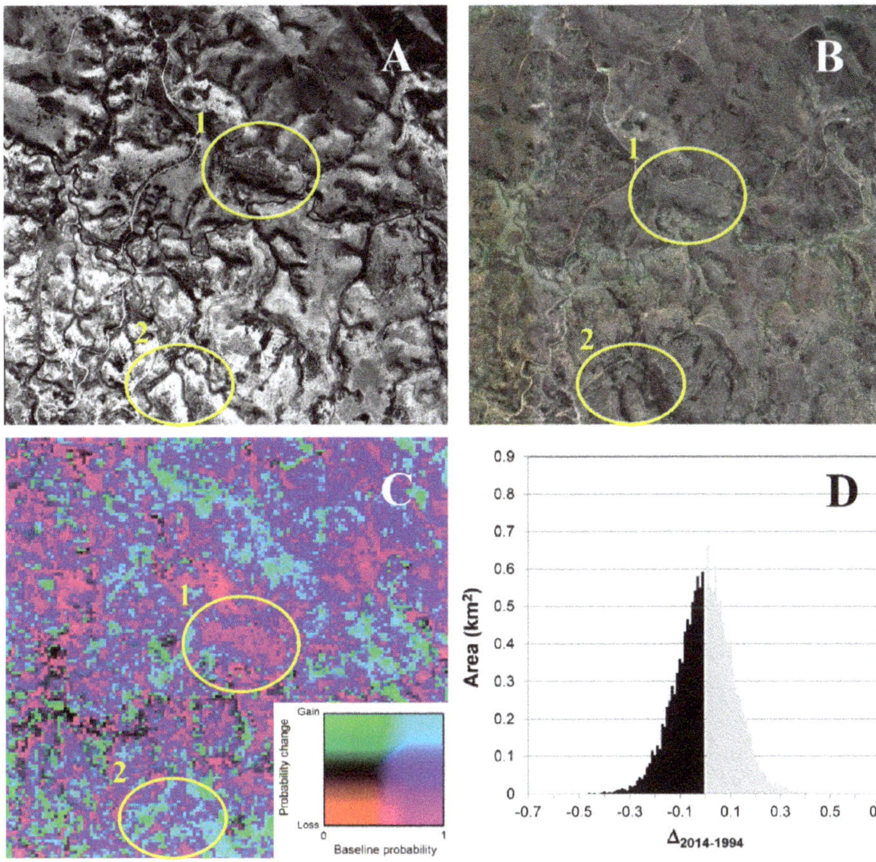

Figure 3. Dynamics of woody cover at Site 01, Ambongabe, near the town of Anjozorobe north of Antananarivo (18°19'S, 47°54'E). (**A**) Aerial photograph taken in 1992. (**B**) Digital Globe image taken in 2014. (**C**) Change map based on Landsat imagery (see Figure 2). (**D**) Histogram depicting the frequency distribution of the delta of woody cover probability values between 2014 and 1994 based on Landsat imagery. Woody cover loss is represented in black, while gain is represented in gray. Ellipses in (A–C) relate to particular woody cover dynamics described in the text. See Figure 1 for the geographic location of Site 01 within the study area.

Of great interest to the present study are areas appearing in the Landsat change map in bright green tones, representing substantial increases in woody cover over the 20 year period. This can only occur when initial woodland probability was low, and is concentrated in the central and western portions of the study area. Three of our field sites exhibit these dynamics.

Near Ambatolampy (Site 03; Figure 5) well-organized plantations were already evident in the 1991 aerial photographs, and the 2013 CNES Astrium image shows these had expanded significantly, resulting in bright green tones in the Landsat change map (Ellipse 1 in Figure 5). Field work confirms these to be eucalyptus plantations. Farther west at the same site is a dense, well-managed pine forest, with some araucaria, sweet gum, and silver wattle. This is a private 48 ha land holding of a family residing in the capital (Ellipse 2 in Figure 5). A ground-level panorama shot in 2003 (Figure S2 in Supplementary Materials) shows the widespread, low-density eucalyptus plantations, as well as a foreground "scrub" of *Helichrysum* spp. and silver wattle.

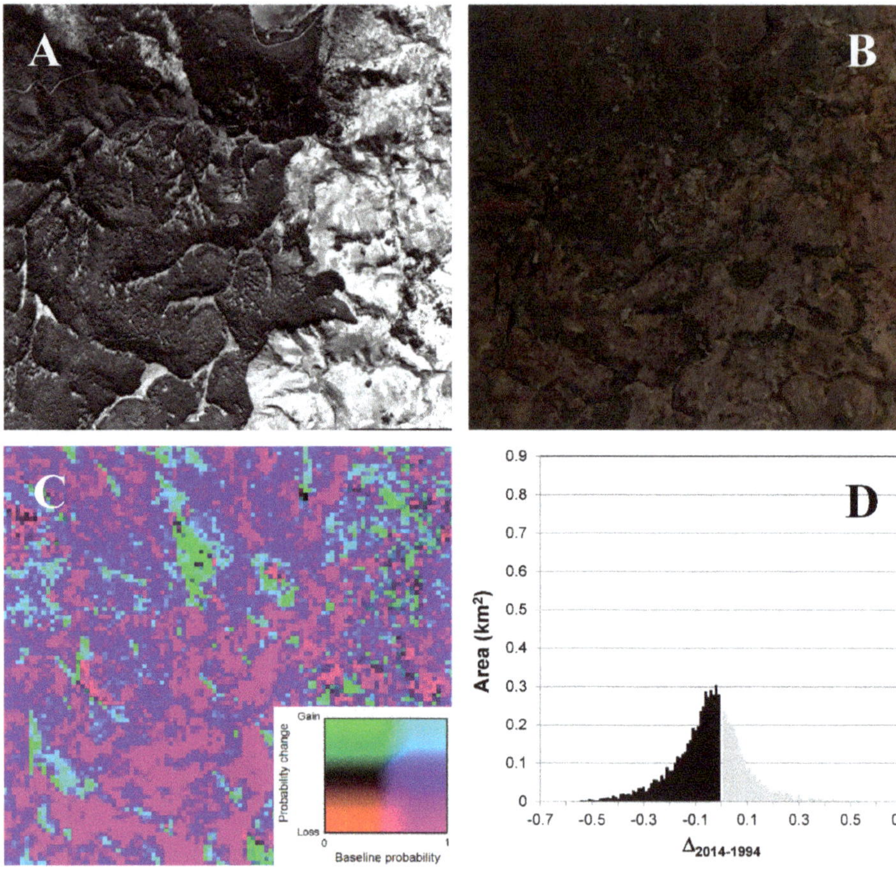

Figure 4. Dynamics of woody cover at Site 02, Ambozontany, south of Fianaratsoa (21°48'S, 47°6'E). (**A**) Aerial photograph taken in 1991. (**B**) Digital Globe image taken in 2014. (**C**) Change map based on Landsat imagery (see Figure 2). (**D**) Histogram depicting the frequency distribution of the delta of woody cover probability values between 2014 and 1994 based on Landsat imagery. Woody cover loss is represented in black, while gain is represented in gray. See Figure 1 for the geographic location of Site 02 within the study area.

Meanwhile Ambohijatovo (Site 04; Figure 6) is a typical intensively cultivated central highland site, marked by eucalyptus woodlots, rice fields, and market vegetables, as shown in a panorama photo taken in 2003 (Figure S3 in Supplementary Materials) from the valley bottom. Comparing the 1991 aerial photograph and the 2013 CNES Astrium image reveals the expansion of a hilltop plantation down to the rice paddies in the valley bottom, once again shown in the Landsat change map in shades of green (Ellipse in Figure 6). Likewise, in the western portion of our Soamahazina site (Site 05; Figure 7), an expansion of tree cover is evident between 1991 and 2013, showing up in the Landsat change map in shades of green. Fieldwork in 2003 documented the development of village woodlots in these cool uplands (1700-2000m) where transport is relatively difficult, but the site is close enough to the town of Antsirabe that charcoal production is economically viable (Ellipse in Figure 7). The haphazard arrangement and sizes of the trees at this site suggest that these woodlots are being expanded in incremental fashion, different from the more regimented rows of even-aged trees seen in other sites.

Finally, it should be noted that there are other "new" tree covers than the spatially dominant eucalyptus and pine woodlots, also responsible for the green tones in the Landsat change map. In particular, Kull [72,73,76] documented a dynamic involving fruit orchards, hedge trees, ornamentals, and other trees in and around settlements. Given their spectral similarities at low stem densities, these cannot easily be distinguished from other woodlots in our remote sensing analysis, but result from different land management processes. Such a landscape element is shown in the panorama photo (Figure S3 in Supplementary Materials) of the Ambohijatovo case study (Site 04), notably on the right hand side (location indicated in Figure 6).

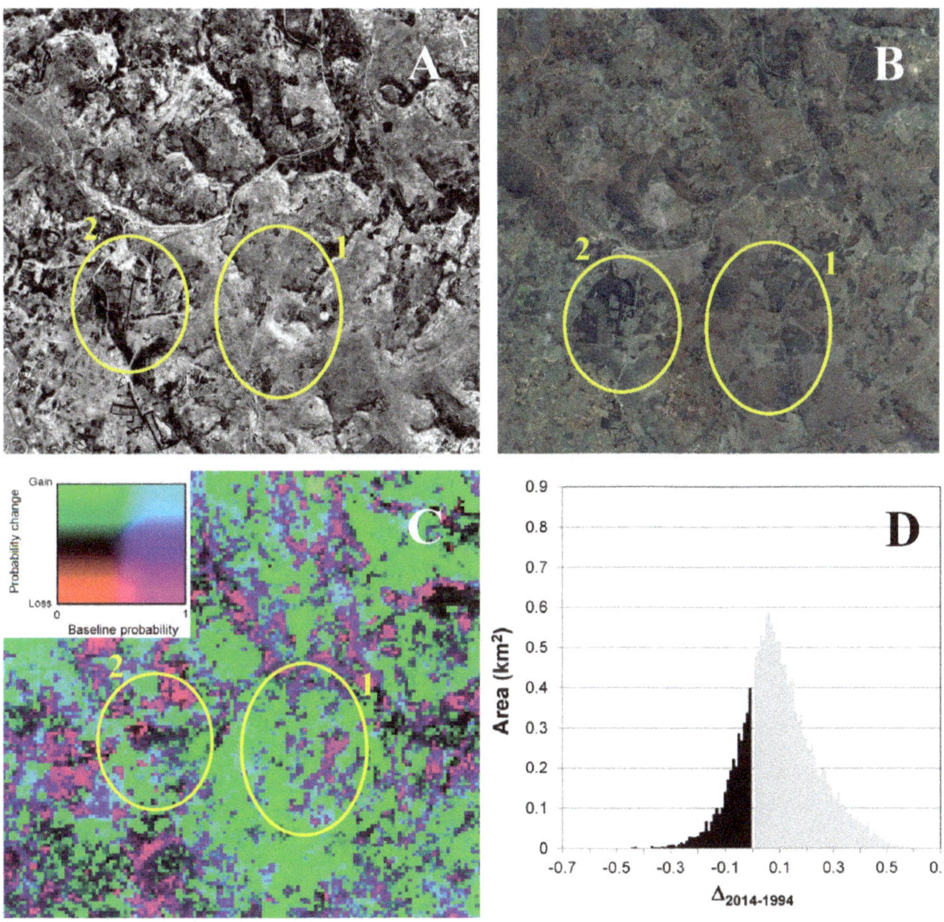

Figure 5. Dynamics of woody cover at Site 03, south of the town of Ambatolampy (19°25'S, 47°25'E). (**A**) Aerial photograph taken in 1991. (**B**) CNES Astrium image taken in 2013. (**C**) Change map based on Landsat imagery (see Figure 2). (**D**) Histogram depicting the frequency distribution of the delta of woody cover probability values between 2014 and 1994 based on Landsat imagery. Woody cover loss is represented in black, while gain is represented in gray. Ellipses in (A–C) relate to particular woody cover dynamics described in the text. See Figure 1 for the geographic location of Site 03 within the study area.

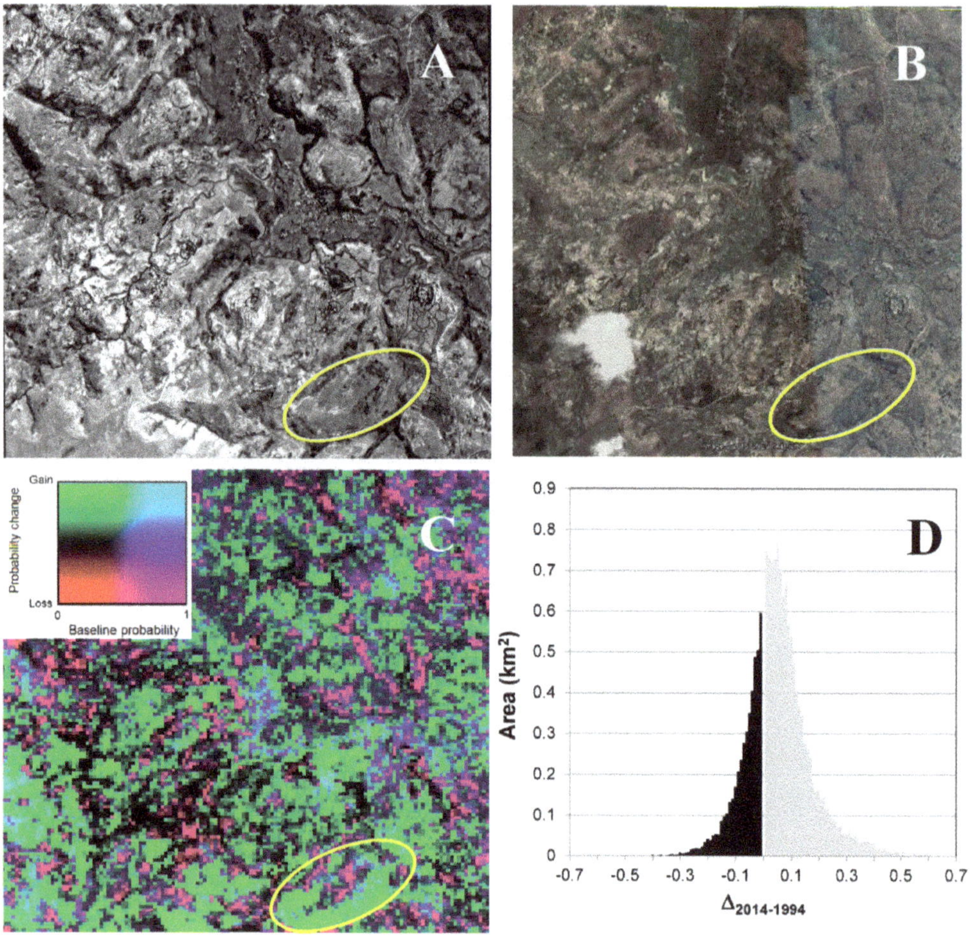

Figure 6. Dynamics of woody cover at Site 04, Ambohijatovo, centered on a rice cultivating valley just south of Anstirabe (19°56'S, 47°0'E). (**A**) Aerial photograph taken in 1991. (**B**) CNES Astrium imagery taken in 2013. (**C**) Change map based on Landsat imagery (see Figure 2). (**D**) Histogram depicting the frequency distribution of the delta of woody cover probability values between 2014 and 1994 based on Landsat imagery. Woody cover loss is represented in black, while gain is represented in gray. Ellipses in (A–C) relate to a particular woody cover dynamic described in the text. See Figure 1 for the geographic location of Site 04 within the study area.

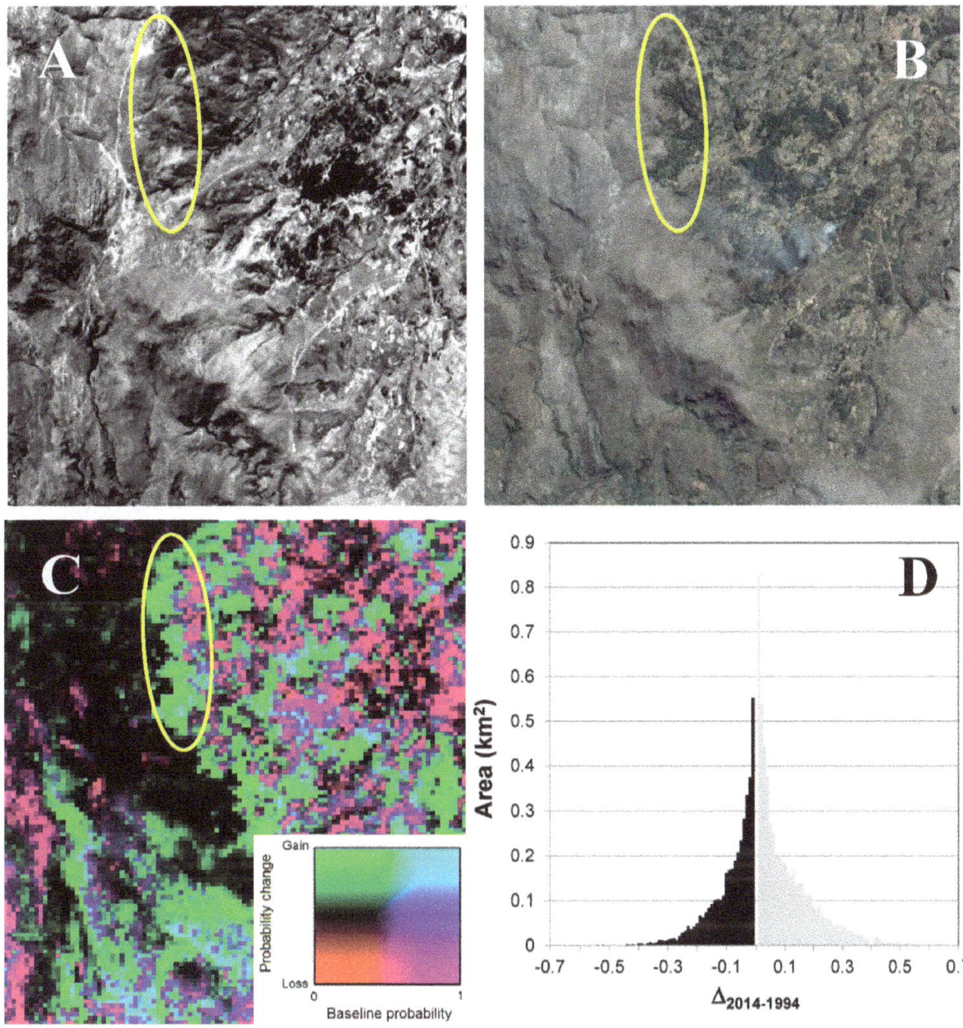

Figure 7. Dynamics of woody cover at Site 05, Soamahazina, an upland area southwest of Antsirabe (20°06'S, 46°54'E). (**A**) Aerial photograph taken in 1991. (**B**) CNES Astrium taken in 2013. (**C**) Change map based on Landsat imagery (see Figure 2). (**D**) Histogram depicting the frequency distribution of the delta of woody cover probability values between 2014 and 1994 based on Landsat imagery. Woody cover loss is represented in black, while gain is represented in gray. Ellipses in (A–C) relate to a particular woody cover dynamic described in the text. See Figure 1 for the geographic location of Site 05 within the study area.

4.2. *Political Ecological Context of Tree Cover Change*

A variety of intertwined endogenous and exogenous factors can be shown to lie behind the tree cover dynamics of the highlands. Six factors emerged from the analysis of the empirical case study sites and the literature: household livelihood strategies, urban market demand, land tenure struggles, dominant environmental ideas and resulting policy, government forestry interventions, and the invasiveness of some of the trees themselves. Each of these factors was already operative at the beginning of the study period. We argue that the conjuncture of these processes spurred the tree cover increase (following [77] and [78]). Further research could determine the precedence of particular factors in different places. We review each factor in turn.

First, a vast array of household needs is met by the trees increasingly covering the highlands. They are utilized in diverse ways, sometimes to meet subsistence needs, sometimes for cash income. Domestic fuel in rural areas comes almost purely from locally grown woody cover. Construction uses include floors, roofs, balconies, railings, stairs, cattle pens, pig sheds, and fence posts. Carpentry uses include beds, chairs, tables, cooking utensils, oxcarts, toys, and much more. Fruit trees and shrubs—from oranges to mangos, from vines to peaches—provide obvious consumable and marketable products, as well as secondary uses of dead branches as fuel wood, and mulberry leaves as silk worm fodder, among many other uses. Other benefits derived from trees include the production of honey based on eucalyptus flowers, the improvement of soil fertility based on acacia nitrogen fixation, and the sale of eucalyptus essential oils [79–81]. It is important to note that population growth in Madagascar is still quite rapid, with the country's population doubling between 1990 and 2013. Meanwhile, economic stagnation in urban areas has actually led to highly unusual urban-to-rural migration [82]. At the same time, the benefits previously derived from the cultivation of uplands have decreased with declining soil quality, while the benefits of grazing cattle have been undermined by increasing cattle rustling [56], which may in turn be interpreted as symptomatic of a weak, neopatrimonial state [83].

Second, urban market demand for wood fuel—particularly charcoal for urban consumers, or even to run industrial installations—as well as for construction wood, drives investment in woodlots [55,75,79,80,84–87]. Several areas of the highlands contain landscapes particularly devoted to the production of charcoal, firewood, and wood products. The most notable are the eucalyptus-dominated hills east and northeast of the capital, in the districts of Manjakandriana and Anjozorobe, where over 70 percent of the land—mostly former grassland—is now treed. At a smaller scale other similar landscapes are found outside other highland cities, such as the high-elevation pine and wattle groves between Antsirabe and Faratsiho [80].

Third, land tenure issues play a contributing role. Since around 1930 and continuing to the present, the plantation of woodlots was recognized by the state as evidence of land appropriation and development, a prerequisite to formal title registration (it should be noted that, in contrast, extant native forests were long ago declared state domain). Thus, urban and rural investors not only benefit from a growing urban market demand for the products from their woodlots, but they can also use the woodlots as part of a strategy to claim title to lands otherwise considered state-owned [55,84,88] or to lay out claims to common lands.

Fourth, it is important to understand the context of prominent environmental discourses in shaping perceptions of the highlands and policy responses. From early colonial days through to the present, scientists, foresters, and government agents have perceived the highlands as "barren" and deforested, and in need of "re-greening." This discourse is based both in a concern over soil degradation and biodiversity loss, as well as more pragmatic concerns with the supply of fuel and construction material [64,72]. Understanding the context of prominent environmental discourses is crucial to understanding government interventions, the next factor.

Fifth, both the colonial and independent governments, together with non-governmental organizations (NGOs), have repeatedly promoted the afforestation of the highlands. Tree planting to create communal forest reserves was at times obligatory and at other times a punishment for crimes [64,89]. The government's *Direction des Eaux et Forêts* (the Forest Service), founded at the dawn of French colonial control, still oversees a number of tree nurseries, forest stations, and forest plantations. Forestry stations, such as Manankazo (near Ankazobe), Manjakatompo and Antsampandrano (on the Ankaratra massif) and Angavokely (east of the capital) oversaw important reforestation projects and are at the center of regions with smallholder woodlots. The Forest Service also embarked on large, industrial plantations of *Pinus* spp, in particular in the *Haute Matsiatra*, northeast and southeast of Fianarantsoa as well as the *Haute Mangoro*, north of Moramanga (outside our study area). In the past two decades, however, the approaches have been more voluntary, sponsored for instance by projects run through local and foreign NGOs, aid agencies, universities, and conservationists. As just one example among many, the overseas wing of the French *Office National des Forêts* sponsors the work of the Malagasy NGO *Tany Meva* to plant trees across a few hundred hectares in a municipality some 30 km outside the capital [90]. At the same time, the state's inability to provide rural security has created disincentives to traditional livestock practices and surplus grain production, thus encouraging a shift in land use favoring trees, while its regulation of the charcoal trade has had mixed results [91] (see also [92] on the issue of corruption in law enforcement in Madagascar generally, and [93] on the specific question of conservation enforcement).

Sixth, and finally, the trees themselves have played a role in expanded tree cover in the highlands, taking advantage of suitably invasible landscapes. In particular, both pines and acacias have been relatively successful in reproducing themselves and dispersing from initial plantations [94], mirroring their behavior elsewhere [95,96].

5. Discussion and Conclusions

5.1. On Pathways and Processes of Forest Transitions and Landscape Change

While the rainforests of the eastern side of Madagascar's highlands continue to be exploited, a dynamic of woodlot rotations and tree planting is transforming the central and western highlands. Based on the state-of-the-art remote sensing evidence, we argue that this represents an incipient, sub-national forest transition. Building on the analysis of factors facilitating this trend, we furthermore argue that this transition follows a "scarcity" pathway pushed and pulled by endogenous and exogenous factors [97].

The land-use transition underway in the highlands of Madagascar shares little with other cases prominent in the forest transition literature. The economic development pathway is not relevant, since neither rural depopulation driven by industrialization nor rising affluence are operative in Madagascar, where the opposite trends are at work [82]. As for the traditional forest scarcity pathway, the few large-scale government-sponsored plantations account for a modest portion of the increased woody cover detected, and many of those were seen to be reverting to agricultural uses. Meanwhile, there are no large-scale private operations like the rubber and palm oil plantations of Southeast Asia. The situation in Madagascar bears some resemblance to the "smallholder stewardship" dynamic in the Ecuadorian Amazon described by [98], in which farmers with abundant land and labor are increasingly allowing certain trees to grow in their pastures. In the Malagasy highlands, though, pasture clearing is typically less labor-intensive than in Amazonia and, while farmers allowing spontaneous regeneration is certainly a part of the story, smallholders are also purposely planting trees that would not otherwise occur. Perhaps more instructive is the Vietnam case presented by Lambin and Meyfroidt [99] and [100], in which smallholders incorporate trees in their landscapes in response to a mix of endogenous and exogenous factors. In that case, the key exogenous factors included market liberalization and land tenure reform, which increased the relative profitability of lowland cultivation, leading farmers to reallocate labor from shifting cultivation, and planting the uplands with trees. In the case of highland Madagascar, market liberalization has not spurred a similar intensification of paddy land, despite massive investment by the United States Agency for International Development and other donors. Rather, it appears that continued uncertainty in rice prices, in concert with the increasing risk of theft—both of grain stores and cattle purchased with the proceeds—have created incentives for farmers to use former grazing land to plant trees. Thus, external factors, including the provision of material, knowledge and incentives to plant trees at the micro level, seem to have combined with the state's inability to provide functioning markets and rural security to spur farmers to plant trees. To our knowledge, this pathway to the forest transition has not been documented elsewhere.

Outside the forest transition literature, the predominant way to understand the long-term evolution of the highlands landscapes has been the Boserupian model of agricultural intensification linked to intertwined growth of population, markets, and property structures [101,102]. For instance, Rakoto Ramiarantsoa [55] has called the style of population-driven agricultural intensification and expansion found at the margins of the central highlands "*mérinisation*," by which he refers to a progressive transformation of open land to an intensively-cultivated cultural landscape typical of the Merina ethnic group in its homeland of Imerina, centered around the capital city. His concept imbues the Boserupian process with a cultural symbolic role. Yet his use of the term does not capture the conquest of the highlands by woodlots of exotic forestry tree species. Indeed, Rakoto Ramiarantsoa calls these eucalyptus-draped landscapes a departure from the Merina model ([55]; p. 299). Given the widespread adoption of these trees into peoples' livelihoods and landscapes over the past century (as he himself documents: e.g., [103]), these trees arguably deserve to be considered as an integral part of current day highland landscapes, whether Imerina or other cultural and regional divisions, like Vakinankaratra or Betsileo.

5.2. On Resilient Livelihoods and Sustainable Environments

What are the implications of the typically patchy, often low density, largely exotic tree cover that we have documented to be continuing to expand over significant parts of the highlands? We have demonstrated that it is central to rural livelihoods, and meets growing urban needs. Few other places in Africa can boast such a supply of locally produced items, instead being reliant on imported materials (sometimes for local manufacture) such as metal doors and window frames, cooking utensils, *etc.* The supply of such household goods perhaps leaves the Malagasy economy on a much more sustainable footing than it would otherwise enjoy. One could argue that this is a renewable wood supply, particularly in contrast with the rest of sub-Saharan Africa, where biomass for fuel (whether wood or charcoal) comes predominantly from natural forests. The extent and importance of private, and especially smallholder woodlots in meeting—and exceeding—subsistence needs, is a question of great importance. Madagascar, along with Rwanda, appears to be at the leading edge in the proportion of urban energy needs being met by such small-scale woodlots [104] (although doubt has recently been cast on the sustainability of Antananarivo's peri-urban eucalyptus plantations, given observations of a shortening of the coppice cycle below the six years recommended for maximum productivity [87]). The present study contributes to our understanding of this important phenomenon.

On the policy front, the pattern of state support for, and indeed investment in, forestry plantations corroborates the findings of Chidumayo and Gumbo [105] and Zulu and Richardson [106] that the provision of woodfuel supplies from plantations has been a long-standing policy goal in many countries, but the success of large, state-sponsored plantations has been mixed. Highland farmers have long been encouraged in their afforestation efforts by the colonial and independent governments of Madagascar, despite the fact that Ghilardi and colleagues [107] report that charcoal supply is often absent from the energy policies of countries that rely heavily on this source of energy.

In some ways, this is a tale of a Machakos Miracle writ large [39]. It highlights the role of endogenous factors, with government support, as opposed to the alternative approach of large government plantations. However, caution is warranted on several counts. First, closer examination of the Machakos case has revealed that, while it may represent a laudable example of environmental rehabilitation in the face of population pressure, not all people involved benefited equally. In that Kenyan case, elite retirees returned to their home areas, bringing with them intellectual and financial capital not available to others who ended up displaced to marginal lands, with negative environmental and social consequences [108]. In the Malagasy highlands, the tree-based dynamics of land control and connections to urban markets likewise are part and parcel of the dynamics of social differentiation between local or urban elites and marginalized members of society, though perhaps without the obvious displacement to marginal lands seen in Machakos [103].

Second, the ecological consequences of expanded woody cover in the highlands remain to be investigated. The trees affect soil chemistry, soil biology, and soil erosion, as well as hydrology, often in complex ways [109,110]. Several of the species common in the new tree cover (*Acacia* spp., *Pinus* spp.) are widely known to be invasive elsewhere, which is one reason for their spread here, and a reason some farmers like them or at least see them as an opportunity. But this also constitutes a potential challenge for the future, since, for instance, pines have invaded endemic *Uapaca* woodlands and montane heaths.

On the more positive side, this new tree cover is unsurprisingly not of central interest to biodiversity conservation planners, yet there are indications that such landscapes are not irrelevant. For instance, some plantation forests or hedgerows may serve as corridors or nursery landscapes, particularly in the more humid forest edge areas [111–114]. In addition, the new forests have implications for Madagascar's contributions to carbon sequestration [115].

5.3. Future Directions

We can only go so far in quantifying and explaining the changes we observe, and in understanding the social and environmental consequences, until further and more detailed field studies are undertaken. Seeking to balance and assess the complex implications of these environmental transformations in the highlands of Madagascar is a challenge; the most important and relevant assessment is probably that of the residents of the highlands themselves, for which these new woody resources constitute a treasure not to be squandered.

Acknowledgments: Air photo analyses and fieldwork by CK were funded by the Province of Quebec (FCAR 2002-SC-79002), the Australian Research Council (DP0666131), the Anne U. White Fund of the AAG, and Monash University.

Author Contributions: WM and AV conceived the research; WM, AV, CK, and CB collected and analyzed the data; WM, AV, and CK wrote the paper.

Conflicts of Interest: The authors declare no conflicts of interest.

References

1. Harper, G.J.; Steininger, M.K.; Tucker, C.J.; Juhn, D.; Hawkins, F. Fifty years of deforestation and forest fragmentation in Madagascar. *Environ. Conserv.* **2007**, *34*, 325–333. [CrossRef]
2. Kull, C.A.; Ibrahim, C.K.; Meredith, T.C. Tropical forest transitions and globalization: NEO-liberalism, migration, tourism, and international conservation agendas. *Soc. Nat. Resour.* **2007**, *20*, 723–737. [CrossRef]
3. Brannstrom, C.; Vadjunec, J.M. Notes for avoiding a missed opportunity in sustainability science: Integrating land change science and political ecology. In *Land Change Science, Political Ecology, and Sustainability Synergies and Divergences*; Brannstrom, C., Vadjunec, J.M., Eds.; Routledge: New York, NY, USA, 2013; pp. 1–23.
4. Mather, A.S.; Needle, C.L. The forest transition: A theoretrical basis. *Area* **1998**, *30*, 117–124. [CrossRef]
5. Rudel, T.K. Is there a forest transition? Deforestation, reforestation, and development. *Rural Sociol.* **1998**, *63*, 533–552. [CrossRef]
6. Foster, A.D.; Rosenzweig, M.R. Economic growth and the rise of forests. *Q. J. Econ.* **2003**, *118*, 601–637. [CrossRef]
7. Kauppi, P.E.; Ausubel, J.H.; Fang, J.; Mather, A.S.; Sedjo, R.A.; Waggoner, P.E. Returning forests analyzed with the forest identity. *Proc. Natl. Acad. Sci. USA* **2006**, *103*, 17574–17579. [CrossRef] [PubMed]
8. Hansen, M.C.; Potapov, P.V.; Moore, R.; Hancher, M.; Turubanova, S.; Tyukavina, A.; Thau, D.; Stehman, S.; Goetz, S.; Loveland, T. High-resolution global maps of 21st-century forest cover change. *Science* **2013**, *342*, 850–853. [CrossRef] [PubMed]
9. Caldas, M.; Walker, R.; Arima, E.; Perz, S.; Aldrich, S.; Simmons, C. Theorizing land cover and land use change: The peasant economy of Amazonian deforestation. *Ann. Assoc. Am. Geogr.* **2007**, *97*, 86–110. [CrossRef]
10. Gutman, G.; Janetos, A.C.; Justice, C.O.; Moran, E.F.; Mustard, J.F.; Rindfuss, R.R.; Skole, D.; Turner, B.L., II; Cochrane, M.A. *Land Change Science: Observing, Monitoring and Understanding Trajectories of Change on the Earth's Surface*; Kluwer Academic Publishers: Dordrecht, The Netherlands, 2004.

11. Lambin, E.; Geist, H.; Rindfuss, R.R. Introduction: Local processes with global impacts. In *Land Use and Cover Change*; Lambin, E., Geist, H., Eds.; Springer: Berlin, Germany, 2006.
12. Rindfuss, R.R.; Walsh, S.J.; Turner, B.L.; Fox, J.; Mishra, V. Developing a science of land change: Challenges and methodological issues. *Proc. Natl Acad. Sci. USA* **2004**, *101*, 13976–13981. [CrossRef] [PubMed]
13. Rudel, T.K.; Coomes, O.T.; Moran, E.; Achard, F.; Angelsen, A.; Xu, J.C.; Lambin, E. Forest transitions: Towards a global understanding of land use change. *Glob. Environ. Change-Hum. Policy Dimens.* **2005**, *15*, 23–31. [CrossRef]
14. Von Thünen, J.H. *Die isolierte Staat in Beziehung auf Landwirtshaft und Nationalökonomie*; Wartenberg, C.M., Translator; Pergamon Press: New York, NY, USA, 1826.
15. Angelsen, A. *Forest Cover Change in Space and Time: Combining the von Thünen and Forest Transition Theories*; World Bank Policy Research Working Paper 4117; World Bank: Washington, DC, USA, 2007.
16. Perz, S.G. Grand theory and context-specificity in the study of forest dynamics: Forest transition theory and other directions. *Prof. Geogr.* **2007**, *59*, 105–114. [CrossRef]
17. Whittaker, R.J.; Grainger, A. Tropical open woodlands special issue—Introduction. *Glob. Ecol. Biogeogr.* **1999**, *8*, 177–178. [CrossRef]
18. Aspinall, R. Modelling land use change with generalized linear models—A multi-model analysis of change between 1860 and 2000 in Gallatin Valley, Montana. *J. Environ. Manag.* **2004**, *72*, 91–103. [CrossRef] [PubMed]
19. Evans, T.P.; Kelley, H. Multi-scale analysis of a household level agent-based model of landcover change. *J. Environ. Manag.* **2004**, *72*, 57–72. [CrossRef] [PubMed]
20. Kok, K. The role of population in understanding Honduran land use patterns. *J. Environ. Manag.* **2004**, *72*, 73–89. [CrossRef] [PubMed]
21. Walsh, S.J.; Evans, T.P.; Welsh, W.F.; Entwisle, B.; Rindfuss, R.R. Scale-dependent relationships between population and environment in northeastern Thailand. *Photogramm. Eng. Remote Sens.* **1999**, *65*, 97–105.
22. Perz, S.G.; Skole, D.L. Secondary forest expansion in the Brazilian Amazon and the refinement of forest transition theory. *Soc. Nat. Resour.* **2003**, *16*, 277–294. [CrossRef]
23. Klooster, D. Forest transitions in Mexico: institutions and forests in a globalized countryside. *Prof. Geogr.* **2003**, *55*, 227–237.
24. Walker, R. Deforestation and economic development. *Can. J. Reg. Sci.* **1993**, *16*, 481–497.
25. Farley, K.A. Grasslands to tree plantations: forest transition in the andes of Ecuador. *Ann. Assoc. Am. Geogr.* **2007**, *97*, 755–771. [CrossRef]
26. Southworth, J.; Nagendra, H.; Cassidy, L. Forest transition pathways in Asia—Studies from Nepal, India, Thailand, and Cambodia. *J. Land Use Sci.* **2012**, *7*, 51–65. [CrossRef]
27. Meyfroidt, P.; Lambin, E.F. Forest transition in Vietnam and Bhutan: Causes and environmental impacts. In *Reforesting Landscapes*; Nagendra, H., Southworth, J., Eds.; Springer: Dordrecht, The Netherlands, 2010; pp. 315–339.
28. Xu, J.; Yang, Y.; Fox, J.; Yang, X. Forest transition, its causes and environmental consequences: Empirical evidence from Yunnan of Southwest China. *Trop. Ecol.* **2007**, *48*, 137.
29. Hecht, S.B.; Morrison, K.D.; Padoch, C. *The Social Lives of Forests: Past, Present, and Future of Woodland Resurgence*; University of Chicago Press: Chicago, IL, USA, 2014.
30. Nagendra, H.; Southworth, J. *Reforesting Landscapes: Linking Pattern and Process*; Springer Science & Business Media: Dordrecht, The Netherlands, 2009.
31. Peet, R.; Watts, M. *Liberation Ecologies*; Routledge: London, UK, 2004.
32. Robbins, P. *Political Ecology: A Critical Introduction*; Blackwell: Oxford, UK, 2004.
33. Batterbury, S.P.J.; Bebbington, A.J. Environmental histories, access to resources and landscape change: An introduction. *Land Degrad. Dev.* **1999**, *10*, 279–289. [CrossRef]
34. Tiffen, M.; Mortimore, M.; Gichuki, F. *More People, Less Erosion*; John Wiley and Sons: Chichester, UK, 1994.
35. Fairhead, J.; Leach, M. *Misreading the African Landscape*; Cambridge University Press: Cambridge, UK, 1996.
36. Leach, M.; Mearns, R. *Lie of the Land: Challenging Received Wisdom on the African Environment*; James Currey: Oxford, UK, 1996.
37. McCann, J.C. The plow and the forest—Narratives of deforestation in Ethiopia, 1840–1992. *Environ. Hist.* **1997**, *2*, 138–159. [CrossRef]
38. Kull, C.A. Deforestation, erosion, and fire: Degradation myths in the environmental history of Madagascar. *Environ. Hist.* **2000**, *6*, 423–450. [CrossRef]

39. Mortimore, M.; Tiffen, M. *Population and Environment in Time Perspective: The Machakos Story*; John Wiley and Sons: Chichester, UK, 1995.
40. Arnold, J.E.M.; DeWees, P. *Farms, Trees and Farmers*; Earthscan: London, UK, 1997.
41. Clinecole, R.A. Manufacturing Forests & Farming Trees in Nigeria. In *The Lie of the Land: Challenging Received Wisdom on the African Environment*; Leach, M., Mearns, R., Eds.; Heineman: Portsmouth, UK, 1996; pp. 122–139.
42. Weston, P.; Hong, R.; Kaboré, C.; Kull, C.A. Farmer-managed natural regeneration enhances rural livelihoods in dryland west Africa. *Environ. Manag.* **2015**, *55*, 1402–1417. [CrossRef] [PubMed]
43. Fortmann, L.; Bruce, J.W. *Whose Trees? Proprietary Dimensions of Forestry*; Westview Press: Boulder, CO, USA, 1998.
44. Rocheleau, D.; Edmunds, D. Women, men and trees: Gender, power and property in forest and agrarian landscapes. *World Dev.* **1997**, *25*, 1351–1371. [CrossRef]
45. Mather, A.S. Recent Asian forest transitions in relation to forest-transition theory. *Int. For. Rev.* **2007**, *9*, 491–502.
46. Dietz, T.; Ostrom, E.; Stern, P.C. The struggle to govern the commons. *Science* **2003**, *302*, 1907–1912. [CrossRef] [PubMed]
47. Moran, E.; Ostrom, E. *Seeing the Forest and the Trees: Human-Environment Interactions in Forest Ecosystems*; MIT Press: Cambridge, MA, USA, 2005.
48. Dufils, J.M. Remaining forest cover. In *The Natural History of Madagascar*; Goodman, S., Benstead, J., Eds.; The University of Chicago Press: Chicago, IL, USA, 2003; pp. 88–96.
49. McConnell, W.; Kull, C.A. Deforestation in Madagascar: Debates over the island's forest cover and challenges of measuring forest change. In *Conservation and Environmental Management in Madagascar*; Scales, I.R., Ed.; Earthscan Conservation and Development Series; Routledge: London, UK, 2014; pp. 67–104.
50. Hansen, M.; DeFries, R.; Townshend, J.; Carroll, M.; Dimiceli, C.; Sohlberg, R. Global percent tree cover at a spatial resolution of 500 meters: First results of the MODIS vegetation continuous fields algorithm. *Earth Interact.* **2003**, *7*, 1–15. [CrossRef]
51. Barrett, C.B. Food marketing liberalization and trader entry: evidence from Madagascar. *World Dev.* **1997**, *25*, 763–777. [CrossRef]
52. Fraslin, J.H. Quel avenir pour les paysans de Madagascar? *Afr. Contemp.* **2002**, *202–203*, 93–110.
53. Minten, B.; Randrianarison, L.; Swinnen, J.F. Global retail chains and poor farmers: Evidence from Madagascar. *World Dev.* **2009**, *37*, 1728–1741. [CrossRef]
54. Minten, B.; Barrett, C.B. Agricultural technology, productivity, and poverty in Madagascar. *World Dev.* **2008**, *36*, 797–822. [CrossRef]
55. Rakoto Ramiarantsoa, H. *Chair de la Terre, Oeil de l'Eau... Paysanneries et Recompositions de Campagnes en Imerina (Madagascar)*; Éditions de l'Orstom: Paris, France, 1995.
56. Fafchamps, M.; Moser, C. Crime, Isolation, and Law Enforcement. *J. Afr. Econ.* **2003**, *12*, 625–671. [CrossRef]
57. Olson, S. Environments as shock absorbers, examples from Madagascar. *Environ. Hist. Rev.* **1988**, *12*, 61–80. [CrossRef]
58. Shackleton, C.; Shackleton, S. The importance of non-timber forest products in rural livelihood security and as safety nets: A review of evidence from South Africa. *S. Afr. J. Sci.* **2004**, *100*, 658–664.
59. Montagne, P.; Bertrand, A. *Kajiala, Tattali, Djekabaara: Valoriser les produits pour mieux conserver les forêts. Tome 1: les expériences nationales Volume 1: Madagascar*; Projet GESFORCOM; Antananarivo, Madagascar, 2012.
60. Gardner, C.J.; Gabriel, F.U.; St John, F.A.; Davies, Z.G. Changing livelihoods and protected area management: A case study of charcoal production in south-west Madagascar. *Oryx* **2014**. [CrossRef]
61. Raison, J.P. *Les hautes terres de Madagascar et leurs confins occidentaux: enracinement et mobilité des sociétés rurales*; Karthala Editions: Paris, France, 1984.
62. République. In *Regions et Developpement: Programmes regionaux et projects locaux (Fianarantsoa)*; République de Madagascar, Ministere de l'Economie at du Plan; Programme des Nations Unies pour le Developpement Projet PNUD/OPS MAG/89/018; Dirasset: Tunis, Tunisia, 1991.
63. Kull, C.A. *Isle of Fire: the Political Ecology of Landscape Burning in Madagascar*; University of Chicago Press: Chicago, IL, USA, 2004.
64. Gade, D.W. Deforestation and its effects in highland Madagascar. *Mt. Res. Dev.* **1996**, *16*, 101–116. [CrossRef]
65. Jolly, A. Madagascar: A world apart. *Natl. Geogr.* **1987**, *171*, 148–183.

66. Unruh, J.D.; McConnell, W.; Rodman, J. Environmental change and adaptation in degraded agro-ecosystems: the case of highland Madagascar. *Area* **2010**, *42*, 359–368. [CrossRef]
67. Jaynes, E.T. Information theory and statistical mechanics. *Phys. Rev.* **1957**, *106*, 620–630. [CrossRef]
68. Phillips, S.J.; Anderson, R.P.; Schapire, R.E. Maximum entropy modeling of species geographic distributions. *Ecol. Model.* **2006**, *190*, 231–259. [CrossRef]
69. FAO. *Global Forest Resources Assessment 2010*; Food and Agriculture Organization of the United Nations: Rome, Italy, 2010.
70. Hanley, J.A.; Mcneil, B.J. The meaning and use of the area under a receiver operating characteristic (ROC) curve. *Radiology* **1982**, *143*, 29–36. [CrossRef] [PubMed]
71. Coppin, P.; Jonckheere, I.; Nackaerts, K.; Muys, B.; Lambin, E. Digital change detection methods in ecosystem monitoring: a review. *Int. J. Remote Sens.* **2004**, *25*, 1565–1596. [CrossRef]
72. Kull, C.A. Air photo evidence of historical land cover change in the highlands: wetlands and grasslands give way to crops and woodlots. *Madag. Conserv. Dev.* **2012**, *7*, 144–152. [CrossRef]
73. Kull, C.A. Leimavo revisited: Agrarian land-use change in the highlands of Madagascar. *Prof. Geogr.* **1998**, *50*, 163–176. [CrossRef]
74. McConnell, W.J.; Sweeney, S.P. Challenges of forest governance in Madagascar. *Geogr. J.* **2005**, *171*, 223–238. [CrossRef]
75. Verhaegen, D.; Randrianjafy, H.; Montagne, P.; Danthu, P.; Rabevohitra, R.; Tassin, J.; Bouvet, J.-M. Historique de l'introduction du genre Eucalyptus à Madagascar. *Bois et Forêt des Tropiques* **2011**, 17–25.
76. Kull, C.A. Historical landscape repeat photography as a tool for land use change research. *Norsk Geografisk Tidsskrift* **2005**, *59*, 253–268. [CrossRef]
77. Ragin, C.C. *The Comparative Method: Moving beyond Qualitative and Quantitative Strategies*; University of California Press: Berkeley, CA, USA, 2014.
78. Rudel, T.K. *Tropical Forests: Paths of Destruction and Regeneration*; Columbia University Press: New York, NY, USA, 2013.
79. Carrière, S.M.; Randriambanona, H. Biodiversité introduite et autochtone: antagonisme ou complémentarité? Le cas de l'eucalyptus à Madagascar. *Bois et Forêts des Tropiques* **2007**, *292*, 5–21.
80. Kull, C.A.; Tassin, J.; Rangan, H. Multifunctional, scrubby, and invasive forests? Wattles in the highlands of Madagascar. *Mt. Res. Dev.* **2007**, *27*, 224–231. [CrossRef]
81. Tassin, J.; Rangan, H.; Kull, C.A. Hybrid improved tree fallows: harnessing invasive woody legumes for agroforestry. *Agrofor. Syst.* **2012**, *84*, 417–428. [CrossRef]
82. The World Bank Group. *Madagascar: Systemic Country Diagnostic*; Report #99197; The World Bank: Washington, DC, USA, 2015.
83. Van der Walle, N. *African Economies and the Politics of Permanent Crisis*; Cambridge University Press: Cambridge, UK, 2001.
84. Rakoto Ramiarantsoa, H. Les boisements d'eucalyptus dans l'est de l'Imerina (Madagascar). In *Terre, Terroir, Territoire: Les Tensions Fonciers*; Blanc-Pamard, C., Cambrézy, L., Eds.; OSTROM: Paris, France, 1995; pp. 83–103.
85. Bertrand, A. La Dynamique Seculaire des Plantations Paysannes d'Eucalyptus sur les Hautes Terres Malgaches. *Afr. Stud. Q.* **1999**, *3*, 61–68.
86. Carrière, S.M.; Andrianotahiananahary, H.; Ranaivoarivelo, N.; Randriamalala, J. Savoirs et Usages des Recrus Post-Agricoles du Pays Betsileo: Valorisation d'une Biodiversite Oubliee a Madagascar. *VertigO – La revue en sciences de l'environnement* **2005**, *16*, 1–15. [CrossRef]
87. Verhaegen, D.; Randrianjafy, H.; Rakotondraoelina, H.; Trendelenburg Rakotonirina, M.-C.; Andriamampianina, N.; Montagne, P.; Rasamindisa, A.; Chaix, G.; Bouillet, J.-P.; Bouvet, J.-M. Eucalyptus robusta pour une production durable de bois énergie à Madagascar: Bilan des connaissances et perspectives. *Bois et Forêts des Tropiques* **2014**, *320*, 15–30.
88. Bertrand, A. Le boisement, le bail, et la législation environnemental à Madagascar: trois articles courts. *Afr. Stud. Q.* **1999**, *3*, 47–59.
89. Gade, D.W.; Perkins-Belgram, A. Woodfuels, reforestation, and ecodevelopment in highland Madagascar. *GeoJournal* **1986**, *12*, 365–374. [CrossRef]

90. Office National de Forets. *Reboisement Communautaire des Hautes Terres, Madagascar*. 2010. Available online: http://www.developpement-durable.gouv.fr/IMG/pdf/Madagascar_ONF_Reboisement.pdf (accessed on 13 November 2015).
91. Minten, B.; Sander, K.; Stifel, D. Forest management and economic rents: Evidence from the charcoal trade in Madagascar. *Energy Sustain. Dev.* **2013**, *17*, 106–115. [CrossRef]
92. Root, H. *Environment for Investment in Madagascar: Institutional Reform for Market Economy*; Center for Institutional Reform and the Informal Sector, University of Maryland at College Park: College Park, MD, USA, 1993.
93. Gore, M.L.; Ratsimbazafy, J.; Lute, M.L. Rethinking corruption in conservation crime: Insights from Madagascar. *Conserv. Lett.* **2013**, *6*, 430–438. [CrossRef]
94. Kull, C.A.; Tassin, J.; Carriere, S.M. Approaching invasive species in Madagascar. *Madag. Conserv. Dev.* **2014**, *9*, 60–70. [CrossRef]
95. Richardson, D.M.; Rejmánek, M. Trees and shrubs as invasive alien species—A global review. *Divers. Distrib.* **2011**, *17*, 788–809. [CrossRef]
96. Van Wilgen, B.W.; Richardson, D.M. Challenges and trade-offs in the management of invasive alien trees. *Biol. Invasions* **2014**, *16*, 721–734. [CrossRef]
97. Laney, R.M. Disaggregating induced intensification for land-change analysis: A case study from Madagascar. *Ann. Assoc. Am. Geogr.* **2002**, *92*, 702–726. [CrossRef]
98. Lerner, A.M.; Rudel, T.K.; Schneider, L.C.; McGroddy, M.; Burbano, D.V.; Mena, C.F. The spontaneous emergence of silvo-pastoral landscapes in the Ecuadorian Amazon: Patterns and processes. *Reg. Environ. Change* **2014**, *15*, 1–11. [CrossRef]
99. Lambin, E.F.; Meyfroidt, P. Land use transitions: Socio-ecological feedback *versus* socio-economic change. *Land Use Policy* **2010**, *27*, 108–118. [CrossRef]
100. Meyfroidt, P. Environmental cognitions, land change and social-ecological feedbacks: Local case studies of forest transition in Vietnam. *Hum. Ecol.* **2013**, *41*, 367–392. [CrossRef]
101. Boserup, E. *The Conditions of Agricultural Progress*; Aldine Publishing Company: Chicago, IL, USA, 1965.
102. Kull, C.A. Saving the land with spades: human landscape transformations in the central highlands. In *Greening the Great Red Island: Madagascar in Nature and Culture*; Kaufmann, J.C., Ed.; Africa Institute of South Africa: Pretoria, South Africa, 2008; pp. 113–135.
103. Blanc-Pamard, C.; Rakoto Ramiarantsoa, H. *Le Terroir et son Double: Tsarahonenana 1966–1992 Madagascar*; Editions de l'Institut de Recherches pour le Developpement: Paris, France, 2000.
104. Owen, M.; van der Plas, R.; Sepp, S. Can there be energy policy in Sub-Saharan Africa without biomass? *Energy Sustain. Dev.* **2013**, *17*, 146–152. [CrossRef]
105. Chidumayo, E.N.; Gumbo, D.J. The environmental impacts of charcoal production in tropical ecosystems of the world: A synthesis. *Energy Sustain. Dev.* **2013**, *17*, 86–94. [CrossRef]
106. Zulu, L.C.; Richardson, R.B. Charcoal, livelihoods, and poverty reduction: Evidence from sub-Saharan Africa. *Energy Sustain. Dev.* **2013**, *17*, 127–137. [CrossRef]
107. Ghilardi, A.; Mwampamba, T.; Dutt, G. What role will charcoal play in the coming decades? Insights from up-to-date findings and reviews. *Energy Sustain. Dev.* **2013**, *2*, 73–74. [CrossRef]
108. Rocheleau, D. More on Machakos. *Environment* **1995**, *37*, 3–5.
109. Bruijnzeel, L.A. Hydrological functions of tropical forests: not seeing the soil for the trees? *Agric. Ecosyst. Environ.* **2004**, *104*, 185–228. [CrossRef]
110. Baohanta, R.; Thioulouse, J.; Ramanankierana, H.; Prin, Y.; Rasolomampianina, R.; Baudoin, E.; Rakotoarimanga, N.; Galiana, A.; Randriambanona, H.; Lebrun, M. Restoring native forest ecosystems after exotic tree plantation in Madagascar: Combination of the local ectotrophic species Leptolena bojeriana and Uapaca bojeri mitigates the negative influence of the exotic species Eucalyptus camaldulensis and Pinus patula. *Biol. Invasions* **2012**, *14*, 2407–2421.
111. Ganzhorn, J.U. A possible role of plantations for primate conservation in Madagascar. *Am. J. Primatol.* **1987**, *12*, 205–215. [CrossRef]
112. Martin, E.A.; Viano, M.; Ratsimisetra, L.; Laloë, F.; Carrière, S.M. Maintenance of bird functional diversity in a traditional agroecosystem of Madagascar. *Agric. Ecosyst. Environ.* **2012**, *149*, 1–9. [CrossRef]

113. Kull, C.A.; Carrière, S.M.; Moreau, S.; Ramiarantsoa, H.R.; Blanc-Pamard, C.; Tassin, J. Melting pots of biodiversity: Tropical smallholder farm landscapes as guarantors of sustainability. *Environ.: Sci. Policy Sustain. Dev.* **2013**, *55*, 6–16. [CrossRef]
114. Gérard, A.; Ganzhorn, J.U.; Kull, C.A.; Carriere, S.M. Possible roles of introduced plants for native vertebrate conservation: The case of Madagascar. *Restor. Ecol.* **2015**, *23*, 768–775. [CrossRef]
115. Razakamanarivo, R.H.; Grinand, C.; Razafindrakoto, M.A.; Bernoux, M.; Albrecht, A. Mapping organic carbon stocks in eucalyptus plantations of the central highlands of Madagascar: A multiple regression approach. *Geoderma* **2011**, *162*, 335–346. [CrossRef]

Section 3:

Smallholder Vulnerability and Adaptation

land

MDPI

Article

Agriculture on the Brink: Climate Change, Labor and Smallholder Farming in Botswana

William G. Moseley

Geography Department, Macalester College, 1600 Grand Avenue, St. Paul, MN 55105-1899, USA; moseley@macalester.edu; Tel.: +1-651-696-6126; Fax: +1-651-696-6116

Academic Editors: Claudia A. Radel and Jacqueline M. Vadjunec
Received: 27 October 2015; Accepted: 13 June 2016; Published: 27 June 2016

Abstract: Botswana is a semi-arid, middle-income African country that imports 90 percent of its food. Despite its relative prosperity, Botswana also suffers from one of the highest measures of income inequality in the world, persistent poverty, and relatively high levels of food insecurity. The objective of this paper is to explore how political economy, climate change and livelihood dynamics are synergistically impacting household food security. The major finding is that the marginalization of smallholder farming in Botswana has as much or more to do with domestic, regional and international political economy as it does with climate change. As such, international efforts to support climate change adaptation in Botswana will have a limited effect on smallholder farming livelihoods and rural food security unless such efforts take account of political economic constraints. Effective support must be based on a grounded understanding of the real drivers of marginalization and food insecurity. One initiative that merits further exploration is the government's backyard gardening initiative, which could be viewed as a pro-poor climate adaptation strategy. The findings of this paper are based on semi-structured interviews with policymakers and surveys with urban, peri-urban and rural households undertaken in 2012 and 2015.

Keywords: agriculture; climate change; farm labor; food security; gardening; livelihoods

---

1. Introduction and Context in the Literature

There is a growing global discourse on the linkages between global climate change and food insecurity. The Intergovernmental Panel on Climate Change (IPCC) Fifth Assessment Report [1] predicts that the impacts of climate change will lead to more flooding, famine, drought and disease, which could have a negative impact on millions of people in the poorest parts of the world, especially Africa. There is also an increasing recognition that climate change impacts various countries, and groups within countries, differently. For example, the climate change vulnerability index [2] is one tool for understanding these differential impacts.[1] It has been used to show how many areas of the Global South, especially the Africa region, are more vulnerable to climate change.

Botswana is likely to suffer greatly from climate change [3]. Most of the country's agriculture is rainfed and occurs in areas of relatively higher precipitation of around 400–600 mm of rainfall per annum (see Figure 1). Different climate change scenarios make varying predictions for the 2000–2050 period. Of the four downscaled global climate models that have been used in Botswana (IPCC Fourth Assessment Report, Commonwealth Scientific & Industrial Research Organisation and Model for Interdisciplinary Research on Climate), most predict a 50–100 mm decline in rainfall in the

---

[1] This index is based on three factors: exposure to extreme climate-related events; the sensitivity of populations, in terms of health, education, agricultural dependence and available infrastructure; and the adaptive capacity of countries to combat the impacts of climate change.

south and southeast parts of the country and an increase in annual maximum temperature from 1.5 °C to 2.5 °C [4]. While these may seem like small changes, they could be quite significant for an area that was already marginal for agriculture. Even though this marginality, or precarity, has long been a fact of life for farmers in areas close to the limits for rainfed agriculture (400 mm per annum for the most drought-tolerant grains), irregular rainfall in recent years has contributed to anemic production [5]. Sorghum and maize are the two primary food crops produced in Botswana, and both are typically grown under rainfed conditions. Sorghum is the more traditional crop, predating colonialism, and needs 450–600 mm of annual precipitation for optimal production [6]. Maize was introduced during the colonial period [7], has expanded rapidly in recent decades, and requires 500–800 mm of rainfall per annum [8]. Were the wettest parts of the country to drop to 500 mm of rainfall per annum, maize would be nearly impossible to grow [9]. Less impacted by seasonal variation in rainfall is irrigated agriculture, which includes some large grain farms in the northeast part of the country (near the town of Kasane), as well as commercial and subsistence horticulture of various sizes in urban, peri-urban and rural areas. Nearly all forms of agriculture in Botswana are dominated by women.

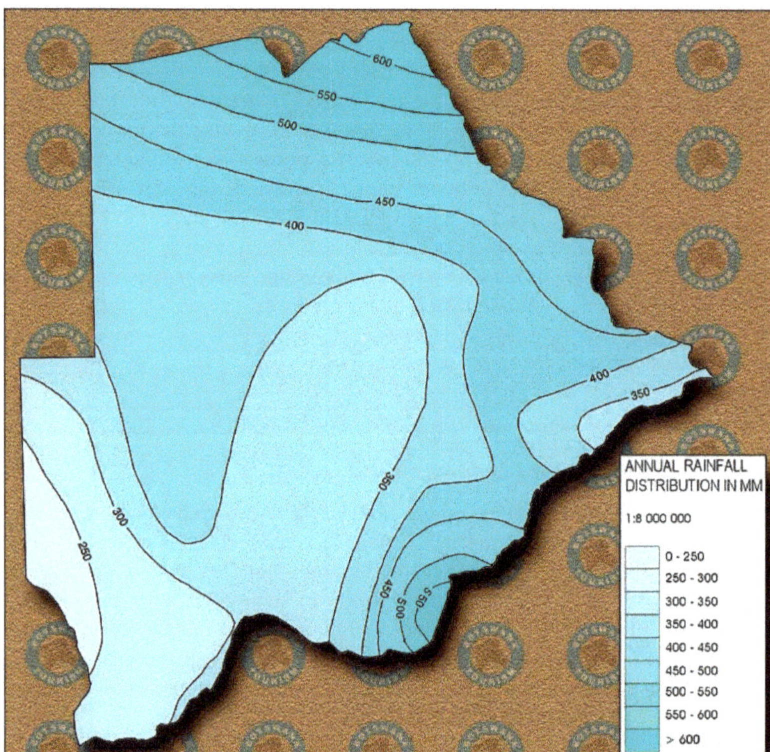

Figure 1. Rainfall in Botswana. Source: Botswana Tourism Organization [10]. Used with permission.

The Government of Botswana is concerned about the impacts of climate change on agriculture, often framing vulnerability as externally driven and best addressed by technology. As early as 2010, the Government began working on a climate change adaptation and mitigation policy, noting that this problem could drive the country back into poverty [11]. President Ian Khama has urged farmers to adopt more innovative approaches to cope with the impacts of climate change, citing the example of technologies developed by Israel [12]. In response, the Government has launched a climate smart agriculture program [13]. It has also been a pioneer among African countries on this issue. For example, Botswana played a leading role in crafting the Gaborone climate change declaration (October 2013), which calls for an African voice in global climate change negotiations. Another example is the 2014 Tripartite Climate Change 2014 Meeting. The COMESA-EAC-SADC[2] Tripartite Programme Management Units of the Programme on Climate Change Adaptation and Mitigation met in Gaborone, Botswana in January, 2014. This group has: "supported countries developing strategies in climate change as well as scaling up climate smart agriculture; and developed vulnerability assessments to help avert and mitigate suffering and vulnerability by providing clear information on populations vulnerable to food insecurity and the effects of shocks including climate change" [14]. Finally, the Southern African region, via the Southern African Development Community (SADC), has programs to support climate smart agriculture, a controversial approach given its frequent reliance on high tech solutions [15], as well as vulnerability assessment programs.

While the impact of a shifting climate could be significant for agriculture and food security in Botswana, political ecology and political agronomy[3] [16] scholars argue that we must understand this situation as refracted through the context of local livelihood dynamics and political economy. In this case, livelihood dynamics refers to the ways in which households interact with the environment, allocate labor and distribute resources within the immediate and extended family. Political economy has more to do with the macroeconomic structures and policies at the state, regional and international level which impinge on household maneuverability and decision-making. These lenses complicate conventional analysis in a couple of ways.

First, according to these scholars, the links between food supply and food security are not direct. For example, even if there is plenty of food on the market, poor households may not be able to access this because of high prices or low incomes, a problem known as entitlement failure [17]. This is particularly relevant in the case of Botswana because food had not been in short supply due to ample imports, but prices spiked in 2007–2008 and 201–2012 because of global market conditions [18,19]. Despite being a middle-income country, Botswana has segments of its population which suffer from the interlinked phenomena of persistent poverty and food insecurity [20–22]. Botswana's Gini index of 63 means that it has one of the most unequal income distributions in the world, along with neighboring South Africa and Namibia. Persistent poverty and food insecurity are accentuated when food prices rise, something which is a particular issue for Botswana given its exposure to global food price fluctuations resulting from a very high dependence on imported food (90 percent in recent years).

Second, cropping and livelihood systems are differentially vulnerable to climatic shocks (such as periodic drought). Furthermore, this vulnerability (read sensitivity and resilience) is often shaped by histories of colonialism or economic integration into the global system. For example, Watts' [23] groundbreaking work in Northern Nigeria showed how famine was not the result of drought and overpopulation, but farming systems made more vulnerable (through the transition to monocrops for cash) by British colonial policy. More recent critiques of the climate change and vulnerability literature by Ribot [24,25] and Bohle et al. [26] make similar points, arguing that vulnerability can have as much or more to do with livelihoods constrained by political economic factors than shifting climates. For example, colonial policies creating a regional maize market in Southern Africa [7], labor constraints [27], or both acting synergistically [28], may be limiting adaptation to climate change. The squeeze or tension that agrarian livelihoods face as they simultaneously cope with fluctuating global markets and shifting weather patterns has been explored by O'Brian and Leichenko [29,30] in India. They dubbed this problem 'double exposure,' or the layering of vulnerabilities to the perils of globalization and climate change.

---

[2] COMESA-EAC-SADC stands for: Common Market for Eastern and Southern Africa—East African Community—Southern African Development Community.

[3] The term 'political agronomy' refers to the study of relationships and processes which link political, economic and social forces and factors to the creation and use of agronomic knowledge and technology [16].

Debates about climate change, vulnerability and political economy intersect with those on gardening in interesting ways. Gardening, as a government or NGO-led initiative, has a mixed legacy in Botswana and Africa more broadly. Some scholars and policymakers have conceptualized gardening and urban agriculture as pro female and as solutions to poverty and food insecurity [31,32]. Others are quite critical of this approach, noting that gardening has been overly romanticized by outsiders, and that it narrowly conceives of hunger as a supply-side problem [33,34]. Gardening also has come under attack in Botswana for not producing the desired results or being a drain on water resources [35]. Rarely, however, has gardening been conceived of as a water efficient alternative to dryland farming or livestock rearing. In the particular case of Botswana, feminist political ecology[4] [36] scholars argue that women's traditional livelihood activities (e.g., poultry raising, gardening and crop agriculture) are often undervalued and disadvantaged relative to male livelihood activities, such as livestock rearing [37,38]. As such, arguments about water scarcity and agricultural efficiency merit examination from a feminist political ecology perspective.

The overarching objective of this paper is to explore how political economy, climate change and livelihood dynamics are synergistically impacting household food security. The major finding is that the marginalization of smallhold farming in Botswana has as much or more to do with domestic, regional and international political economy as it does with climate change. As such, international efforts to support climate change adaptation in Botswana will have a limited effect on smallholder farming livelihoods and rural food security unless such efforts take account of political economic constraints. Effective support must be based on a grounded understanding of the real drivers of marginalization and food insecurity. One initiative that merits further exploration is the government's backyard gardening initiative, which could be viewed as a pro-poor climate adaptation strategy.

2. Methods

Botswana is 60 percent urban, urbanizing at a rate of 2.5 percent per annum. The top destination for migrants leaving the rural areas is the capital city, Gaborone, with a population of almost 250,000. Many rural migrants move in with existing family members in urban areas, rent rooms in the city's poorer neighborhoods, or settle in peri-urban areas around the city [37]. While rural households are, on average, poorer than urban households, those in urban areas tend to acquire a larger portion of their food via the market. This makes poor urban households especially vulnerable to fluctuating prices for basic foodstuffs. However, given the historically tight connections between urban and rural households [39], the impacts of fluctuating food prices on urban households may impact remittance flows from urban to rural areas.

The findings in this paper are based on qualitative, semi-structured interviews with policymakers, as well as 158 questionnaire surveys with urban, peri-urban and rural households in Southeastern Botswana in 2012. A small set of follow-up, semi-structured interviews were also conducted with government officials and households in 2015. The details of these interviews and surveys are outlined below.

The author interviewed representatives of a number of offices involved with different dimensions of food security and agriculture in Botswana in 2012 and 2015. These included the Ministry of Agriculture's poverty eradication and horticulture units, the famine early warning unit, the Southern African Development Community (SADC) early warning system, and various non-profit agencies. These qualitative interviews focused on the government's food security strategy, how it is being implemented, and whether or not this strategy is evolving given increasingly volatile global food

[4] Feminist political ecology explores the role of gender in the political ecological landscape. These scholars see gender roles as socially constructed, yet important to consider when examining human-environment interactions. Such scholars have explored resource use, agrarian reconstruction and rural-urban transformation, among other topics [36].

prices and global climate change scenarios. In some cases, questions were also asked about the government's backyard gardening initiative.

Household questionnaire surveys were conducted in three poorer neighborhoods in Gaborone: Old Naledi, White City, Broadhurst (see Figure 2), and one peri-urban area: Tlokweng. The urban neighborhoods selected for sampling cannot be considered representative of the city more broadly (as they likely lean towards being poorer). These areas were chosen, in part, so that comparisons could be made with a survey undertaken four years earlier by African Food Security Urban Network (AFSUN) [20]. The author and two research assistants systematically selected 89 households for semi-structured interviews in these urban neighborhoods by moving across these areas from east to west and knocking on the door of every fourth house for a potential interview. Representatives of these households were interviewed if they played an active role in household budgeting and food purchases. A similar selection process was undertaken for 30 households in the peri-urban village of Tlokweng. The urban and peri-urban household interviews focused on: income and expenditures; percent income spent on food; level of food insecurity; the changes in food status since 2008; the extent to which high food prices are a problem; coping strategies employed by households to deal with food shortfalls; and the potential impacts on extended family in rural areas of coping strategies in urban areas.

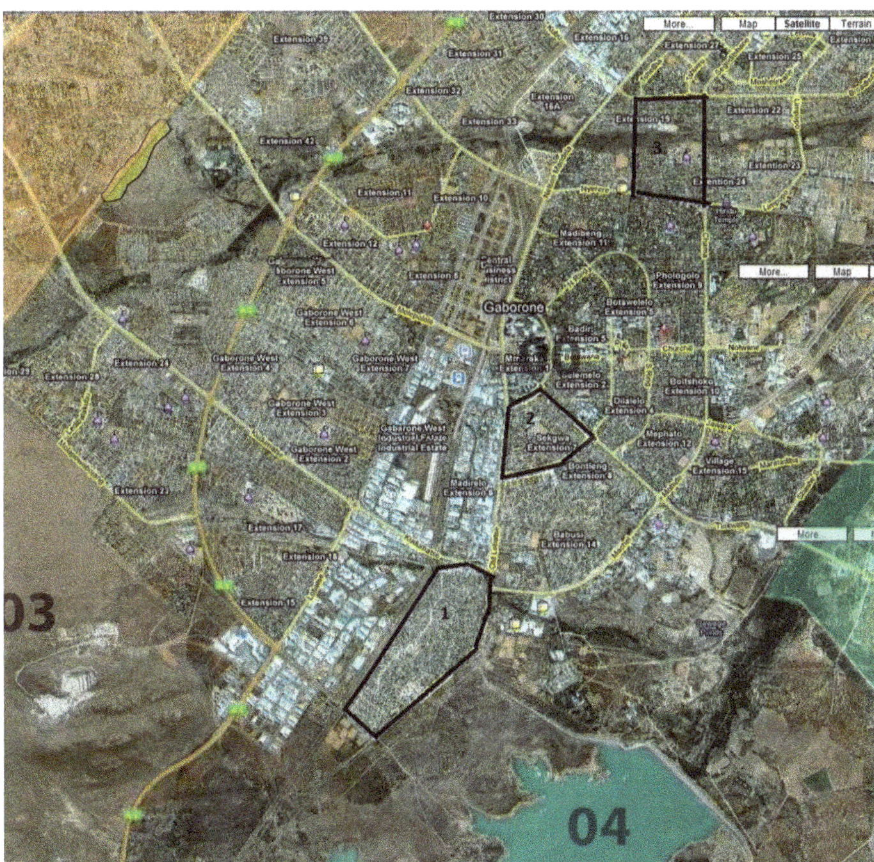

Figure 2. Map of Urban Survey Locations (outlined in black). Survey neighborhoods as follows 1. Old Naledi, 2. White City and 3. Broadhurst. Source: Base photo from Google imagery and used within fair use guidelines.

An additional 39 questionnaire surveys were undertaken in two rural communities outside of Gaborone: Molepolole and Thamaga. These are relatively large villages within one hour's drive of Gaborone. They could be considered representative of rural Southeastern Botswana, but not the country as a whole. Households were selected in a similar manner: moving from east to west across communities, systematic selection by knocking on the door of every fourth house for a potential interview, and representatives of these households interviewed if they played an active role in household budgeting and food purchases. The author asked a similar set of questions to those posed to urban households. The goal of these interviews was to examine how coping strategies in a period of high food prices may differ from those of urban households. The author also explored how food and cash transfers with urban-based family members may be changing.

The main measure the author used to assess household food insecurity in surveys was the Household Food Insecurity Access Scale (see Table 1). This measure was developed by the United Nations Food and Agriculture Organization (FAO) [40] and asks a series of questions, gradually increasing in severity, about a household's state of anxiety over food and access to meals, as well as actual experienced food scarcity, over the past four weeks.

Table 1. Household food insecurity access scale.

No.	Occurrence Questions
1.	In the past four weeks, did you worry that your household would not have enough food?
2.	In the past four weeks, were you or any household member not able to eat the kinds of foods you preferred because of a lack of resources?
3.	In the past four weeks, did you or any household member have to eat a limited variety of foods due to lack of resources?
4.	In the past four weeks, did you or any household member have to eat some foods that you really did not want to eat because of a lack of resources to obtain other types of food?
5.	In the past four weeks, did you or any household member have to eat a smaller meal than you felt you needed because there was not enough food?
6.	In the past four weeks, did you or any household member have to eat fewer meals in a day because there was not enough food?
7.	In the past four weeks, was there ever no food to eat of any kind in your household because of a lack of resources to get food?
8.	In the past four weeks, did you or any household member go to sleep at night hungry because there was not enough food?
9.	In the past four weeks, did you or any household member go a whole day and night without eating anything because there was not enough food?

Note: Each question is scored for 0–3, with 0 = never, 1 = rarely (once or twice in past four weeks), 2 = sometimes (three to ten times in past four weeks); and 3 = often (more than ten times in past four weeks). The household food insecurity access score, or HFIA, (1 = food secure, 2 = mild food insecurity, 3 = moderate food insecurity and 4 = severe food insecurity) is tabulated based on the scores to the nine questions in Table 1. A score of 1 to 4 is based on the following: HFIA category = 1 if [(Q1a = 0 or Q1a = 1) and Q2 = 0 and Q3 = 0 and Q4 = 0 and Q5 = 0 and Q6 = 0 and Q7 = 0 and Q8 = 0 and Q9 = 0]; HFIA category = 2 if [(Q1a = 2 or Q1a = 3 or Q2a = 1 or Q2a = 2 or Q2a = 3 or Q3a = 1 or Q4a = 1) and Q5 = 0 and Q6 = 0 and Q7 = 0 and Q8 = 0 and Q9 = 0] HFIA category = 3 if [(Q3a = 2 or Q3a = 3 or Q4a = 2 or Q4a = 3 or Q5a = 1 or Q5a = 2 or Q6a = 1 or Q6a = 2) and Q7 = 0 and Q8 = 0 and Q9 = 0] HFIA category = 4 if [Q5a = 3 or Q6a = 3 or Q7a = 1 or Q7a = 2 or Q7a = 3 or Q8a = 1 or Q8a = 2 or Q8a = 3 or Q9a = 1 or Q9a = 2 or Q9a = 3].

Thirty follow-up, semi-structured interviews with peri-urban and rural households in 2015 focused more narrowly on gardening. These included 10 peri-urban households interviewed in 2012, plus 20 new rural households in different villages than 2012. All representatives of households interviewed were engaged in gardening, some with the government's backyard gardening initiative and others on their own initiative. Questions in these shorter interviews focused on the degree to which gardening may or may not be contributing to food security, dietary diversity or income generation.

3. Results

The results of this research are organized into two sections. The first is the empirical findings related to household food security. The second is an analysis of policy and macroeconomic factors driving food insecurity. In particular, the second section discusses the political economy of the livestock, diamond and ecotourism sectors and their role in driving inequality, as well as the precarity of smallholder production in Botswana today.

3.1. *Food Insecurity*

In Botswana, 30 percent of the general population lives below a $2/day poverty threshold [20]. The mean per capita income of the households surveyed in Gaborone was $3.55/day, whereas that for peri-urban households was $5.56/day, and that for rural households was $1.67. Household interview data revealed challenging food security circumstances for the poorest households (see Table 2).

Table 2. Food security findings.

Sample Size	Demographic	Mean per Capita Income per Day		Percent Income Spent on Food	Food Security	
		Pula	USD		HFIA Score	Qualitative
89	All Urban	25.5	$3.55	35%	2.8	Moderate
30	Urban lower tercile	1.71	$.24	38%	2.9	Moderate
30	Urban middle tercile	16.9	$2.35	35%	2.9	Moderate
29	Urban upper tercile	59.14	$8.21	31%	2.5	Moderate
30	Peri-urban	40.1	$5.56	26%	2.8	Moderate
39	Rural	12	$1.67	61%	3.2	Moderate

Note: The household food insecurity access score (HFIA) is tabulated based on the scores to the nine questions in Table 1: 1 = food secure, 2 = mild food insecurity, 3 = moderate food insecurity and 4 = severe food insecurity.

In Gaborone, the poorest tercile of urban respondents had a per capita income of $0.24 per day and a household food insecurity access (HFIA) score of 2.9, indicating moderate food insecurity (see note following Table 1 for a description of how question responses were converted to HFIA scores). Amongst the urban households surveyed, the low and middle income categories spend 38 percent and 35 percent of their income on food (as compared to 31 percent for wealthiest third). The average for the urban group was 35 percent. In total, 12 percent of the urban respondents were experiencing moderate food insecurity and 44 percent were experiencing severe food insecurity (see Table 3). Interviews revealed that urban residents have a very limited ability to cope with fluctuating global food prices. Some respondents in the Old Naledi neighborhood (area #1 in Figure 2) did discuss the spatial or locational advantage of their neighborhood in terms of coping. They mentioned that it was close enough to the city center to minimize the cost of commuting, yet also within walking distance of the Gaborone dam area to the southeast where firewood could be collected. Historically, residents from this neighborhood also salvaged food scraps from a dump in the Gaborone dam area [41]. The most common strategy for coping came in the form of reducing remittances to rural areas. This is important because it means that rural areas may be bearing the brunt of belt tightening in urban areas in response to fluctuating food prices.

Table 3. Proportion of households facing food insecurity.

Sample Size	Demographic	Food Secure	Mild Food Insecurity	Moderate Food Insecurity	Severe Food Insecurity
89	Urban	24%	20%	12%	44%
30	Peri-Urban	27%	7%	23%	43%
39	Rural	18%	6%	19%	57%

In the survey, the proportion of households facing food insecurity was highest in rural areas (see Table 3). Here some 76 percent of households in the sample were moderately or severely food insecure, as compared to 56 percent in the urban sample and 66 percent in the peri-urban sample.

The rural sample was composed of 39 households. Sixty-one percent of the rural respondents were female, with an average age of 47 and an average household size of 4.7. Average assets for these rural households were: 7 cows, 10.6 goats, 5 sheep, and 13.6 chickens. These mean asset figures mask lower numbers for poorer and more food insecure households. Thirty eight percent of rural inhabitants surveyed identified as farmers, of which 87 percent were women. The major crops raised were maize, sorghum and cowpeas. The minor ones included: watermelon, sweet reed and pumpkins. The typical farm was five hectares in size and all benefited from government support for seeds and plowing.

Of the farmers interviewed, 20.5 percent reported that weather impacted their household's access to food. Furthermore, 67 percent of these farmers cultivate less than they did five years ago. When asked why they farmed less, two-thirds indicated that it was because of labor constraints. The second most common reason for farming less was weather, mentioned by 20 percent.

In terms of cropping strategies, very few farmers indicated that they alter what they grow in response to market signals. As such, farmers do not necessarily grow more maize, sorghum or cowpeas when the prices for those crops are high. One of the most surprising findings was that farmers are growing considerably less sorghum than they did 10 years ago. This was surprising because six of the 10 years prior to 2012 (the year of the survey) were drought years, and sorghum is generally considered to be much more drought tolerant than maize. Informants reported that sorghum is mainly being replaced with maize. According to interviewees, the advantages of maize (even though it is more susceptible to drought) are that it is less labor intensive, it matures in a shorter time period and it is not subject to predation by birds to the same extent as sorghum.

Finally, of the sample of 30 peri-urban households (all from Tlokweng, a large peri-urban village next to Gaborone to the east), ten were involved in urban gardening. The majority of these gardeners were female and their average age was 44. Urban gardening in Botswana is largely supported by the Ministry of Agriculture's Backyard Gardening initiative, part of President Ian Khama's signature Poverty Eradication Programme. This program targets households and individuals living below the poverty line and generally frames "backyard gardening" as a female activity. Beneficiaries of this program receive shade netting, rain barrels to collect roof water, drip irrigation tubing, seeds, tools and extension services (see Figure 3). Based on interviews with these ten gardeners, the major constraint they face is access to affordable water. Follow-up interviews in 2015 (including backyard and commercial gardeners) revealed that access to affordable water was the key variable impacting food security and economic success [42]. This program has been criticized in the local press for being impractical due to the high cost of water, as well as for not being a priority for some recipients [3,43].

3.2. *Agriculture, Poverty and Botswana's Political Economy*

Understanding Botswana's political economy is critical for gaining insights into the major drivers of poverty, inequality and food insecurity in the country. While agriculture was once Botswana's most important sector, accounting for 40 percent of GDP at independence, it now accounts for only 1.8 percent of GDP [44]. Not unlike many African countries [45], Botswana adhered to a policy of food self-sufficiency in the first few decades following independence. In the early 1980s, however, Botswana transitioned from a policy of food self-sufficiency to one of food security [21]. This shift, from trying to produce as much food as possible, to importing increasing quantities of food, was consistent with neoliberal economic reforms of the era which emphasized specializing in one's comparative advantage and trade for other goods [45,46].

Figure 3. The backyard gardening initiative in Botswana (photo by author).

Today Botswana imports 90 percent of its food overall, and crop agriculture (not to be confused with livestock rearing) is in a more marginal position than ever in the country's political economy. Despite this fragility and marginality, crop agriculture still remains an important activity for the poor and women. Small farmers (defined as those who cultivate 20 hectares or less) work 80 percent of planted area and produce 38 percent of the harvest. Agriculture employs 30 percent of the total workforce and 35 percent of women [47].

As agriculture has stagnated and other sectors of the economy have grown, many rural Botswanans purchase more of their food or have moved to the city in search of employment [48,49]. The government of Botswana does provide a fairly robust safety net for poor urban and rural dwellers—especially when compared to other African countries. In rural areas, the key welfare support comes in the form of drought relief whenever an official dry year has been declared (more than half of the last ten years). In urban areas, there is a significant public works program, known as Ipelegeng in Setswana, which employs those who fall below the poverty line, and are able-bodied, in activities, such as street sweeping and trash collection. Botswana's current president, Ian Khama, has made poverty eradication a major policy priority. Despite these efforts to address poverty, and inter-related food insecurity, it is structural shifts in Botswana's economy over the past 50 years that explain the growing gap between Botswana's rich and poor. The three main components of the country's economy are cattle, diamonds and high-end ecotourism. Below, each of these components is discuss briefly, followed by an articulation of how a comprehension of the political economy of these three sectors allows us to understand the precarity of smallholder production in Botswana today.

3.2.1. Cattle

The livestock industry is Botswana's original export activity and accounts for 80 percent of the agricultural sector. While livestock represented 85 percent of exports at independence (largely on the hoof to South Africa), this had shrunk to two percent by 2006 as other sectors of the economy grew. There has been significant public investment in the livestock sector since independence, including the sinking of thousands of boreholes, land tenure reform favoring large cattle ranchers [50], and the establishment of the Botswana Meat Commission (BMC). The BMC regulates the livestock industry, provides veterinary services and disease control, runs or oversees slaughter houses, and facilitates the export of meat to lucrative European Union markets [51]. What is critical to understand is that these public investments have largely benefitted wealthy men who own the majority of the nation's livestock herd. Even urban elites continue to invest surplus wealth in cattle. The national herd grew ten-fold from beginning of 20th century, reaching three million in 1978. It then declined to two million by 1990 because of drought and is now back up to three million today.

3.2.2. Diamonds

Botswana is the largest producer of gem-quality diamonds in the world (as opposed to industrial diamonds). After discovering diamonds a few short years after independence, Botswana shrewdly negotiated with DeBeers for a 50/50 split of revenues [52]. This industry has been incredibly well managed, and the Government of Botswana now owns 15 percent of DeBeers today. It is diamond revenues which largely fund the government, including education and health care for all citizens. While this wealth is shared by all citizens, it tends to be wealthier, better educated citizens who hold the highest paying civil service positions [44]. The government's stake in the diamond industry has allowed it to bankroll services, employment and contracts in cities and towns that have fueled urbanization in Botswana and drawn labor away from agriculture [44,49].

3.2.3. Eco-Tourism

Last, but not least, Botswana is known for its high-end, low-volume eco-tourism sector. After looking at the eco-tourism industry in neighboring African countries, Botswana decided early on that it would focus on the wealthy end of the market to maximize revenues and limit potential environmental degradation related to high-volume tourism [53]. Most eco-tourism is located in the northern parts of the country in and around the Okavango Delta. Here the government leases out concessions to tourism operators for extremely high prices. Unfortunately, the highest paying jobs in this sector are held by wealthy investors, many of whom are naturalized Botswana citizens (often former white South Africans and Zimbabweans). Most employment in this sector is composed of low paying, service sector jobs.

The structure of the Botswana economy, dominated by the cattle industry, diamond mining and high-end eco-tourism, has tended to deepen inequality and done little to alleviate poverty and related food insecurity. This structure has also served to pull labor and resources away from crop farming, a sector involving more women and poorer members of society.

4. Discussion

Climate change has had minimal direct impact on the food security of the majority rural households surveyed in this study. First and foremost, with agriculture becoming less and less of a factor in Botswana's rural food economies (largely due to structural changes in the economy), production declines related to poor rainfall only impact a relatively small source of food for most rural households. Furthermore, rainfall declines, as an exogenous shock, only explain part of the crop losses experienced in dry years. Labor constraints have forced changes in Botswana's farming systems (namely a shift from sorghum to maize production) that have exacerbated losses in dry years because maize is less drought tolerant than sorghum.

While this only impacts a fraction of rural households, declining rainfall does constrain the government's backyard gardening initiative because gardens often receive minimal rainfall in the December to March 'rainy season' and roof top rain barrels are only minimally replenished during this period. Absent sufficient rainfall, most urban and peri-urban backyard gardeners cannot afford to cover the cost of municipal tap water with their meager returns from gardening. Given an on-going water shortage, most municipalities also frown on the use of tap water for gardening purposes. More rural gardeners with access to borehole water typically face less of a constraint.

A more likely and significant impact of climate change is on the overall price of grain in regional and global markets (which is an issue for Botswana which imports 90 percent of its food). That said, global food price fluctuations to date have not been tied to weather related supply shocks—but more clearly to increasing energy prices. A stronger argument could be made for importance of regional climate change, especially that affecting South Africa, because it supplies the majority of Botswana's food. This has been especially evident in 2016 with an El Niño-related drought devastating crop production across Southern Africa [54].

The major direct constraint on rural food security in Botswana is increasingly high and variable food prices—both directly and indirectly—coupled with constrained income-generating opportunities for rural households. The problem is that the vast majority of Botswana's rural households are not food self-sufficient and purchase a large proportion of their food. This means that rural households (like urban households) are often equally affected by high and fluctuating food prices. Rural households are also affected indirectly by high food prices, as interviews with rural and urban informants suggested that rural households receive fewer remittances from urban relatives when food prices go up. As discussed previously, the structure of the Botswana economy, dominated by the cattle industry, diamond mining and high-end eco-tourism, has tended to deepen inequality and done little to alleviate poverty and related food insecurity. This structure has also served to pull labor and resources away from crop farming, a sector involving more women and poorer members of society. In terms of moving forward, it may be worth thinking about backyard gardening in different terms than just poverty alleviation, but as a climate proof, small-scale production initiative. While Southeastern Botswana used to benefit from some of the better rainfall in the country (an area of the country where the nation's capital and majority of the population are located today), the spatial distribution of rainfall seems to be shifting away from this region [3]. Although groundwater reserves are not sufficient to support the irrigation of extensive row crop cultivation, an argument could be made for small, intensive horticulture production, which is more frugal with water resources and produces higher value crops. As such, the government of Botswana may want to consider framing backyard gardening as a climate change adaptation that is worthy of international support. The initiative could also be improved upon. For example, while most backyard gardening and commercial horticulture in Botswana has largely focused on European vegetable production to date, there may be other types of vegetables or fruits (such as traditional beans whose leaves are consumed), which require less water. The targeting of recipients for government backyard garden support could also be strengthened, as some of the current beneficiaries did not see this as their first priority for a livelihood activity).

Detractors may argue that it is inefficient for a water-constrained country like Botswana to produce more of its own food when it can import food from more water-rich countries (essentially importing what some have referred to as virtual water) [55]. The Ricardian argument suggests that Botswana should focus on those goods for which the country has a comparative advantage (namely cattle, diamonds, and high-end ecotourism) [19]. There are at least two counter arguments to this line of thinking. First, backyard gardening is a pro-poor activity (rather than investing in the core elements of Botswana's economy that may deepen inequality) and subsistence food production protects the poorest of the poor against food price fluctuations. Small-scale, irrigated backyard gardening is also quite different from traditional irrigation schemes that have often marginalized the poor [56]. Second, given the fact that South Africa produces the vast majority of Botswana's food, and South Africa increasingly faces similar climate change threats [57], the environmental comparative advantage argument in favor of South African foodstuffs is somewhat weak.

A key challenge for making small-scale, irrigated gardening a viable, pro-poor adaptation in Botswana will be overcoming the politics and power dynamics involved in water allocation in southeastern Botswana. With rainfall declining in the most densely populated area of the country, the powerful livestock industry has tightened its grip on borehole water resources in this region. Furthermore, municipal water, which increasingly comes to Gaborone via the north-south water carrier (a huge pipeline bringing water from the north to the south of the country) [58] because the Gaborone dam no longer fills, is mainly allocated for domestic use. Critics have pointed out that these domestic uses also include watering golf courses, as well as lawns in wealthy Gaborone suburbs, such as Phakalane (arguably frivolous uses that far exceed the quantities that might be used in vegetable gardens). Reorienting water use over the long term will be key to developing more viable livelihoods for the poor.

5. Conclusions

Despite the growing concern about the impact of climate change on livelihoods and food security in Sub-Saharan Africa, there are larger structural issues driving the coupled problems of poverty and food insecurity in Botswana. Poor urban households in Botswana's capital, Gaborone, are challenged by increasingly volatile food prices. Facing a limited ability to cope, these households often reduce remittances to rural households as a way to address rising food prices. This creates a double burden for rural households, who are rarely food self-sufficient, buy a large proportion of their own food, and are squeezed by declining remittances from family members in the city. As such, it is the structure of the economy and periodically high food prices that are the main drivers of food insecurity in both urban and rural areas, not climate change. While rainfed agriculture is an important source of food for some rural households, labor constraints mean that this approach to agriculture is increasingly vulnerable to declining rainfall regimes as households grow less labor intensive crops that also happen to be less drought tolerant. While the government's backyard gardening initiative was only moderately successful according to official accounts, and recently discontinued amidst criticism, small-scale, irrigated horticulture may represent a form of climate change adaptation that deserves further experimentation. However, given that horticulture, and women's farming more broadly, is marginalized within Botswana's economy in terms of access to water, addressing this constraint is a prerequisite for women's horticultural success.

Acknowledgments: The fieldwork for this project was supported by a Wallace Grant from Macalester College in 2012, as well as the Womadix Foundation in 2015. The author is grateful for the advice and support of colleagues and friends in the Environmental Science Department at the University of Botswana, especially Thando Gwebu and Isaac Mazonde. He also wishes to thank two graduate students at the University of Botswana, Abrahm Botlhale and Chasha Masabase, for their work conducting interviews for this project in 2012, and to Macalester student Rachel Fehr for her assistance in 2015. The author finally wishes to express his appreciation for the constructive advice of two anonymous reviewers as well as for input from the Special Issue editors, Claudia A. Radel and Jacqueline M. Vadjunec.

Conflicts of Interest: The author declares no conflict of interest.

References

1. Core Writing Team; Pachauri, R.K.; Meyer, L.A. *Climate Change 2014: Synthesis Report. Contribution of Working Groups I, II and III to the Fifth Assessment Report of the Intergovernmental Panel on Climate Change*; Intergovernmental Panel on Climate Change (IPCC): Geneva, Switzerland, 2014.
2. De Sherbinin, A. Climate change hotspots mapping: What have we learned? *Clim. Chang.* **2014**, *123*, 23–37. [CrossRef]
3. Batisani, N.; Yarnal, B. Rainfall variability and trends in semi-arid Botswana: Implications for climate change adaptation policy. *Appl. Geogr.* **2010**, *30*, 483–489. [CrossRef]
4. Zhou, P.; Simbini, T.; Ramokgotlwane, G.; Hachigonta, S.; Sibanda, L.M.; Thomas, T.S. Botswana. In *Southern African Agriculture and Climate Change: A Comprehensive Analysis*; Hachigonta, S., Nelson, G.C., Thomas, T.S., Sibanda, L.M., Eds.; International Food Policy Research Institute: Washington, DC, USA, 2013; pp. 41–70.
5. Temoso, O.; Villano, R.A.; Hadley, D. Agricultural productivity, efficiency and growth in a semi-arid country: A case study of Botswana. *Afr. J. Agric. Resour. Econ.* **2015**, *10*, 192–206.
6. United Nationas Food and Agriculture Organization. Crop Water Information: Sorghum. Available online: http://www.fao.org/nr/water/cropinfo_sorghum.html (accessed on 8 February 2016).
7. McCann, J. *Maize and Grace: Africa's Encounter with a New World Crop, 1500–2000*; Harvard University Press: Cambridge, MA, USA, 2007.
8. United Nations Food and Agriculture Organization. Crop Water Information: Maize. Available online: http://www.fao.org/nr/water/cropinfo_maize.html (accessed on 8 February 2016).
9. Moalafhi, D.B.; Tsheko, R.; Atlhopheng, J.R.; Odirile, P.T.; Masike, S. Implications of climate change on water resources in Botswana. *Adv. J. Phys. Sci.* **2012**, *1*, 4–13.

10. Botswana Tourism Organization. Annual Rainfall Map. Available online: http://www.botswanatourism.us/experience_botswana/annual_rainfall_map.html (accessed 21 June 2016).
11. Government of Botswana. Botswana Committed to Tackling Climate Change. 2010. Available online: http://www.gov.bw/en/News/Botswana-committed-to-tackling-climate-change-/ (accessed on 8 February 2016).
12. Kayawe, B. Hunger Heralds Climate Change's Arrival in Botswana. 2015. Available online: http://www.ipsnews.net/2015/11/hunger-heralds-climate-changes-arrival-in-botswana/ (accessed on 6 February 2016).
13. Government of Botswana. Botswana Climate-Smart Agriculture Program. 2015. Available online: http://canafrica.com/wp-content/uploads/2015/05/Botswana-CSA-Program.pdf (accessed on 8 February 2016).
14. Southern African Development Community (SADC). News: Tripartite Climate Change 2014 Meeting. 2014. Available online: http://www.sadc.int/news-events/news/tripartite-climate-change-2014-meeting/ (accessed on 21 June 2016).
15. Neufeldt, H.; Jahn, M.; Campbell, B.; Beddington, J.; DeClerck, F.; de Pinto, A.; Gulledge, J.; Hellin, J.; Herrero, M.; Jarvis, A.; et al. Beyond climate-smart agriculture: Toward safe operating spaces for global food systems. *Agric. Food Secur.* **2013**, *2*. [CrossRef]
16. Sumberg, J.; Thompson, J. *Contested Agronomy: Agricultural Research in a Changing World*; Routledge: New York, NY, USA, 2012.
17. Sen, A. *Poverty and Famines*; Clarendon: Oxford, UK, 1981.
18. Kebakile, P. Rising Global Food Prices: Causes and Implications for Botswana. BiDPA Briefing. September 2008. Available online: http:/www.bidpa.bw (accessed on 21 June 2016).
19. Moseley, W.G. Africa's future? Botswana's growth with hunger. *Al Jazeera-English*, 8 May 2012.
20. Frayne, B.; Pendleton, W.; Crush, J.; Acquah, B.; Battersby-Lennard, J.; Bras, E.; Chiweza, A.; Dlamini, T.; Fincham, R.; Kroll, F. *The State of Urban Food Insecurity in Southern Africa*; Urban Food Security Series No. 2; Queen's University: Kingston, ON, Canada; AFSUN: Cape Town, South Africa, 2010.
21. Lado, C. Environmental and socio-economic factors behind food security policy strategies in Botswana. *Dev. South. Afr.* **2001**, *18*, 141–168. [CrossRef]
22. Riley, L.; Legwegoh, A. Comparative urban food geographies in Blantyre and Gaborone. *Afr. Geogr. Rev.* **2014**, *33*, 52–66. [CrossRef]
23. Watts, M. On the poverty of theory: Natural hazards research in context. In *Interpretations of Calamity*; Hewitt, K., Ed.; Allen and Unwin: Boston, MA, USA, 1983; pp. 231–262.
24. Ribot, J. The causal structure of vulnerability: Its application to climate impact analysis. *GeoJournal* **1995**, *35*, 119–122. [CrossRef]
25. Ribot, J. Vulnerability does not fall from the sky: Toward multiscale, pro-poor climate policy. In *Social Dimensions of Climate Change*; Mearns, R., Norton, A., Eds.; The World Bank: Washington, DC, USA, 2010; pp. 47–74.
26. Bohle, H.G.; Downing, T.E.; Watts, M.J. Climate change and social vulnerability: Toward a sociology and geography of food insecurity. *Glob. Environ. Chang.* **1994**, *4*, 37–48. [CrossRef]
27. Bryceson, D. The scramble in Africa: Reorienting rural livelihoods. *World Dev.* **2002**, *30*, 725–739. [CrossRef]
28. Moseley, W.G. *Paradoxical Constraints to Agricultural Intensification in Malawi: The Interplay between Labor, Land and Policy*; Georgia Department of Geography Discussion Paper Series. No. 00-1; University of Georgia: Athens, GA, USA, 2000.
29. O'Brien, K.L.; Leichenko, R. Double exposure: Assessing the impacts of climate change within the context of economic globalization. *Glob. Environ. Chang.* **2000**, *10*, 221–232. [CrossRef]
30. O'Brien, K.L.; Leichenko, R. Winners and losers in the context of global change. *Ann. Assoc. Am. Geogr.* **2003**, *93*, 89–103. [CrossRef]
31. Binns, T.; Nel, E. Beyond the development impasse: The role of local economic development and community self-reliance in rural South Africa. *J. Mod. Afr. Stud.* **1999**, *37*, 389–408. [CrossRef]
32. Mkwambisi, D.D.; Frase, E.D.G.; Dougill, A.J. Urban agriculture and poverty reduction: Evaluating how food production in cities contributes to food security, employment and income in Malawi. *J. Int. Dev.* **2011**, *23*, 181–203. [CrossRef]
33. Battersby, J. Beyond the food desert: Finding ways to speak about urban food security in South Africa. *Geogr. Ann. Ser. B Hum. Geogr.* **2012**, *94*, 141–159. [CrossRef]
34. Battersby, J.; Marshak, M. Growing communities: Integrating the social and economic benefits of urban agriculture in Cape Town. *Urban Forum* **2013**, *24*, 447–461. [CrossRef]

35. Kebadiretse, F. Batswana dismiss backyard gardening as "a big joke.". *Mmegi Online*. 23 May 2011. Available online: http://www.mmegi.bw/index.php?sid=1&aid=1058&dir=2011/May/Monday23#comments (accessed on 6 February 2016).
36. Rocheleau, D.E.; Thomas-Slayter, B.; Wangari, E. *Feminist Political Ecology: Global Perspectives and Local Experience*; Routledge: London, UK, 1997.
37. Hovorka, A.J. Urban agriculture: Addressing practical and strategic gender needs. *Dev. Pract.* **2006**, *6*, 51–61. [CrossRef]
38. Hovorka, A.J. The No. 1 Ladies' Poultry Farm: A feminist political ecology of urban agriculture in Botswana. *Gend. Place Cult.* **2006**, *13*, 207–225. [CrossRef]
39. Lesetedi, G.N. Urban-rural linkages as an urban survival strategy among urban dwellers in Botswana: The case of Broadhurst residents. *J. Political Ecol.* **2003**, *10*, 37–46.
40. Coates, J.; Swindale, A.; Bilinsky, P. *Household Food Insecurity Access Scale (HFIAS) for Measurement of Food Access: Indicator Guide*; U.S. Agency for International Development: Washington, DC, USA, 2007.
41. Rankokwane, B.; Gwebu, T. Characteristics, threats and opportunities of landfill scavenging: The case of Gaborone-Botswana. *GeoJournal* **2006**, *65*, 151–163. [CrossRef]
42. Fehr, R. *Exploring the Role of Horticulture in Alleviating Food Insecurity among Women in Botswana*; Geography Honors Projects, Paper 48; Macalester College: St. Paul, MN, USA, 2016.
43. Kajevu, Z. Poverty alleviation projects under scrutiny. *Sunday Standard*, 14 August 2013.
44. Hillbom, E. Diamonds or development? A structural assessment of Botswana's forty years of success. *J. Mod. Afr. Stud.* **2008**, *46*, 191–214. [CrossRef]
45. Moseley, W.G.; Schnurr, M.; Bezner Kerr, R. Interrogating the technocratic (neoliberal) agenda for agricultural development and hunger alleviation in Africa. *Afr. Geogr. Rev.* **2015**, *34*, 1–7. [CrossRef]
46. Moseley, W.G.; Carney, J.; Becker, L. Neoliberal policy, rural livelihoods and urban food security in west Africa: A comparative study of the Gambia, Côte d'Ivoire and Mali. *Proc. Natl. Acad. Sci. USA* **2010**, *107*, 5774–5779. [CrossRef] [PubMed]
47. Food and Agricultural Organization of the United Nations (FAO). *FAO Statistical Yearbook 2013*; FAO: Rome, Italy, 2013.
48. Twyman, C.; Sporton, D.; Thomas, D. 'Where is the life in farming?': The viability of smallholder farming on the margins of the Kalahari, Southern Africa. *Geoforum* **2004**, *35*, 69–85. [CrossRef]
49. Cavrie, B.; Mosha, A.; Keiner, M. Chapter 3: Gaborone: An emerging urban oasis. In *From Understanding to Action: Sustainable Urban Development in Medium Sized Cities in Africa and Latin America*; Keiner, M., Zegras, C., Schmid, W., Salmeron, D., Eds.; Springer: New York, NY, USA, 2004; pp. 41–73.
50. Peters, P. Embedded systems and rooted models: The grazing systems of Botswana and the commons debate. In *The Question of the Commons*; McCay, B., Acheson, J., Eds.; The University of Arizona Press: Tuscon, AZ, USA, 1987; pp. 171–194.
51. Darkoh, M.B.K.; Mbaiwa, J.E. Globalisation and the livestock industry in Botswana. *Singap. J. Trop. Geogr.* **2002**, *23*, 149–166. [CrossRef]
52. Samatar, A. *An African Miracle: State and Class Leadership and Colonial Legacy in Botswana Development*; Heinneman: Portsmouth, NH, USA, 1999.
53. Swatuk, L.A. From "Project" to "Context": Community based natural resource management in Botswana. *Glob. Environ. Politics* **2005**, *5*, 95–124. [CrossRef]
54. Fragoso, A.D. Africa is Going through Serious Drought and El Niño is Making it Worse. *Climate Progress*, 21 January 2016.
55. Turton, A.R. Precipitation, people, pipelines and power in southern Africa: Towards a 'virtual water'-based political ecology discourse. In *Political Ecology: Science, Myth and Power*; Stott, P., Sullivan, S., Eds.; Arnold: London, UK, 2000; pp. 132–153.
56. Kates, R. Cautionary tales: Adaptation and the global poor. *Clim. Chang.* **2000**, *45*, 5–17. [CrossRef]
57. Kings, S. Subsistence farmers in a hot spot. *Mail & Guardian*, 29 May 2015.
58. Sejabosigo, J. Botswana: Govt addresses water supply issues. *AllAfrica.com*, 5 March 2015.

land

MDPI

Article

Fluid Waters and Rigid Livelihoods in the Okavango Delta of Botswana

Brian King [1,*], Jamie E. Shinn [2], Kelley A. Crews [3] and Kenneth R. Young [3]

1 Department of Geography, The Pennsylvania State University, University Park, PA 16802, USA
2 Department of Geography, Texas A&M University, College Station, TX 77843, USA; jamieshinn@tamu.edu
3 Department of Geography and the Environment, University of Texas at Austin, Austin, TX 78712, USA; kelley@utexas.edu (K.A.C.); kryoung@austin.utexas.edu (K.R.Y.)
* Correspondence: king@psu.edu; Tel.: +1-814-865-1628; Fax: +1-814-863-7943

Academic Editors: Claudia A. Radel and Jacqueline M. Vadjunec
Received: 17 August 2015; Accepted: 31 May 2016; Published: 11 June 2016

Abstract: Current and future impacts of climate change include increasing variability in a number of biophysical processes, such as temperature, precipitation, and flooding. The Intergovernmental Panel on Climate Change (IPCC) has suggested that Southern Africa is particularly vulnerable to the anticipated impacts from global climate change and that social and ecological systems in the region will be disrupted and likely transformed in future decades. This article engages with current research within geography and cognate disciplines on the possibilities for responsive livelihoods within socio-ecological systems experiencing biophysical change. The paper draws from an ongoing research project that is evaluating perceptions of environmental change, specifically of precipitation and flooding dynamics, in order to understand social responses. We report on the findings from qualitative interviewing conducted in 2010 and 2011 in the communities of Etsha 1, Etsha 6, and Etsha 13 within the Okavango Delta of Botswana. While flooding and precipitation patterns have been dynamic and spatially differentiated, some livelihood systems have proven rigid in their capacity to enable adaptive responses. We assert this demonstrates the need for detailed research on livelihood dynamics to support adjustments to biophysical variability within socio-ecological systems experiencing change.

Keywords: climate change adaptation; livelihood; governance; socio-ecological system; Botswana

---

1. Introduction

The Fourth Assessment Report of the Intergovernmental Panel on Climate Change (IPCC) (AR4) detailed current and future impacts of climate change, including increasing variability in a number of biophysical processes, specifically temperature, precipitation, and flooding [1,2]. The recently released Fifth Assessment Report (AR5) of the IPCC confirms these findings, with added confidence in the likelihood of climate-driven disruptions to human and environment systems [2]. Working Group II of the IPCC has projected that Southern Africa is one of several areas particularly vulnerable to global climate change and that livelihood systems in the region will likely be disrupted and transformed in future decades [3]. The predictive models for Southern Africa are not uniform in their assessments of these changes, with some studies predicting wetter seasons while others anticipate drier months than those at the present time [4]. While the implications for institutional systems and livelihood decision-making are stark, research has been limited in addressing the specific factors shaping perceptions of environmental variability and potential for effective responses. A key factor is the governance regime, which involves the individuals, organizations, rules, and traditions that help shape environmental perceptions and livelihood responses. Yet the specific relationships between livelihoods and governance regimes are not well understood, particularly within settings experiencing socio-environmental variability. This complicates the possibilities for adaptive responses that are sensitive to local context and sustainable in the face of future perturbations.

This article engages with current research on climate change adaptation while focusing upon the possibilities for responsive livelihoods within socio-ecological systems experiencing biophysical change [5,6]. We begin with a review of some of the recent scholarship on environmental perceptions and environmental governance. This is followed by a discussion of the case study and methodology, which relied upon semi-structured interviewing and livelihood mapping conducted in May 2010 and from May to June 2011. The paper reports findings from an ongoing research project in the Okavango Delta of Botswana that is evaluating perceptions of environmental change, specifically of precipitation and flooding dynamics, in order to understand livelihood responses to variability. The timing of this research is particularly opportune because the study region experienced annual flooding from 2009 to 2012 that was higher than the previous decade. The geomorphological interactions of the Okavango Delta's landscape features and spatio-temporally variable flooding dynamics have produced intra-regional and intra-village impacts including the permanent displacement of residents within the study area. We find that while environmental perceptions vary within the region, adjustments to biophysical variability remain tied to livelihood systems that can constrain adaptive responses. We conclude by suggesting that this research helps demonstrate the need for future research on climate change adaptation within the natural and social sciences.

2. Environmental Perceptions and Governing for Adaptation

The ways in which human populations understand and adjust to environmental variability has been a topic of interest within geography and other social science disciplines for decades. Environmental dynamics can be variable and unpredictable, and generate local perceptions that influence decision-making in multifaceted ways. For example, drylands have precipitation regimes that are both temporally and spatially variable. Agriculturalists living in these types of variable environments must make decisions based on a level of uncertainty and risk, including assessing when, where, and which crop varieties to plant during a growing season. As climate change leads to greater variability in the drivers of agrarian systems, agriculturalists will need to make decisions amidst higher levels of uncertainty. Other livelihood practices, such as livestock management and natural resource collection, are similarly impacted by dynamic environmental change. People perceive risk through an integration of analytical and experiential processing systems [7,8]. Marx *et al.* [8] argue that where most communication about climate risk assumes people process information analytically, there is a need for greater attention to the role of experiential processing in how people perceive risk. In turn, an expanded focus on experiential processing recognizes that decision-making is in part based on a person's or group's past experiences of similar situations, including the emotions (positive and negative) associated with those experiences. To this end, there is an existing body of work focused on how to foster more effective communication between climate scientists and local populations living in variable environments [8–10].

A larger body of case-study specific work mainly focusing on precipitation patterns considers the differences between meteorological data and environmental perceptions. Some studies reveal at least some congruency between environmental perceptions and scientific interpretations of rainfall patterns [11,12]. However, a larger number of studies detail inconsistencies between perceptions and regional climate data [9,13,14]. Importantly, many of these studies reveal that inconsistencies are not the result of "incorrect" interpretation, but rather result from these groups paying attention to different variables in terms of rainfall and its impacts. For example, a number of studies have shown that while agriculturalists are concerned with the timing of rain (e.g., how well precipitation aligns with a growing season), scientists tend to be concerned with the overall quantity of rain in a given time period (e.g., a year) (examples include [9,11,12,14]). Further, some studies suggest that it is not just the quantity and quality of rainfall that impacts environmental perceptions, but also a set of social and ecological variables in a given context [13–15]. Meze-Hausken ([14]; p. 30) noted that "peoples' perceptions about climate are in fact a combination of various environmental aspects ... any analysis of subjective observations about weather and climate requires a deeper investigation of the socioeconomic, cultural and environmental conditions experienced by the affected people." Taken together, these studies show that perceptions of environmental variability are likely the result not only of current experiences, but result from complicated socio-environmental relationships that are rooted in historical experiences with biophysical variability.

Understanding perceptions of environmental change has proven important "at a time when biophysical and social environmental conditions are seen as increasingly volatile, unpredictable, uncontrollable, and interconnected" ([16]; p. 616). The possibilities for climate change adaptation exist within decision-making processes that extend across multiple spatial and temporal scales. While environmental governance has been a unifying theme for a variety of academic disciplines and research fields, there are divergent understandings of what this concept actually entails. One useful conceptualization is of the governance regime, which is understood as the collections of institutions and stakeholders operating across spatial and temporal scales that intersect in making the decisions of how environmental systems are perceived and managed. Governance systems include the individuals, organizations, rules, and traditions that help shape how decisions are made, power is exercised or resisted, and commitment to collective objectives achieved [5,17–20]. The governance systems under particular frames of study are themselves embedded within political and economic relationships that make them vulnerable to perturbations or stresses. The increase in scholarship on resilience or adaptive governance systems [21] reflects the concern that some of these systems are not well positioned to respond to environmental change [22]. For example, in some of this work global climate change and associated weather-related hazards are presented not as problems with the natural resource base but as "governance failures" ([23]; p. 354). Under circumstances of social and ecological fluctuation and change, the governance regimes therefore prove critical in providing flexibility for livelihoods to respond to changing circumstances, while offering resources and capacity building to anticipate and respond to future pressures.

While environmental variability, environmental perception, governance, and livelihood production are central to the adaptive capacity of individuals and households in the Global South, more research is needed to provide explicit empirical detail as to their relationships. This is particularly important within contexts that are experiencing environmental change, to identify the specific ways in which environmental variability is understood, in addition to whether livelihood systems can support adaptive responses. As detailed in the next section, the Okavango Delta of Botswana is characterized by variability in precipitation and flooding dynamics that necessitate dynamic responses at multiple decision-making scales. As such, it is a unique setting for examining the relationships between livelihood systems and biophysical variability.

3. Case Study and Methodology

The Okavango Delta is not actually a delta but is an inland alluvial fan stemming from the panhandle into the distal portions. This wetland system is notable for its location within a savanna system in the Kalahari Desert, with its waters being fed by local precipitation and upstream precipitation in the Angolan highlands. The relative contribution and amounts of these two inputs vary temporally and are thought to respond to the shifting of the Southern Atlantic Oscillation (SAO) [24]. The Delta is characterized by high spatial and temporal variability with regards to both precipitation and flooding. Water levels in the fan vary considerably seasonally and spatially each year, covering anywhere from 8000 to 12,000 km^2 [24]. The floods arrive into the distal reaches of the Delta an average of four months after the peak of the local dry season (winter) and after the local rainy season [4,25]. The Okavango Delta is additionally characterized not only by strong seasonal cycles, but also by longer term (interannual and quasi-decadal) variability [26], meaning that sites within the Delta may experience high and differential variability in proximity to, amount of, and timing of the arrival of the water [27].

The Delta is comprised of permanent, seasonal, and occasional swamps that flood typically every 2–3 years [28]. The fluvial geomorphological processes of this system move sand such that some rechannelization occurs in the fan portion of the Delta, which at times is unpredictable [29]. The Delta is comprised of interspersed wetlands, grasslands, savanna and woodland, including riparian, interfluves. As such it offers not only a variety of habitats and resources but serves as a source of water for both wildlife and anthropogenic uses, creating a complex and shifting socio-ecological system driven by irregular annual flooding. The Okavango Delta was declared a wetland of international importance in 1996 by the Ramsar Treaty (site #879) and is globally significant as one of the world's largest wetlands. Land use in the region traditionally ranges from open access communal areas with subsistence agriculture and some livestock husbandry, to hunting and photography tourism concessions, game reserves and national parks, as well as settlements and villages along the periphery of the Delta. A recent and controversial national ban on wildlife hunting was implemented in January 2014 and is altering the assignment and use of a number of concession areas. The diverse management zones of the Delta are loosely controlled through veterinary fences and staffed checkpoints established by the national government intended to regulate the movement of animals and humans, in large part a response to earlier wildlife-cattle disease transmission. An outbreak of contagious bovine pleuropneumonia (CBPP), also known as cattle lung disease, occurred in the region in 1996, leading to the eradication of all cattle in Ngamiland in northern Botswana. In lieu of cattle killed, the Government of Botswana offered residents the option of replacement cattle or a cash reimbursement.

The Okavango Delta is one of Botswana's leading economic sectors, with tourism serving as one of the country's top three industries along with the cattle industry and mining. Many different ethnicities occupy the region including the long-term resident Bayei and Basarwa, with more recent immigrants from outside the country including Hambukushu, Herero and many Tswana-descended tribes. The study region for this project is predominately occupied by Hambukushu people, who were refugees from Angola during the country's civil conflict in the early 1970s, along with a small Bayei population, who moved to the villages after the Hambukushu settled there. The Delta is also a major source of resource extraction for local communities living on its periphery. The population of the Ngamiland West District, including the Etsha villages where this research was completed, is estimated at 64,864 [30]. There is a total of 175,631 people living in the entire North-West District, which encompasses the entire Delta and other areas of Northwestern Botswana. As detailed in the next section, it is well established that many residents of the Okavango Delta depend upon a diversified set of natural resource-based livelihoods [31–33], and that many of these livelihoods depend upon the precipitation or flooding patterns of the Delta [34].

This paper reports on research conducted in May 2010 and from May to June 2011 in the northwest of Botswana in the Okavango Delta (Figure 1). Semi-structured interviews were conducted along a twenty kilometer transect in the villages of Etsha 1, Etsha 6, and Etsha 13 (see Figure 1). A total of 82 semi-structured household interviews were conducted with 18 interviews in Etsha 1, 37 interviews in Etsha 6, and 27 interviews in Etsha 13. The number of interviews was conducted to reflect the different population sizes of the villages, and respondents were selected randomly using a spatially stratified sampling procedure that was designed to ensure geographic representation from the village. The increase in flooding caused differential impacts within the region, particularly within Etsha 13, where some village residents were displaced due to the floods. In order to understand these dynamics, seven of the Etsha 13 interviews were intentionally targeted. Semi-structured interviews of households were undertaken by the authors in English, Setswana, and Hambukushu with assistance from local interpreters. Interviews concentrated upon household demographics and history, use of natural resources and collection strategies, and perceptions of environmental variability within the area. Additionally, interviews probed individual and household perceptions of how variations in flooding and precipitation impact the collection and availability of natural resources. Text-based analysis of the interviews was conducted to identify themes related to environmental change and livelihood practices. These semi-structured interviews were complemented with three livelihood mapping exercises with select informants from each of the villages that provided spatially explicit detail on the location of livelihood practices and ways in which environmental change intersects with and reworks them. The mapping exercises included participant observation of resource collection and farming practices, the identification of key natural resource zones, and Global Positioning System (GPS) point collection for spatial analysis. This method was intended to build upon recent research intent upon spatializing livelihood systems to examine the ways in which livelihood systems and spatial processes are interlinked and co-produced [35].

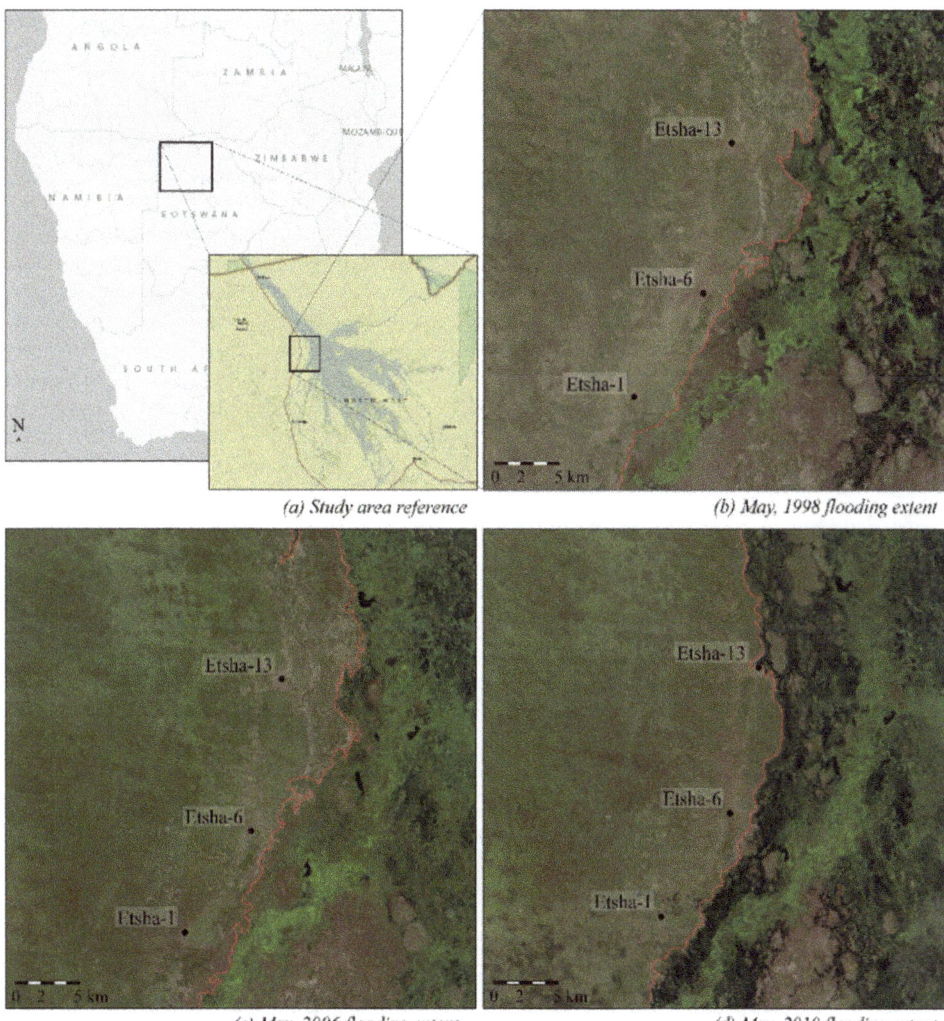

Figure 1. Map of the study region with the variance in the spatial location of flooding indicated by the red line along the edge of the floodwaters. (**a**) Study area reference; (**b**) May, 1998 flooding extent; (**c**) May, 2006 flooding extent; (**d**) May, 2010 flooding extent.

In order to understand the institutional context shaping livelihood systems, semi-structured interviews were conducted with tribal authority officials, including the village chiefs, and governmental representatives from the Department of Wildlife and National Parks and the Tawana Land Board. These interviews concentrated upon the rules governing resource access in the region and how they are enforced. Similarly, the household semi-structured interviews asked detailed questions about not only the spatiality of natural resource collection, but also the rules shaping access regimes and the institutions responsible for administering them. These interviews concentrated upon the four key elements of environmental governance as identified by Pahl-Wostl (2009), particularly the role of formal and informal institutions, interactions between state and non-state actors, administrative boundaries, and bureaucratic hierarchies. As detailed in the next section, livelihoods are shaped by dynamic ecological systems, thereby challenging institutions to be responsive in ways that support livelihood systems while reducing vulnerability to future change.

4. Livelihoods in the Okavango Delta

Given the dynamic environmental conditions of the Okavango Delta, livelihood systems are either tightly calibrated to these biophysical dynamics or forced to make adjustments in response to biophysical variability [36]. One study of five villages located throughout the Delta found that households participate in an average of three livelihood activities [33]. One agricultural practice in the region is *molapo* (floodplain) farming (Figure 2), which is a traditional form of agriculture dating back at least several hundred years and that takes advantage of the unique flooding dynamics of the Delta [37]. As the floodwaters recede at the end of the wet season (roughly around August-September, though this varies spatially and temporally), *molapo* farmers plant seeds in the relatively nutrient-rich, moist soil left behind by the waters [37]. *Molapo* soils are preferred by some residents to those in the dryland outside of the floodplain because they are seen as more fertile and they depend less on variable precipitation patterns because of their proximity to water. Maize, a staple grain in the region, is grown in nearly all *molapo* fields, along with beans, melons, squash, and sorghum [37]. There is risk associated with *molapo* farming due to crop loss resulting from years of high levels of flooding, insecure land tenure, and damage from livestock and wild animals. When damage to *molapo* fields occurs, residents in some regions switch to dryland farming, while people in other areas prefer to try to continue to *molapo* farm [34].

Figure 2. *Molapo* field in Etsha 1. Credit: Brian King.

Livestock husbandry is practiced throughout the Okavango Delta. Motsholapheko *et al.* [33] found that livestock husbandry was the most common livelihood practice in four of the villages.

Another study in five different villages of the Okavango Delta (including Etsha 6, one of the study sites for this project) found that livestock husbandry was ranked as the third most common livelihood practice after arable farming and formal employment [34]. In one study, some households complained that increased flooding led to less pasture for their cattle when grazing areas were submerged in water for long periods of time [33]. Another study found that 33% of residents in 22 villages surveyed use the Delta to water their livestock [34]. Thus, livestock husbandry is widely practiced in the region and has direct relationships with variable precipitation (and associated variability in groundwater depth and quality) and flooding patterns.

The collection of non-timber plant products is also integral to livelihoods in the Okavango Delta. These resources include reeds, thatch grass, palm, water lilies, and wild fruits [38]. Wild fruits are collected in both wetland and dryland areas; however, there is little information on their importance to household food security, or on the status of their availability. Water lilies are also collected as a food source, and it has been recently reported that they are not readily available due to both an increase in demand and shifting flooding dynamics [31]. Palms and grasses are used for fibers for basket weaving, which is a source of income for women in the region. Leaves and roots of various other species are used to create dyes for the baskets. Reeds and thatch grasses are also used in the construction of homes, roofs, and fences (Figure 3).

Figure 3. Transporting reeds by *mokoro*, a traditional dugout canoe, in Etsha 13. Credit: Brian King.

Lastly, fishing has been reported as a livelihood practice in the study region. One study found that 65% of the population of Northern Ngamiland depends on fishing as one source of food or for income generated from sale ([39], cited in [31]). However, a competing study found that fishing is less essential to local livelihoods [32]. Taken together, the collection and use of natural resources serves as a critical component to livelihood production within the Okavango Delta, including in the study region. What is particularly important to emphasize is that because this environmental setting is dynamic, the implications of changing precipitation and flooding for social systems can be significant for residents. This has proven to be the case in recent years with flooding in the study region reaching much higher levels than in the previous decade. Beginning in 2008, there was an expansion of floodwaters in the Delta that was driven by a number of factors. This continued in the years following, including the two years during which this research was conducted. Because much of the existing scholarship cited above was conducted prior to the onset of increased flooding in 2008, the research reported here was designed to evaluate how shifting aquatic ecologies are shaping perceptions of environmental change and resulting livelihood decision-making within the Okavango Delta.

5. "We Do Not Understand How and When It Rains": Perceptions of Environmental Change in the Etshas

Many of the households in the three villages engage in at least one form of subsistence agriculture, either in the form of *molapo* or dryland farming. Given this, people's lives are intimately tied to precipitation patterns, largely in the case of dryland farming, and to the floods, primarily in the case of *molapo* farming. Respondents in all three villages reported that precipitation has become increasingly variable over time. People did not always describe changes in rainfall in the same way, with some people reporting more rain in recent years, while some people reported less. However, the majority of respondents indicated that precipitation patterns were *less predictable* than in the past, with negative implications for their ability to grow crops. One respondent described these changes as a "new-fashion" kind of rain. Respondents often detailed that the rains now come in the "wrong" season when temperatures are high. Many interviewees described their dryland crops getting "scorched" in the sun, when the rain unexpectedly stopped for long periods of time. Some respondents also indicated that their crops rotted due to too much rain. For example, in Etsha 13, 30% of these respondents mentioned that too much rain caused their crops to rot, whereas 37% mentioned precipitation changes causing their crops to get "scorched." Some respondents even discussed crops in the same field being scorched and then rotting in the same season. This indicates that at times there was more rain than expected, and at times there was less than expected, or simply no rain at all.

There were high percentages of people in all three villages who perceive changes in precipitation (Table 1), with most of the comments concerning the unpredictability of rainfall. One resident of Etsha 6 described how the unpredictability in precipitation impacted his household's ability to dryland farm in 2011, describing how first it rained, then it stopped and the crops wilted. After that, the family plowed a second time to replace the lost crops, but they had to settle upon a smaller area. Of particular note, 22% of respondents in Etsha 13 indicated that changes in precipitation were leading to food insecurity for their households, compared to 0% of respondents in Etsha 1 and 11% of respondents in Etsha 6. While precipitation data from this region is not available for comparison with local perceptions, precipitation gauges at the Maun airport, which is located roughly 200 kilometers southeast of the study region, show variability over the time periods of study [40]. These data should be interpreted with some caution because precipitation can be extremely localized; however, they support the patterns articulated by our respondents.

Table 1. Perceptions of changes in precipitation.

	Etsha 1 (n = 18)	Etsha 6 (n = 37)	Etsha 13 (n = 27)
% of households reporting changes in precipitation	88%	97%	81%
% of households reporting unpredictable precipitation	72%	81%	81%

While respondents in the three villages had somewhat similar perceptions of changes in precipitation, there was greater variation in views of flooding dynamics (Table 2). This can be interpreted as a function of each of the villages along the transect having a distinct spatial relationship with flooding, given their disparate proximity to the floodwaters. Etsha 1 is situated furthest from the floodwaters, with Etsha 6 slightly closer, and Etsha 13 located on the edge of the water during high flood seasons. Additionally, residents in Etsha 13 have built homes in the secondary floodplain thereby increasing vulnerability to flooding. This is not to say that residents of Etsha 1 have no relationship to changing flood patterns. As one resident said, "It is different than in the past. Water was not there at all and now it is everywhere." There was a general sense that the spatial distribution of water is changing near Etsha 1, but less discussion of direct impacts on the village, as compared to interviews conducted in the other two villages. Residents also talked about the flooding as something happening elsewhere, such as further away from their homes, than did village members in Etsha 6 or Etsha 13. Therefore, a key difference is that households in Etsha 1 were less likely to be directly impacted by flooding than households in Etsha 6 or Etsha 13. Distinct from Etsha 1, residents of Etsha 6 and Etsha 13 articulated a clear knowledge of spatial history of flooding near the village. Over half of the people interviewed in Etsha 6 mentioned that in the past, floods came directly to the village; 49% of those interviewed in Etsha 6 mentioned that water had risen closer to the village once again.

Table 2. Perceptions of changes in flooding and related livelihood impacts.

	Etsha 1 (n = 18)	Etsha 6 (n = 37)	Etsha 13 (n = 27)
% of households reporting changes in flooding	83%	86%	88%
% of households reporting loss of *molapo* fields (complete or partial)	50%	35%	59%
% of households reporting increased difficulty in reed and grass collection	44%	45%	66%

Residents perceive these changing flooding patterns as impacting livelihoods in the study area (Table 2). A high percentage of respondents in all three villages mentioned complete or partial loss of a *molapo* field, and commented about how reed and grass collection was getting more difficult due to higher waters. Some of these residents indicated that reed and grass collection would be made easier in the time of higher water if they had access to a *mokoro* (traditional canoe) to allow for harvesting. Other people indicated that they were able to collect reeds and grasses, but the season is later than in the past, because they have to wait for the water to go back down. One interviewee explained that if he waits for the water to go down before reeds are collected, it can be time to go back to the fields and he prefers to farm in the drylands rather than collect reeds because farming is their primary source of food. As the next section details, the seasonality of permitted collection of certain resources is one of the ways in which livelihood patterns are proving inflexible to biophysical change.

A number of findings on perceptions of environmental variability emerge from the semi-structured interviews in the study region. First, changes in precipitation and flooding are perceived as increasing the difficulty for subsistence agriculture in the three villages. Variability in precipitation is challenging dryland farming, and for households that depend on both dryland and *molapo* agriculture, there is a double impact from changes in flooding and precipitation variability. One respondent indicated that there is not as much food as in the past, because then they had plenty of rainfall and the water was closer and higher. A number of people expressed the view that floods will remain too high to practice *molapo* farming in the future, indicating a perception of permanent loss of this livelihood practice. However, many households mentioned that they would continue to farm, particularly in the dryland area, even if it yielded fewer crops in the future.

Second, results from this study indicate that residents experience increased difficulty in accessing reeds and grasses due to higher floodwaters near all three villages. This represents a livelihood challenge because of the importance of these materials for household construction and in the basket making industry, which is a lucrative source of income for some women in the region. Many people discussed higher waters leading to hazards during collection, either in the form of dangerous animals like crocodiles and hippopotamuses, or from drowning. As a result, a number of people mentioned the increased need for *mekoros* (plural of *mokoro*) to make collection easier. However, few families own *mekoros*, and they are difficult to build and expensive to purchase. Also of note, some interviewees indicated that one benefit of the higher floods is that once the water goes down and they can be collected, reeds and grasses are believed to be stronger than in the past.

Finally, this research speaks to important spatial relationships between community location and perceptions of environmental change. This is especially the case in Etsha 13, where many residents have been displaced from their homes and relocated to a dryland area, known as the New Stands, located a couple of kilometers from the village center [41]. In 2009, 2010, and 2011, floodwaters reached all the way to the village for the first time in recent decades. As a result, not only did many people lose *molapo* fields, but also some residents were flooded out of their homes entirely. While some people emphasized that life is better in the New Stands because they are out of the reach of future floods, others raised a number of concerns. Included was a lack of ability to build new homes (many residents were still living in government issued tents at the time of the interviews), lack of access to electricity, and further distances from standpipes, which are sources of potable water. Some people displaced by the floods previously possessed private standpipes and electricity connections but could not afford to install them in their new homes. Finally, some people simply preferred to live closer to the water and did not want to live in the dryland area. Therefore, while the relocation afforded some safety from floodwaters, it also posed a number of new difficulties for residents.

6. Rigid Livelihoods within Dynamic Environmental Systems

Patterns of biophysical change within the study region necessitate closer scrutiny to the decision-making processes that enable and constrain resource access and livelihood production for residents. There is a diversity of actors and institutions that contribute in shaping livelihood systems within the Etsha villages. At the national level, the Department of Wildlife and National Parks in the Ministry of Environment, Wildlife and Tourism sets rules for hunting of wildlife in the Delta and gives permits for fishing that specify seasons and the size and type of the nets that can be used. The Tawana Land Board is responsible for allocating plots of land for settlement and has been particularly active within Etsha 13 due to the displacement of residents from the floods and subsequent relocation efforts to the New Stands. Almost all land in the vicinity of the Etsha villages is considered communally owned tribal land, which is managed and allocated to citizens by the Tawana Land Board. Regional Land Boards in Botswana are part of the federal government system, as outlined in the Tribal Land Act of 1968. The Tawana Land Board provides 99-year leases on tribal lands to Botswana citizens for residential, arable, and borehole purposes. At the regional level in northern Botswana, there are officers working from the towns of Maun or Gumare who provide extension support to farmers or provide other types of development interventions. Within the villages themselves, traditional authorities, including local chiefs and headmen, adjudicate on local issues and, within the context of this study, provide guidelines on the location and timing of resource collection and *molapo* farming. The tribal authority office is based in Etsha 6, with the other villages having their own headmen that work within this institutional structure under the supervision of the chief. These governance institutions should not be interpreted as being fixed to particular locations; rather, they overlap and there is a flow of information and power among them. As such, residents must navigate multiple agencies in securing land or acquiring permission for specific forms of resource collection. For example, while the Land Board designates the dryland fields, residents noted that the chief provides the list of farmers and therefore is integral to the process of land allocation.

The institutional context is critical in either constraining or enabling livelihood flexibility to deal with variabilities in the biophysical system [22,23]. In the Etsha villages, local residents take their cues from a mix of traditional authorities, local governmental agencies, and regional or national agencies, as well as from cultural practices and biases, knowledge bases, and household and community histories. Different ethnic groups, whether Hambukushu or Bayei, evidence distinct livelihood practices that are tied to precipitation and flooding dynamics. As one informant from the village of Etsha 13 explained, "The Bayei prefer to plow at the river because our staple food is maize. When we arrived here the river was not flooded. It is getting smaller now because the flowing is changing, especially on the edges." The informal institutions that govern these landscapes can be localized to the particular *molapo* plot or floodplain island that supports seasonal fishing communities. Livelihood systems are therefore tightly calibrated to perceptions of environmental change on a monthly, seasonal, annual, and decadal scale. When asked about future preparations for precipitation, one resident from Etsha 13 stated: "We are not doing any preparation because the rain is a new rain. We do not understand how and when it rains. For next year, we might do the preparation and unfortunately the rain will come late."

Another example of how livelihood systems are shaped by environmental change is through the variability in flooding patterns, which is reflected in the seasonality of natural resource collection. Residents explained that permits are needed for the collection of plant fibers for basket making, in addition to wild plants that are used for medicinal purposes. One fisherperson indicated that a permit is given by the Department of Wildlife and National Parks and costs roughly 200 pula (US$35 at the time of fieldwork in 2011) for three years. During the interview, the man shared the logs he keeps of his daily catches that he turns in to the local office, explaining that this helps in getting future permits. He insisted that the fishing season takes place from March until December 31st and that there is no fishing in the intervening months. In addition to fishing permits, permits are needed to collect and sell reeds and grasses commercially but not for household use. There is a season for reed and grass collection that is advocated by local tribal authority officials working in coordination with the national government. Residents shared that they are informed by the chief and government to collect these materials in July and August after the "seeds have dropped." Other people talked about how the water needed to go down before reed collection could commence. One resident in Etsha 6 expressed displeasure with this, explaining that reeds are a natural product and people should be free to go where they want to collect them, noting that in the past she used to collect reeds and grasses whenever she wanted but now the national government has imposed rules limiting this access. Another resident in Etsha 6 emphasized the importance of having a *mokoro* to collect these materials. As she explained, the Bayei collect reeds and grasses in August, but with *mekoros* you can collect earlier, even in June. If you do not have a *mokoro* you need to wait for the water to go down, and then there might not be any reeds to collect by the time you arrive. This same informant explained that reed collection is a woman's responsibility and that women were taught when to collect. There would be meetings at the Kgotla (the traditional political center of the village) and then they would negotiate as a group for resource access. She noted that now people go to the meetings but they do not listen, and they collect when they want to.

One interview conducted with an Etsha 1 resident provides an example of how environmental change shapes livelihood decision-making. He was one of the residents in Etsha 13 displaced by the increased flooding that began in 2008, although it is worth noting that his move to Etsha 1 in that year was voluntary, done proactively before the waters destroyed the usefulness of his *molapo* field. This respondent was able to acquire rights from the Land Board to farm that same year, and was able to harvest enough thatch and reeds to make his new house and compound fencing. Unfortunately, at that time the rising waters had begun to affect floodplain lands further to the south, meaning that he lost much of his harvest in the next two years. Acquiring land rights outside of the secondary floodplain is apparently a longer process, one he had still not been able to complete, but that he was pursuing in order to have a less productive but non-floodable area to practice dryland farming of millet, maize, and sorghum. This respondent is an example of how local residents can be resourceful, using multiple strategies for growing crops, acquiring the thatch and reeds needed for building a new house compound, and proactively reducing losses while preparing for new opportunities. At the same time, he continued to face institutional constraints to acquiring new land that challenged his capacity to adapt his livelihood system to environmental variability.

Interviews in Etsha 13 revealed particularly strong views on the rules constraining livelihood flexibility. These views were intensified by the displacement of residents to the New Stands in the dryland area. One interview was completed with a family that was being encouraged by the government to move to the dryland area, because it was believed their house would likely be in the path of future floods. The male respondent explained that it will be difficult because no one in the household is working and they do not have enough money for a new plot. Previously, they had the money from the 1996 cattle eradication, but presently they did not have money to build new structures. When asked about resource collection, one family member explained that it was not possible to collect grass or reeds during early June and finish in July because the water level will not go down by that time. He explained that "if you go collect during July, you will get punished." When asked if the government will change the season because of changes to the water, he said that if they do, it will take a very long time. When asked if he thought the rules should change, he noted that "the government should not set rules on how we collect the reeds and grasses, especially the time for collection ... Maybe your house can burn in January, you have to wait to June. It makes it difficult to build houses."

7. Conclusions

This paper details the myriad ways in which variability, specifically of precipitation and flooding dynamics, in the study region of the Okavango Delta influences environmental perceptions and livelihood responses. Changes in environmental dynamics are challenging for residents because they reveal how adaptive responses are constrained due to rigidities within broader governance regimes. Residents in all three of the villages viewed precipitation as unpredictable and novel in a way that decreased the viability of farming, particularly within the drylands in the region. Flooding patterns were similarly difficult to predict, which most likely reflected the spatial variability of the flooding. Even with these dynamic environmental patterns, residents continue to rely upon a diverse set of natural resources and livelihood practices that are interlinked with biophysical dynamics. Attending to the interconnections and tensions between environmental change and livelihoods is much needed to understand the capacity of social and biophysical systems to respond to change. Residents within the Etsha villages experience and perceive environmental variability in precipitation and flooding in distinct ways that are related to residence location, while also being shaped by natural resource needs. The Etsha villages provide an illustration of how rigid livelihood systems can limit the ability of social actors to adapt to environmental changes. One example of this is the seasonality of reed collection that, while explained by governmental officials is intended to ensure sustainable extraction, constrains the ability of community members to collect materials at critical periods. Thus, even though the shifting flooding dynamics challenge the collection of reeds and thatch grasses, the rules for collection, including seasonality, are not as fluid as biophysical processes. In a similar way, agricultural production is directly impacted by variability in the extent, location, and duration of floodwaters, which results in uneven impacts for residents within the study region.

While scholarship on environmental governance and resilience often emphasizes social dynamics that allow for adaptation to environmental change, some studies underemphasize how these patterns are simultaneously produced by livelihood production systems, environmental perceptions, and biophysical processes. At a time of reported increasing climate variability, scholarship on environmental governance must attend not only to the rules that shape socio-environmental relationships but also environmental perceptions and the ways in which livelihood systems are able to respond. If livelihood systems remain rigid in response to biophysical change they will be unable to successfully adapt and may generate tensions with existing governance regimes. This has material import for the ways in which residents are able to access natural resources critical for livelihood production, in addition to land for *molapo* and dryland farming. Future scholarship and policy on climate change adaptation must attend to the ways in which institutional options for adaptation can be simultaneously produced by the dynamic and fluid interactions between social and ecological systems.

Acknowledgments: The research that informs this article was supported by the United States National Science Foundation (BCS/GSS-0964596) and BCS/GSS Research Experiences for Undergraduates (REU) Supplement Award. Support from the Frederick S. Pardee Center for the Study of the Longer-Range Future at Boston University helped the first author attend an environmental governance workshop that clarified many of the ideas in this paper. We are grateful to Fuata John, Japhet John, and Kentse Madise for their outstanding work as research assistants. Allison White and Evan Griffin were extremely helpful in conducting the interviews in 2011. We also thank Thomas Christiansen for making Figure 1. Finally, we appreciate the two anonymous reviewers and special issue editors whose feedback helped strengthen an earlier version of this manuscript.

Author Contributions: All four authors contributed to the research design and collection of data in the Etsha region in 2011. Brian King undertook the preparation of the manuscript with invaluable support from Jamie E. Shinn. Kelley A. Crews and Kenneth R. Young contributed to the writing of the paper and assisted with revisions during the review process.

Conflicts of Interest: The authors declare no conflict of interest.

References

1. Solomon, S.; Qin, D.; Manning, M.; Chen, Z.; Marquis, M.; Averyt, K.B.; Tignor, M.; Miller, H.L., Jr. *Climate Change 2007: The Physical Science Basis. Contribution of Working Group I to the Fourth Assessment Report of the IPCC*; Cambridge University Press: Cambridge, UK; New York, NY, USA, 2007.
2. IPCC. Summary for Policymakers. In *Climate Change 2013: The Physical Science Basis. Contribution of Working Group I to the Fifth Assessment Report of the Intergovernmental Panel on Climate Change*; Stocker, T.F., Qin, D., Plattner, G.-K., Tignor, M., Allen, S.K., Boschung, J., Nauels, A., Xia, Y., Bex, V., Midgley, P.M., Eds.; Cambridge University Press: Cambridge, UK; New York, NY, USA, 2013; pp. 3–29.
3. Niang, I.; Ruppel, O.C.; Abdrabo, M.A.; Essel, A.; Lennard, C.; Padgham, J.; Urquhart, P. Africa. In *Climate Change 2014: Impacts, Adaptation, and Vulnerability. Part B: Regional Aspects. Contribution of Working Group II to the Fifth Assessment Report of the Intergovernmental Panel on Climate Change*; Barros, V.R., Field, C.B., Dokken, D.J., Mastrandrea, M.D., Mach, K.J., Bilir, T.E., Chatterjee, M., Ebi, K.L., Estrada, Y.O., Genova, R.C., *et al.*, Eds.; Cambridge University Press: Cambridge, UK; New York, NY, USA, 2014; Chapter 22; pp. 1199–1266.
4. Murray-Hudson, M.; Wolski, P.; Ringrose, S. Scenarios of the impact of local and upstream changes in climate and water use on hydro-ecology in the Okavango Delta, Botswana. *J. Hydrol.* **2006**, *331*, 73–84. [CrossRef]
5. Ostrom, E. Frameworks and theories of environmental change. *Glob. Environ. Change* **2008**, *18*, 249–252. [CrossRef]
6. Ostrom, E. A general framework for analyzing sustainability of social-ecological systems. *Science* **2009**, *325*, 419–422. [CrossRef] [PubMed]
7. Slovic, P.; Finucane, M.L.; Peters, E.; MacGregor, D.G. Risk as analysis and risk as feelings: Some thoughts about affect, reason, risk, and rationality. *Risk Anal.* **2004**, *24*, 311–322. [CrossRef] [PubMed]
8. Marx, S.M.; Weber, E.U.; Orlove, B.S.; Leiserowitz, A.; Krantz, D.H.; Roncoli, C.; Phillips, J. Communication and mental processes: Experiential and analytic processing of uncertain climate information. *Glob. Environ. Change* **2007**, *17*, 47–58. [CrossRef]
9. Roncoli, C.; Ingram, K.; Kirshen, P. Reading the rains: Local knowledge and rainfall forecasting in Burkina Faso. *Soc. Nat. Resour.* **2002**, *15*, 409–427. [CrossRef]
10. Valdivia, C.; Seth, A.; Gilles, J.L.; García, M.; Jiménez, E.; Cusicanqui, J.; Navia, F.; Yucra, E. Adapting to climate change in Andean ecosystems: Landscapes, capitals, and perceptions shaping rural livelihood strategies and linking knowledge systems. *Ann. Assoc. Am. Geogr.* **2010**, *100*, 818–834. [CrossRef]
11. Ovuka, M.; Lindqvist, S. Rainfall variability in Murang'a District, Kenya: Meteorological data and farmers' perception. *Geogr. Ann.: Ser. A Physic. Geogr.* **2000**, *82*, 107–119. [CrossRef]
12. Gamble, D.W.; Campbell, D.; Allen, T.L.; Barker, D.; Curtis, S.; McGregor, D.; Popke, J. Climate change, drought, and Jamaican agriculture: Local knowledge and the climate record. *Ann. Assoc. Am. Geogr.* **2010**, *100*, 880–893. [CrossRef]
13. Dahlberg, A.C. Vegetation diversity and change in relation to land use, soil and rainfall—A case study from North-East District, Botswana. *J. Arid Environ.* **2000**, *44*, 19–40. [CrossRef]
14. Meze-Hausken, E. Contrasting climate variability and meteorological drought with perceived drought and climate change in northern Ethiopia. *Clim. Res.* **2004**, *27*, 19–31. [CrossRef]
15. Vedwan, N. Culture, climate and the environment: Local knowledge and perception of climate change among apple growers in Northwestern India. *J. Ecol. Anthropol.* **2006**, *10*, 4–18. [CrossRef]
16. Turner, M.D. Political ecology I: An alliance with resilience? *Prog. Hum. Geogr.* **2014**, *4*, 616–623. [CrossRef]
17. Ostrom, E. *Governing the Commons: The Evolution of Institutions for Collective Action*; Cambridge University Press: New York, NY, USA, 1990.
18. Folke, C.; Hahn, T.; Olsson, P.; Norberg, J. Adaptive governance of social-ecological systems. *Annu. Rev. Environ. Resour.* **2005**, *30*, 441–473. [CrossRef]
19. Lemos, M.C.; Agrawal, A. Environmental governance. *Ann. Rev. Environ. Resour.* **2006**, *31*, 297–325. [CrossRef]
20. Young, O.R. Effectiveness of international environmental regimes: Existing knowledge, cutting-edge themes, and research strategies. *Proc. Natl. Acad. Sci. USA* **2011**, *108*, 19853–19860. [CrossRef] [PubMed]
21. Nelson, D.R.; Adger, W.N.; Brown, K. Adaptation to environmental change: Contributions of a resiliency framework. *Ann. Rev. Environ. Resour.* **2007**, *32*, 395–419. [CrossRef]

22. Moss, R.H.; Meehl, G.A.; Lemos, M.C.; Smith, J.B.; Arnold, J.R.; Arnott, J.C.; Behar, D.; Brasseur, G.P.; Broomell, S.B.; Busalacchi, A.J.; *et al.* Hell and high water: Practice-relevant adaptation science. *Science* **2013**, *342*, 696–698. [CrossRef] [PubMed]
23. Pahl-Wostl, C. A conceptual framework for analyzing adaptive capacity and multi-level learning processes in resource governance regimes. *Glob. Environ. Change* **2009**, *19*, 354–365. [CrossRef]
24. Heinl, M.; Neuenschwander, A.; Sliva, J.; Vanderpost, C. Interactions between fire and flooding in a southern African floodplain system (Okavango Delta, Botswana). *Landsc. Ecol.* **2006**, *21*, 699–709. [CrossRef]
25. Gumbricht, T.; McCarthy, T.S.; Merry, C.L. The topography of the Okavango Delta, Botswana, and its tectonic and sedimentological implications. *S. Afr. J. Geol.* **2001**, *104*, 243–264. [CrossRef]
26. Neuenschwander, A.L.; Crews, K.A. Disturbance, management, and landscape dynamics: Wavelet analysis of vegetation indices in the lower Okavango Delta. *Photogramm. Eng. Remote Sens.* **2008**, *74*, 753–764. [CrossRef]
27. Ellery, W.N.; Dahlberg, A.C.; Strydom, R.; Neal, M.J.; Jackson, J. Diversion of water flow from a floodplain wetland stream: an analysis of geomorphological setting and hydrological and ecological consequences. *J. Environ. Manag.* **2003**, *68*, 51–71. [CrossRef]
28. McCarthy, T.S.; Cooper, G.R.J.; Tyson, P.D.; Ellery, W.N. Seasonal flooding in the Okavango Delta, Botswana—Recent history and future prospects. *S. Afr. J. Sci.* **2000**, *96*, 25–33.
29. Wolski, P.; Todd, M.C.; Murray-Hudson, M.A.; Tadross, M. Multidecadal variability in hydro-climate of Okavango river system, southwest Africa, in the past and under changing climate. *J. Hydrol.* **2012**, *475*, 294–305. [CrossRef]
30. CSO (Botswana Central Statistics Office). *Botswana Population and Housing Census*; Central Statistics Office: Gabarone, Botswana, 2011.
31. Kgathi, D.L.; Mmopelwa, G.; Mosepele, K. Natural resources assessment in the Okavango Delta, Botswana: Case studies of some key resources. *Nat. Resour. Forum* **2005**, *29*, 70–81. [CrossRef]
32. Mbaiwa, J.E.; Darkoh, M.B.K. Sustainable development and natural resource competition and conflicts in the Okavango Delta. *Botsw. Notes Rec.* **2005**, *37*, 40–60.
33. Motsholapheko, M.R.; Kgathi, D.L.; Vanderpost, C. Rural livelihoods and household adaptation to extreme flooding in the Okavango Delta, Botswana. *Phys. Chem. Earth Parts A/B/C* **2011**, *36*, 984–995. [CrossRef]
34. Kgathi, D.L.; Ngwenya, B.N.; Wilk, J. Shocks and rural livelihoods in the Okavango Delta, Botswana. *Dev. S. Afr.* **2007**, *24*, 289–308. [CrossRef]
35. King, B. Spatialising livelihoods: Resource access and livelihood spaces in South Africa. *Trans. Inst. Br. Geogr.* **2011**, *36*, 297–313. [CrossRef]
36. Meyer, T.; Bendsen, H. The dynamics of land use systems in Ngamiland: Changing livelihood options and strategies. In *Environmental Monitoring of Tropical and Subtropical Wetlands*; Bernard, T., Mosepele, K., Ramberg, L., Eds.; HOORC Report Series No. 1; HOORC: Maun, Botswana, 2003; pp. 278–307.
37. Magole, L.; Thapelo, K. The impact of extreme flooding of the Okavango River on the livelihood of the *molapo* farming community of Tubu village, Ngamiland Sub-district, Botswana. *Botsw. Notes Rec.* **2005**, *37*, 125–137.
38. Mbaiwa, J.E.; Ngwenya, B.N.; Kgathi, D.L. Contending with unequal and privileged access to natural resources and land in the Okavango Delta, Botswana. *Singap. J. Trop. Geogr.* **2008**, *29*, 155–172. [CrossRef]
39. Mosepele, K. *Preliminary Description of the Okavango Delta fishery*; Unpublished technical report. Fisheries Section, Ministry of Agriculture: Gaborone, Botswana, 2001.
40. Wolski, P. Okavango Delta Monitoring and Forecasting. 2014. Available online: http://okavangodata.ub.bw/ori/monitoring/rainfall/# (accessed on 3 September 2014).
41. Shinn, J.E.; King, B.; Young, K.R.; Crews, K.A. Variable adaptations: Micro-politics of environmental displacement in the Okavango Delta, Botswana. *Geoforum* **2014**, *57*, 21–29. [CrossRef]

land

MDPI

Article

Changing Livelihoods and Landscapes in the Rural Eastern Cape, South Africa: Past Influences and Future Trajectories

Sheona Shackleton [1,*] and Marty Luckert [2]

1 Department of Environmental Science, Rhodes University, Grahamstown 6140, South Africa
2 Department of Resource Economics and Environmental Sociology, University of Alberta, 515 General Services Building, Edmonton, AB T6G 2H1, Canada; marty.luckert@ualberta.ca
* Correspondence: s.shackleton@ru.ac.za; Tel.: +27-46-603-7009

Academic Editors: Claudia A. Radel and Jacqueline M. Vadjunec
Received: 30 June 2015; Accepted: 5 November 2015; Published: 19 November 2015

Abstract: This paper seeks to understand the drivers and pathways of local livelihood change and the prospects for transformation towards a more sustainable future. Data are used from several studies, and a participatory social learning process, which formed part of a larger project in two sites in the Eastern Cape, South Africa. Secondary information from a wealth of related work is used to place our results within the historic context and more general trends in the country. Findings indicate that livelihoods in the rural Eastern Cape are on new trajectories. Agricultural production has declined markedly, at a time when the need for diversification of livelihoods and food security seems to be at a premium. This decline is driven by a suite of drivers that interact with, and are influenced by, other changes and stresses affecting local livelihoods. We distil out the factors, ranging from historical processes to national policies and local dynamics, that hamper peoples' motivation and ability to respond to locally identified vulnerabilities and, which, when taken together, could drive households into a trap. We end by considering the transformations required to help local people evade traps and progress towards a more promising future in a context of increasing uncertainty.

Keywords: livelihoods; farming; traps; transformation; South Africa; landscapes; poverty; change; multiple stressors; Eastern Cape

1. Introduction

Many parts of the world are facing accelerated change, rising insecurity and vulnerability, and greater uncertainty brought about by a multitude of interacting drivers and shocks ranging from global economic downturns to growing corruption and weak governance, escalating poverty and food insecurity, failing health systems under pressure from epidemics such as HIV/AIDS, and extreme weather events [1,2]. At the same time, the longevity of the goods and services provided by the planet's ecosystems is questioned [3]. These global risks and changes interact with, and may be compounded by, country-level structural, economic and political processes that create and sustain inequities. The effects of these global-and national-level interactions are also felt in localized contextual dynamics, manifesting in institutional breakdown, crime, landscape modification through changing land uses, declining agricultural production, land degradation and biodiversity loss [4].

A case in point is the agrarian change in the Eastern Cape, South Africa that has been taking place over several decades. The present day landscape has been largely shaped by discriminatory apartheid policies that placed most of the land in the hands of white commercial farmers and systematically undermined peasant production amongst black farmers in the Bantustans [5]. However, despite this history, and up until relatively recently, farming continued to provide an important source of livelihood and food security for black rural families. Many people in the rural communal areas of the country identified themselves as "farmers", undertaking both arable and livestock farming. However, more recent work, especially in the Eastern Cape, has suggested that farming, particularly extensive arable production is declining and either shifting to more intensive cultivation of home gardens or being abandoned altogether [5–14]. The trends also suggest a reduction in livestock ownership and numbers, although this change is not as striking [13–15]. Such adjustments in agricultural production are influenced by, and interact with, many other trends at a variety of scales. For instance, under global climate change, the region is projected to become hotter and drier, facing more droughts and, ironically, more severe and frequent floods [16]. These climatic changes are superimposed on a suite of other stresses impacting local livelihoods and human well-being [17,18]. Short-term shocks, such as death and illness in the household, job loss, hunger, violence and crime, often arise as a consequence of longer-term "background" shifts, which, we argue, are fundamentally altering local social–ecological systems. Examples of these "slow variable" background changes include the systemic effects of HIV/AIDS on household structure and assets [19–21], urbanization and changes in demographics, household structure and relations, and values.

For many of the progressive changes observed (such as agricultural decline), no clear trajectories into the future are evident, with short-term stressors working in synchrony with longer-term change to create feedback loops that are potentially leading to increasing vulnerability and food insecurity [22,23]. We, therefore, argue that the legacy of deprivation and neglect in South Africa's rural communal areas, combined with more recent changes that decrease flexibility, diversification and local agency, as well as exposure to increased risk, may lock some households and communities into situations that prevent innovation and response. Such situations are often viewed as being "stuck" or in a trap [24]. It is in these contexts that the possibility of transformation becomes crucial. New policies and actions will need to be put in place to protect and support vulnerable people, and innovative practices or different "ways of doing" need to be developed and implemented. This approach requires recognition and understanding of the changes taking place, the drivers of these changes, and the various factors that may hinder people's ability to accommodate or respond to change.

In this paper we consider changes in ways of life in two marginalized rural areas of the Eastern Cape and investigate how interactions amongst these changes and their drivers emerge and impact on livelihoods and current and future vulnerability. To assemble a holistic and composite picture of the complexity, dynamics and heterogeneity of local livelihoods, and to tease out the multiple factors across scales that drive the observed trends, we synthesize findings from several specific sub-studies conducted in the same sites, and then situate these findings within the relevant literature. Our aim is to investigate changes in livelihoods and landscapes in the Eastern Cape and consider whether some households, through combinations of shocks and long-term stressors, may be trapped in situations that prevent them from responding to vulnerabilities and, thus, to securing resilient livelihoods. We specifically seek to understand how historical processes, national policies and contemporary local dynamics have affected current livelihood portfolios and the capacity to adapt to change, particularly in relation to future climate change. We argue that with an improved understanding of these processes, transformative policies and strategies may be better developed to avoid traps and facilitate more promising futures.

In the next section, we describe our research approach and its theoretical underpinnings. We then present key findings, which identify large changes in the nature and structure of livelihoods and landscapes that have occurred in our study areas, and suggest reasons behind these changes drawing on both our own research and other studies and sources. Next, we consider a number of features of the broader political and socio-economic contexts that have also contributed to, and resulted from, these livelihood changes. We conclude with policy considerations regarding transformations and potential future trajectories that could assist rural Eastern Cape households evade traps and make choices regarding their livelihood strategies, including engagement in arable production.

2. Study Sites

We selected two sites for the project: Gatyana in Mbashe Local Municipality and Lesseyton in Lukanji Local Municipality in the Eastern Cape Province, South Africa (Figure 1). The study sites were chosen to correspond to a gradient of increasing rainfall and decreasing accessibility to urban amenities and markets. Each site consists of multiple villages, with the boundaries of these being less clear-cut in the Gatyana site due to the scattered nature of homesteads. The inland site, Lesseyton, is 20 km from the city of Queenstown. It falls at the intersection of the Grassland and Thicket biomes [25] and has a mean annual precipitation of between 350 and 500 mm. Homesteads are built in a typical village layout and generally have only a small garden space for vegetable cultivation. However, there is access to communal land, which is used primarily for grazing by livestock owners. The coastal site, Gatyana, is more isolated and is about 30 km from the small town of Willowvale. It still lags behind in infrastructure such as tarred roads, and has marginal local markets and poor transport systems. Most homesteads are widely dispersed with large plots of land (half a hectare or more). The site crosses two major biomes, the Indian Ocean Coastal Belt Forests and Savanna, and receives a mean annual rainfall of between 950 and 1100 mm [25].

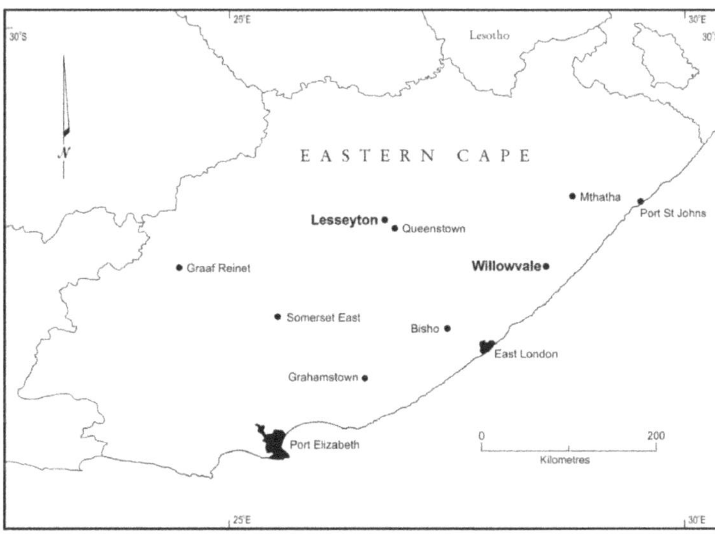

Figure 1. Map of the Eastern Cape, South Africa showing study sites.

3. Approach and Methods

To construct an integrated and cohesive understanding of the complex livelihood and landscape changes in our study sites, and to analyze the drivers behind these changes, we draw on and synthesize findings from a mix of separate sub-studies (five in total from Honors undergraduate to PhD level) conducted as part of a larger project funded by the International Development Research Centre (IDRC) [26–31]. Using livelihoods and assets-based approaches (guided by the sustainable livelihoods framework and the concepts of vulnerability and multiple stressors), this project sought to explore how climate change, together with several other livelihood stressors including HIV/AIDS, affect vulnerability, food security, livelihood strategies and choices, adaptive capacity and coping, adaptive and transformational responses amongst different types of households.

The data presented in this paper are mainly derived from unpublished Masters theses and draw heavily on the baseline household survey used by several students. This survey was administered in 2011 and 2012 to 170 randomly-selected households at each site (total n = 340). Questions related to livelihood and income sources were repeated an additional two times at six monthly intervals to capture seasonal differences. The data presented here are averaged across the three rounds. Data collected included information on household demographics, health and HIV/AIDS proxy indicators, assets, welfare perceptions, shocks and responses, livelihood activities and household income in cash and kind [30]. We interviewed heads of households or the next most senior adult. Men often left the interviews to the women in the household. For some analyses, we disaggregated the data by gender structure (based on the sex of household head and adult members) and wealth categories (see [32]). These data provided a holistic picture of household income shares, livelihood activities and community and household assets.

In addition to this quantitative data, we use qualitative data from a study by Clarke [26] on responses to climate change and HIV/AIDS amongst vulnerable rural households, in which she undertook 20 life histories. Five main "themes" were used to guide the interviews: (1) general household information such as who lived there, and what household members did for a living; (2) the village's history, and any changes that occurred over the period of the household's residence; (3) the respondent's personal life story from growing up with their parents to the present, focusing on the hardships they experienced and the specifics of how they and their family coped with these hardships; (4) the respondent's main concerns for the future; and (5) the types of support the respondent felt were needed in their community. Each interview was primarily conducted and recorded in *isiXhosa* and later transcribed and translated into English through a translator. The findings were systematically analyzed using a software package for textual analysis called MAXQDA. A coding system was created using the following categories: vulnerability (changes, shocks and stressors experienced), responses (responses made towards shocks and stressors), assistance (help received), health (shocks and stressors related to health), climate, farming, and natural-resource use (climate-related changes, shocks and stressors experienced), and future concerns (future shocks and stressors interviewees were concerned about, as well as what support they felt was needed). The text retrieval option allowed specific categories and study sites to be analyzed separately. These broader categories gave rise to more specific themes such as unemployment, social-grant dependency, and crime-related shocks and stressors. Here, we extract representative quotes from the life histories as supporting evidence for the changes and vulnerabilities emerging at the local level. These provide insight into local, lived experiences of vulnerability and change.

Participatory mental maps were constructed at both sites to record the stressors believed by local men and women to be creating vulnerability, and their perceptions of the linkages between these [26]. We held four small focus groups (4–8 participants each) with men and women separately, with each group creating their own mental map. Mental maps are "qualitative representations of a system consisting of variables and the causal relationships between them" [33]. For our study, these took the form of a spider-gram drawn on large sheets of paper, with key stressors being linked via directional lines. HIV/AIDS was suggested to participants as an initial item, to which they added the various cause and effects of this disease. Respondents then added further stressors and drivers of change and began connecting these with what was already on the map, until they felt they had exhausted all the key factors contributing to their vulnerability. All links and decisions were discussed within the focus groups. These maps provide a rich interpretation of the local drivers of vulnerability and their impacts on people's lives. In addition, to obtain the perspectives of the youth on local vulnerabilities and concerns, participatory workshops were held with mixed groups of grade 11 learners in each site (20 learners in Lesseyton and 53 in Gatyana). These scholars discussed and tabulated their viewpoints related to the past, present and future of their area.

For data on agrarian change, we draw on the published results of Shackleton *et al.* [14], which investigated farmers' opinions of landscape change, ecological succession in abandoned fields, and changes in land-cover categories. In addition, we report on unpublished research undertaken in 2013 and 2014 by Shackleton and colleagues on arable production in the Eastern Cape. To assess broad trends in landscape use and land-cover change, 1961 aerial photographs were compared to

2009 photographs by means of randomly selected sample plots (50 m × 50 m). Eight hundred forty-four plots were sampled on the 1961 photographs and 949 on the 2009 photographs, representing approximately 5 percent of the study area. Each plot was categorized by its dominant land use or land cover, *i.e.*, abandoned garden, abandoned field, current field, current garden, forest, woodland, and grassland. In May and June 2012, a household questionnaire was administered to 31 current and 50 former field cultivators to solicit views on changes in land-use practices, including causes and what happens to abandoned fields. A snowball sampling approach was adopted to identify households in each category. To obtain an adequate sample, 38 villages within the wider Gatyana area were surveyed. For the more recent research, we conducted follow-up, in-depth interviews with the same cultivators in October 2013. Some had already ceased cultivating their fields. Furthermore, we undertook a separate survey of 90 randomly-selected households in two villages in the study site regarding cultivation of home gardens. Key data collected related to cultivation and garden details, the reasons for gardening, what garden inputs were used (for example manure, fertilizers, compost and labor), methods of plowing, seasonal determinants, changes/trends in gardening, challenges involved, estimations on yields and sale of crops.

These research results are complemented by the outputs of a parallel, participatory social learning and knowledge exchange process at both study sites, which provided in-depth insights on key research themes [34]. This process, which took place over three years, involved mixed groups of 10–15 people and commenced at the same time as the household survey. We aimed to build problem-solving capacity and strengthen local practices that had evolved in response to multiple stressors. In some of the foundational meetings, we asked participants to share stories of vulnerability and changes in their communities and lives based on their first-hand experiences. Participant conceptualizations of vulnerability and change covered shocks and stressors (drought, illness), "community problems" (truancy amongst the youth, crime, rape of women and girls) and structural aspects such as poverty, poor education and unemployment. The entire process is documented in a handbook [34].

Also important for our analysis are other complementary research findings from the Eastern Cape and various critical commentary papers on higher-level trends and processes in the country. We use this literature to support and contextualize our findings and to explore underlying drivers.

4. Conceptual and Analytical Lens

Our analytical approach to considering past influences and future trajectories is informed by the concepts of transformation and traps that have been developed in the social–ecological systems literature. In this literature, transformation is generally defined as "the fundamental alteration of the nature of a system once the current ecological, social or economic conditions become untenable or undesirable" [35] or a "profound, substantial and irreversible change" [36]. While transformation may be spontaneous or planned, the process often implies conversion to a better or more beneficial state [37,38]. Consequently, it can take on a normative and deliberative dimension. The normative nature of transformation can lead to ethical concerns related to what constitutes a more beneficial state, for whom, at what scale and who determines this. In the climate change literature, the IPCC uses much the same definition, though alterations in various sub-systems are emphasized, including value systems; regulatory, legislative or bureaucratic regimes; and financial institutions and technological or biological systems. Other climate change researchers prefer the term "transformational adaptation", where the distinction between adaptation and transformation is less clear cut, with transformation being related to the scale, intensity and novelty of adaptations made [39–41]. Transformation is said to occur within social–ecological systems when principal members of these systems change core functions through collective action and institutional change (coordinated and uncoordinated action) and from the accumulation of numerous smaller, often behavioral, changes made by individuals. Transformation can also occur through changes in policies and governance, institutions, livelihoods, landscapes, farming systems, locations or identity [36,40]. From this perspective, transformation is, thus, both similar to, and different from, adaptation. In contrast to adaptation, which refers to small, incremental responses to risk and change that keep a system in the same state, transformation is conceptualized as being required when human-environmental systems or components of these are over-whelmed or trapped and a complete shift is required [39]. However, transformation is rarely a discrete event—it often emerges messily over space and time through a series of smaller transformations [36]. We argue in reality the distinctions between these concepts become blurred when faced with the variety and complexity of conditions influencing livelihoods and their development, particularly given the lack of clarity and the uncertainty regarding future development directions and pathways and the normative nature of transformation.

Transformation may be hindered by traps, but, at the same time, transformation is generally required to escape traps. Traps are seen as negative states and are often the outcome of multiple and accumulative barriers to action that prevent or block the ability of people to transform the way they live. Several types of traps are recognized in the socio-economic ecological resilience literature: poverty traps, rigidity traps, and locked-in traps [24]. These traps convey different meanings for different disciplines, but simply put, they imply a situation of "being stuck", often in a dysfunctional or degraded state, and being unable to respond to shocks and changes. In the resilience community, a poverty trap embodies the loss of options to develop or deal with change [24], and represents a condition that inhibits positive responses to a stimulus. Such a trap requires external support to escape [24]. Similarly, rigidity traps can equally prevent a shift towards upwardly mobile trajectories; they represent a system in which the members of organizations and their institutions become highly connected, rigid and inflexible and resist change [24]. This rigidity constrains the ability to respond to new problems and opportunities [24]. Locked-in traps are similar [42]. Walker and Salt [43] argue that to avoid being trapped on an unacceptable trajectory that is difficult to escape, it may be necessary to facilitate almost continual transformational change. This depends on creating new opportunities that, in turn, are dependent on internal sources of resilience and adaptive capacity, and various forms of external support [44]. Various structural (legacy of Apartheid), cognitive (forms of denial), normative and cultural (conformity, traditions, place), and behavioral factors can create path dependence and keep people (or certain groups of people) and systems in these traps. Path dependence occurs when the legacy of past events, states, decisions and actions affect current dynamics and future changes though reinforcing feedback mechanisms and entrenchment of particular self-reproducing institutional arrangements [42,45]. This makes it difficult to divert the system to a new course [46]. In our case, the notion of path dependence helps us to understand the long-term consequences of separate development (under the Apartheid regime) in our study sites.

The starting point for our analysis (below) is to describe the changes in livelihoods and landscapes that have occurred in the rural Eastern Cape. However, surrounding this change, our research identified important past and current contexts that have created several types of interlinked shocks, stressors and other factors that could act as barriers to livelihood improvement, both now and in the future, for certain groups of people. All of these are closely associated and often mutually reinforcing, together creating a situation that resembles aspects of both poverty and locked-in traps. However, some of the changes also have positive dimensions that need to be considered and weighed up against any negative aspects, with social grants being a particular case in point. The primary changes are described below with supporting evidence from both our own research and the literature.

5. Results and Discussion

5.1. Changing Livelihoods and Landscapes

5.1.1. Shifting Livelihood Portfolios

Land-based income (from crops, livestock, and natural resources) makes up, on average, only 15–25 percent of livelihoods in the Lesseyton and Gatyana sites, respectively (see Figure 2). The cash income received from employment (formal, small businesses, casual labor and remittances) is of a similar magnitude, but reversed, with the more rural site receiving 16 percent of their livelihoods from employment activities, while households in Lesseyton receive 29 percent of their income from these employment sectors. The overwhelming source of income for both sites comes from social grants that make up over half of income.

The snapshot of income shares (see Figure 2) is the result of ongoing processes, detailed below, which have caused land-based and employment incomes to decrease, while reliance on social grants has grown (also see [7]). These changes represent a narrowing of livelihood activities/options and a reduction in the use of local landscape resources as a basis for livelihoods and as safety nets needed in response to shocks. In the next three sections, we consider trends within each of these sources of livelihood and the important contexts and underlying causes for such changes through analysis of the broader literature.

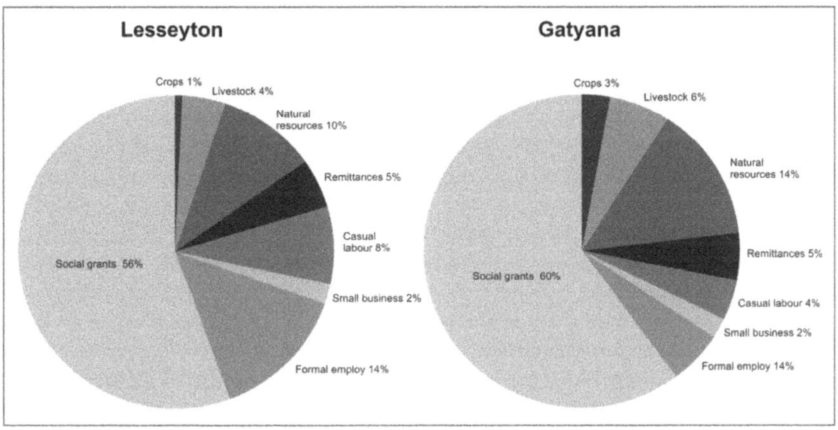

Figure 2. Average income shares from different livelihood sources across all households (n = 340) in Lesseyton and Gatyana (Source: Stadler 2012, unpublished thesis [30]).

5.1.2. Abandonment of Fields and a Decline in Livestock Production

Results from our study sites show that cultivation forms, on average, only between 1 percent and 3 percent of total household income in Lesseyton and Gatyana respectively (Figure 2). This is a substantial decline relative to the past (*i.e.*, from greater than 20 percent of income in 1950) [12], and is reflected in changes in the landscape, especially the increasing area of abandoned fields. This field abandonment and decline in arable farming for both cash and subsistence purposes has been happening over several decades and can be viewed as a social–ecological transformation in itself (Table 1) [12,14,28]. In the Gatyana study site, landscape change over a 50-year period involved mainly a decline in cultivated fields and an increase in woodlands, primarily due to the establishment of the thorny, pioneer species *Acacia karoo* in old lands (Table 1). The encroachment of woody vegetation into these areas acts as a barrier to their future cultivation.

Table 1. Landscape change in Gatyana over a 50-year period.

Aerial Photography Analysis of the Gatyana Landscape over the past 50 Years (1961–2009) has Shown:	
• Reduction in cultivated field cover from 12.5% to 2.7% • Increase in abandoned field cover from 1.5% to 6.9% • Reduction in grassland of 22.5%	• Increase in woodland from 13.9% to 28.8% • Increase in forests by 5% • Increased abandonment of home gardens, but less than abandonment of fields

Source: Shackleton *et al.* 2013 [14].

Many researchers have argued that this move away from agriculture started in the 1940s (or even earlier as discussed by Hebinck and van Averbeke [47]) with the concentration and relocation of people into the Bantustans, and a deliberative process of "depeasantization". This was expedited through the creation of "labor pools" to supply the mining industry and the marginalization of black farmers, while white farmers were heavily subsidized [8,47]. At the same time, the policy of so called "Betterment" that placed people in villages and zoned land for particular uses, served to separate and distance households from their fields and contribute further to their disuse [7,8,14,47,48]. However, some farming still continued, facilitated by income from jobs in the mines, which in those days could mainly be sent home because accommodation (in hostels) and transport were provided by the mining companies [47,49].

The greatest levels of field abandonment have occurred in recent years, with Shackleton *et al.* [14] finding that most of the past farmers interviewed in Gatyana had stopped farming on distant fields, on average, about 18.5 years ago. This timing coincided with the democratic transition and an increase in social welfare [5,50], as well as changes in the process of "formal" migration of mine workers, which has now ceased [50]. Other drivers and reasons given for the decline in agriculture include: lack of access to markets; exhausted soils; unpredictable weather; lack of labor; poverty and lack of access to credit; infrastructure decline, especially fencing and roads; destruction of crops by cattle and wild animals, related to both the former and to the loss of child labor for herding; a decline in farming knowledge; and institutional factors that prevented the emergence of a land rental market [5,6,14,47]. Regarding the latter, Kepe and Tessaro [5] argue that a lack of land rights in the rural communal areas of South Africa has limited the implementation of agricultural programs devised by the state because *de facto* owners simply do not trust anyone else to cultivate their land in the absence of secure tenure. Livestock production has also declined (attributed locally to disease) and become more asymmetrical, although to a lesser extent than arable farming [6]. Indeed, the lack of cattle for draught has been given as another reason for the decline in cropping [6,14,48,51].

Despite the substantial decline in field cultivation, households are still planting so called "home gardens" (fields attached to homesteads), especially in areas like Gatyana where residences are scattered in the landscape and plots can be up to one hectare in size (in Lesseyton this is not the case, and there is little home gardening). Andrew and Fox [6] contend that in the coastal areas of the former Transkei (known as the Wild Coast and of which Gatyana is part), there has, in fact, been a shift from field to homestead cultivation. Our research into the extent and use of gardens shows that most households have gardens, although they may not cultivate them annually. Women (63 percent) are mainly responsibility for managing these home gardens and for making cultivation decisions. Constraints to gardening mentioned included financial (mentioned by about 40 percent of respondents), biophysical (weather, pests, water, soil fertility, fencing and draught power, with at least one of these being mentioned by most respondents) and lack of labor and disinterest among the youth (46 percent). When asked about trends, the majority of households interviewed felt that garden cultivation was declining, with 46 percent saying the greatest decline had been in the last fifteen years. This was also reflected in the land use change data (Table 1).

Given the declines in extensive arable agriculture and to a lesser extent home gardening, an important question is whether this land use trajectory is narrowing options for the future and placing people in a state of greater vulnerability, including blocking positive transformation in relation to future food security. The current high reliance on purchased food and other goods was perceived by the majority of vulnerable people to be detrimental due to an increased need for scarce cash (Table 2) [26], and could become a more serious problem as food prices rise with the impacts of climate change [52]. Moreover, for households in isolated rural areas such as Gatyana, local supplies are costly and the transport to distant markets expensive. This further stretches meager cash resources. The waning in the perceived value of farming and homestead gardening also results in the potential loss of a set of other related benefits. These include communal ethics relating to farming, and other social and cultural benefits [6], reinforced ownership and entitlement rights [51], the supplementation towards incomes and food security, and provision of a safety net [12,53,54]. Gardening, for instance, has become a very individualistic endeavor, with not a single one of the households interviewed mentioning making use of work parties. Hebinck [5] and Andrew and Fox [6] propose that these trends have led to a disconnect between people and the land and the loss of a farming identity, with the prospects for the future suggesting that the youth are unlikely to be interested in farming [7,14]. Regarding the latter, Hebinck and Van Averbeke [47] describe a scenario of rural residents, especially the youth, being in a mental state of "waiting to leave" the moment an urban opportunity arises, and as drifting between urban and rural settings (also mentioned by Dubbeld [49]). Bank [7] also discusses how new opportunities for women (in state-supported off-farm commodity groups), as well as their increased migration to towns and cities, have severed them from their traditional farming and gardening role. These shifts, together with the woody invasion and reforestation of old lands and grasslands [14], constitute a strong disincentive for future agricultural production in the area.

Despite this downward trend in production, cultivation has not totally disappeared; there are still a few dedicated cultivators (Table 3), and about a third of gardening households in Gatyana mentioned that crop cultivation is an important part of *Xhosa* culture and identity and is likely to continue. Some 15 percent believe cultivation will enter a phase of expansion again in the future. A survey conducted for the Mbhashe Local Municipality IDP (2012–2017) [55] showed that out of 31 wards, only seven did not list an agricultural-related need, suggesting that residents of the region are still interested in continuing to farm. However, even if some of the farming challenges could be overcome, and there was effort to support agricultural development in the region, the next challenge for farming is climate change. It has already been mentioned that the climate future for southern Africa, including South Africa, will be increasingly severe. Work in the study areas has shown that people are perceiving changes in weather patterns, which include shifts in the start of the rains to later in the season and more frequent dry periods. These changes were attributed to factors such as "God's punishment for the way society is going", and were seen as a risk particularly by more well-off farming and livestock-owning households [30]. Such risk perception could serve as a further disincentive for farming. If agriculture is to be reinvigorated in these areas, it will be necessary to take into account shifts in seasons, rising temperatures (2–4 °C), and more extreme events such as droughts and floods.

Table 2. Reliance on purchased rather than on collected natural resources and cultivated food amongst vulnerable households in Gatyana and Lesseyton and perceptions of the impacts of this reliance (n = 100).

Site	Does Your Household Rely More on Purchased Goods than in the Past?		Has This Made It Harder or Easier for Your Household?		
	% yes	% no	% Easier	% Harder	% Unsure
Gatyana (n = 50)	72	28	16	72	17
Lesseyton (n = 50)	69	29	23	60	12
Some stated reasons			Gardening wastes money Shop gives credit Can't rely on gardening Crops don't grow Too old to garden	Transport expensive High costs in town Shops far Pay for load in taxi No other option	

Source: Clarke 2012, unpublished thesis [26].

5.1.3. Unemployment, Declining Remittances and Less Cash

Remittances now only form some five percent of total household income sources in our study areas (Figure 2), whereas they used to be one of the main sources of cash income, especially for farm investment, for rural households [7,47]. Remittances have decreased with declining employment opportunities nationally in several sectors and in the cities, especially for the less-connected and less-educated rural poor [5,57]. Marais [58] estimated that between 1994 and 2000, more than 500,000 jobs were lost in the country. Further factors include de-industrialization in some of the smaller towns in the Eastern Cape [7,58] and the process of more-permanent relocation to urban areas, which leaves little money to send home [47]. The impact of HIV/AIDS on the productive-age classes of the population has also reduced remittances, as ill people return home for care or to die [54]. Bank and Minkley ([59]; p. 24) make the point that " . . . Evidence suggests that the rates and real value of rural remittances have been declining steadily since the 1980s," and go on to suggest that "The main dynamic in the post-apartheid era is that rural labor—once so eagerly desired—has become a burden to the state and an irrelevance to capital." The decline in remittances has had the effect of squeezing and stretching the cash obtained from grants and can also be linked to the decline in agriculture as mentioned previously, possibly further trapping people. Connected to a decline in remittances is a change in values related to how income from employment should be spent. Kepe and Tessaro ([5]; p. 271) mention that "Even if there are people who still hold jobs, nowadays migrant workers buy cars instead of cattle that should be used for plowing." In this case, a change in one source of cash income has resulted in cascading effects on livelihood diversification and land use in terms of both cultivation and animal husbandry.

Table 3. Narratives from interviews with households still cultivating fields indicating that there is still life in arable farming in the rural Eastern Cape (n = 19).

Why Are You Farming?	What Do You Think about the Future of Farming?
"No one forces me to farm, it is something I enjoy doing, I love it and have been farming since I was a child. I have never worked but in farming." "I don't buy everything, for example I grind maize here at home and don't buy it in town. Farming saves a lot of money". "I feel very good about being a farmer, I wish I would plant more if I had resources and if I was healthy." "I started farming from 1941 to today, it's something I have been doing all my life." "I'm passionate about farming chickens and crops. Farming guarantees food, you cannot experience hunger when farming."	"There is little interest from the younger generation in farming." "We are still young and now we are negotiating with government to revitalize our fields, we are really pushing for them to come and support us." "My eldest child will definitely continue with farming." "My children and grandchildren will continue with farming because we raised them well regarding the importance of farming. They were raised to respect farming." "There is hope of farming in the future here, people do talk about wanting to plant their fields again, and their hope is on government."

Source: Shackleton and Hebinck (unpublished data) [56].

5.1.4. Social Grants: Window of Opportunity or a Barrier to Transformation?

Social grants form the largest percentage share of livelihood income in both study sites and are sometimes the only source of income for households (Figure 2, Table 4). This contribution has become increasingly important as other income sources have declined and as social protection has increased. Post 1994, in an effort to alleviate poverty and redistribute wealth, old-age pensions in South Africa were equalized (*i.e.*, black and white pensioners now receive the same amounts) and social grants were extended to include a child-support grant, initially for children of up to the age of seven but increased in 2010 to children up to 18 years old [60]. Furthermore, in 2007 the eligibility age of men for a pension was dropped from 65 to 60, the same as for women. While the pension amount is significant and reflects the minimum wage, the child grants are small. These changes have increased the availability of state support to the poorest South Africans. Some 15.2 million people receive grants.

Grants tend to be pooled within the household and are critical in contributing to food security, children's education and reducing the negative impacts of HIV/AIDS [27,60] (Table 4), but they have also created high levels of reliance on them (forming more than half of household income, Figure 2) and there has been some diversion of benefits from the intended beneficiaries (something acknowledged by local people themselves). Many life-history interviewees stressed how they depended on grants as their only source of cash, and that these payments often did not stretch beyond supplying everyday needs (Table 4). This dependence on grants, however, cannot be separated from the apartheid history of the country and its disempowering effects on poor people's self-determination, confidence, and self-sufficiency, as well as the marginalization of the areas in which they live.

Table 4. Narratives from life histories indicating dependency on social welfare grants (n = 20).

"I worry about how I will survive if the government stops the old age grant."
"There are six people living in this house. Me, my two daughters, my brother and grandchildren. We live on the pension that I get."
"I cannot afford to build a decent dwelling for my family due to financial constraints. And to raise children on a social grant is very hard for me. It was better when my husband was alive."
"I am not employed; I do not have a job. I live on the grant I get. I live with my grandchildren. Nobody in this house is employed."
"Life is much harder now. There is no employment for the youth. They depend on my pension for survival. Everything they want they expect me to provide."
"If my grandmother were to pass on I would have no-one to support me, as I depend on her old age grant for surviving."

Source: Clarke, 2012, unpublished thesis [26].

While various authors have suggested that grants can provide a window of opportunity, this does not seem to have been fully mobilized in our study areas, except by more educated households who have greater means of using social grants to invest in improved livelihoods [27]. Hagen-Zanker *et al.* [60] found that pensions can help adults within the pensioner's household seek employment by providing the resources for job-hunting. However, these results would only apply to individuals who are competitive in the job market, especially those with education (as revealed by Ndlovu [27]). Other research has shown the child grants have increased the likelihood of mothers participating in the labor market [61], but probably not in rural areas where there are few local employment opportunities. Other positive aspects of grants have been to reduce risky livelihood activities such as transactional sex, contribute to health care, and support other livelihood strategies.

Recent in-depth, ethnographic work by Dubbeld [49] in the village of Glendale in Kwazulu Natal, South Africa provides a more nuanced perspective on the role of social grants. His work suggests that not everyone sees social grants as a blessing and seldom as a substitute for wage labor. Young men in particular are not eligible for the current types of grants, but at the same time have limited job opportunities. Thus, the grant system puts money in the hands of some but not others. Interestingly, in our study sites it was only young males, especially those living alone, who were food insecure [32]. Dubbeld [49] further writes about how some of his informants perceived grants as "socially corrosive," destroying family values, morals and traditions such as marriage, absolving men of their parental responsibilities and contributing to disrespect of elders by the youth.

The situation in our study sites is complex and multidimensional, and there are no straight-forward answers regarding the impacts of social welfare and how it is affecting future trajectories. Fay [9], working in a neighboring site to Gatyana, found a correlation between the mean number of social grants per household and the presence or absence of field cultivation, with households with more child grants less likely to practice cultivation. Kepe and Tessaro ([5]; p. 271), also working in the Eastern Cape, explain how informants "unanimously agreed that with the prevalence of these grants, many people would rather use the monthly income from the grant to purchase food items rather than cultivate crops." Certainly, this trend seems to make sense, as purchasing food is much less risky and time consuming than investing in cultivation where the outcomes are unpredictable, especially as people are beginning to perceive and experience greater uncertainty in the climate. At the same time, growing food locally could reduce food insecurity into the future as previously highlighted.

The notion of a "cycle of dependency," described by Kofinas and Chapin [44] as a situation where individuals, households and even communities move away from traditional livelihood practices into systems with disincentives for self-sufficiency, is too simplistic in our context. While dependencies on external support could increase vulnerabilities to change, through depressing innovation and productive activity and narrowing the livelihood portfolio, these negative results may be outweighed by the importance of these grants in providing a safety net, especially for the most vulnerable (women, orphans, the elderly and infirm, and those with low levels of education and opportunity). However, the grants are neither permanent nor secure and are lost when the pensioner passes away or the child grows up, potentially plunging households back into highly vulnerable states. Moreover, our results suggest that, given people's situations and other stressors in their lives such as HIV/AIDS, these grants are able to do little more than keep households within the status quo, preventing them from falling deeper into poverty. They cannot be the basis of investment for the future [49]. Other commentators have also recognized that the grant system alone is unable to change rural people's lives [58] and claim that the South African government's social security system is not designed to lift a significant proportion of individuals out of poverty. For such a result to occur, social grants would likely need to co-exist with other complementary transformations such as in education [27] and/or reductions in the effects of HIV/AIDS, and/or be more universally distributed. Consequently, the role of social grants in facilitating transformation, rather than merely helping people cope and survive, will need further investigation [26,30]. Indeed, at the moment, grants are unlikely to play a role in reinvigorating agriculture.

5.2. *Broader Contexts and Changes Influencing Landscape and Livelihood Change*

The changes in livelihoods described above have occurred within the milieu of a number of broader contexts, changes and stressors that characterize our study sites. These include the legacy of apartheid, HIV/AIDS impacts, changing values, demographic changes, unsustainable and maladaptive strategies such as crime, changing demographics and the erosion of social capital. The mental maps in Figure 3 illustrate the multiple stressors and changes recognized by men and women from the study sites and the links between these, including to farming activities; whereas Table 5 reflects the perspectives of the youth.

5.2.1. The Legacy of Apartheid and the Immovability of Path Dependence

While apartheid ended two decades ago, inequality and poverty in South Africa have worsened [54,62,63]. This reflects the legacy of the Bantustan system, with former "homelands" in rural areas remaining under-developed in comparison with former "white" South African areas and cities [57]. Consequently, the communal areas of the Eastern Cape continue to represent some of the poorest and most vulnerable regions in the country. The residual effects of poor education, poor service delivery, lack of land rights, the undermining of agriculture and self-sufficiency, and the creation of a labor pool can still be seen and continues to impact on people's livelihood options today. What is more, many of these deprivations are being perpetuated by the current government (Table 5). National transformation has not yet trickled down to the rural communal areas, and they still face neglect at multiple levels: "More than a decade into our successful democracy, the Eastern Cape Province remains trapped in structural poverty that shows in all aspects of its demographic, health and socioeconomic profile" ([64]; p. 15). This sentiment is recognized and reiterated at the local level as captured in Table 5, Figure 3 and the following quote from one of the in-depth interviews from our study: "The so-called freedom is for the educated people. As for the uneducated, it has brought them nothing." The quality of rural education is in a dismal state, as highlighted by students in our study areas (Table 5). Moreover, local government is failing in many of its delivery mandates (Table 5). Land rights in these areas are still in the hands of the state despite many promises to resolve the land question [5].

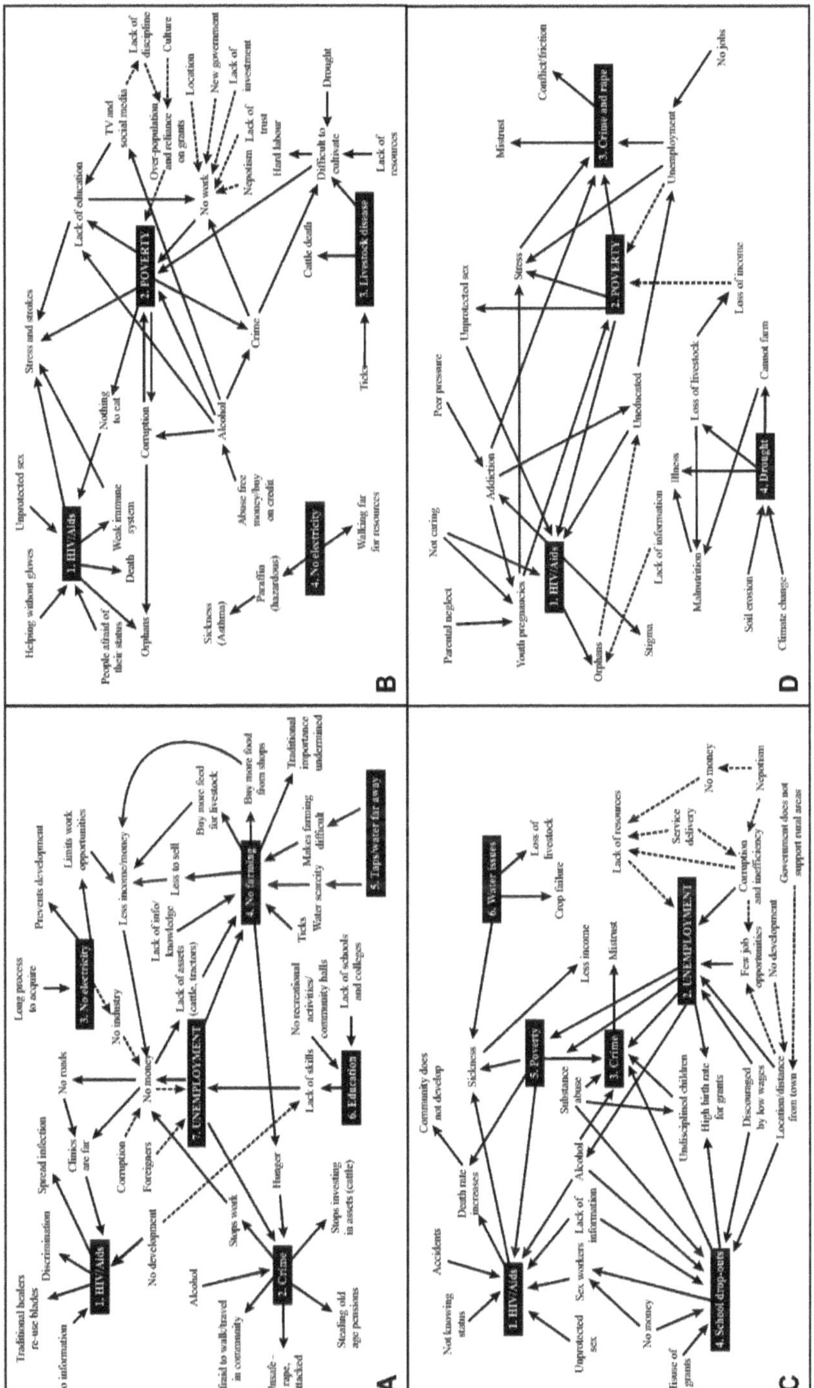

Figure 3. Participatory mental modeling spider-diagrams of perceived stressors in the community and their interconnections; done by (**A**) men in Gatyana, (**B**) women in Gatyana, (**C**) men in Lesseyton and (**D**) women in Lesseyton. Numbered items in the black boxes indicate the order of discussion of the major livelihood challenges/vulnerabilities identified; arrows pointing towards these boxes can be thought of as causes while those pointing away are effects, and dotted arrows indicate relationships revealed from a separate problem tree exercise. (Source: Stadler 2012, unpublished thesis [30]).

Table 5. Past, present and future problems and path dependency as discussed by the youth (Grade 11 learners) in Lesseyton and Gatyana (n = 20).

Lesseyton	Problems	Perceived Causes	Responses
Past	No electricity, not educated, low income	Service delivery, low education level, low income	Look for employment, sell livestock
Present	Not enough water, unemployment, health problems, crime—murder, rape, unhealthy environment (litter, animals), substance abuse, lack of information—distance, teenage pregnancy	Lack of education, theft of taps, dumping in the bush - unclear access rights to dumping site, drugs and alcohol linked to lack of recreation, peer pressure, media, ignorance, unemployment	Rainwater tanks, sell vegetables, sew clothes and other self-employment, prostitution, community punish criminals, contraception, use school library, people clean up area
Future	The same problems as present, but worse	Youth abusing drugs and alcohol, setting bad example for peers, leaders are corrupt and selfish	Knowledge sharing (guest speakers, evening classes, community library, adults should start this), leaders should be more educated so they take issues seriously, recreational activities, more police and police stations, more community skills-based programs, environmental regulations, elder supervision for youth, employ more nurses, street committees
Gatyana	**Problems**	**Perceived Causes**	**Responses**
Past	No education, transport, forced marriage, no rights	Children working, doing domestic duties, no money for transport, wanting money for marriages, Apartheid	
Present	Inadequate school and recreational facilities, teachers, no electricity, service delivery and roads, walk far for water, livestock theft, violence and crime, farming less, people are judgmental, lack of information, e.g., for bursaries	Corruption in school, no money, corruption in municipality, no water tanks and dams, poverty leads to theft, substance abuse, lack of education	Electricity Stronger and more policing No bail for arrests—harsher consequences More trained/qualified teachers Government should provide agricultural support Government should regulate child grants
Future	Violence and crime, no farming, no proper schools, teenage pregnancy increase	Government failing to deliver, lack of discipline and democracy, no consequences for crime	

Source: Clarke 2012, unpublished thesis [26].

The lingering effects of the entrenched policies of apartheid will thus continue to form a barrier to transformation in the previous Bantustans for a long time (see Figure 3), and new policies, for example, that aim to put traditional authority structures back into power have the potential to reinforce the "separateness" of these areas, disempower some groups of people (e.g., women) and heighten social differentiation [65]. Furthermore, education is one of the most important assets to extract households out of poverty [66] but seems to be in a trap of its own in the rural Eastern Cape. Ndlovu's [31] work showed that with education, social grants can be used more effectively to build household resilience. These continued effects of the former apartheid system are supported by Adato *et al.* ([62]; p. 227) who claim that "The legacy of apartheid is an economy in which social exclusion and poverty continue to interact in a mutually self-sustaining fashion" and is likely to persist well into the future. This characteristic, referred to as path dependence [38], we believe it is one of the fundamental aspects to be addressed to avoid traps and support transformation in the rural Eastern Cape. Indeed, if people in our study sites could move beyond living hand to mouth, it is possible that some would invest in revitalizing cultivation, as our research has shown that it is often the more wealthy (a minority) who are continuing to farm [14].

5.2.2. HIV/AIDS: Long-Term, Insidious and Asset-Eroding Impacts

In the rural areas in which we were working about half of households have been impacted in some way by HIV/AIDS (Table 6). This prevalence of HIV is supported by qualitative data, where many informants mentioned the loss of family members to HIV/AIDS and the increasing incidence in the community (see Table 7). With South Africa considered as having one of the highest HIV/AIDS prevalence rates in the world, this disease has had major negative impacts on households and communities across the country; a situation that was further exacerbated by the government's lack of action and slow roll out of antiretroviral drugs. HIV/AIDS is a long-term stressor on livelihoods and not only a shock as frequently framed; its impacts are often gradual, incremental and uneven within communities and regions. Impacted households have higher and more specific dietary needs, and face the dual pressure of less productivity and income, coupled with increasing health-related expenses, often resulting in a downward spiral of increasing stress and food insecurity and the erosion of assets as these are sold to pay for healthcare (Table 7). At the household and community level, the deaths of adults equate to a loss of labor and skills, an upset of knowledge transfer systems, loss of professionals and a rise in orphans. Home-based care also diverts household labor from economically or agriculturally productive activities [67].

The effects of HIV/AIDS tend to be longer lasting than other causes of mortality for two reasons. Firstly, because so many families are affected, traditional coping strategies (such as support from relatives) have been eroded [22,68], which can affect responses to other shocks. Secondly, with the significantly higher death rates of prime-aged adults the number of orphans has grown [22]. These orphans are most often either taken in by relatives, usually grandparents, adding to the strain on household resources, or they form their own child-headed households, where they frequently experience a lack of supervision, stunted growth and poor socialization [69]. Many of these orphans have dropped out of school, and so the effects will last for the rest of their lifetime, if not generations. In the absence of adult caregivers (a major concern in the communities where we worked; Figure 3), these children are particularly vulnerable to exploitation, abuse and HIV infection themselves. Many authors have made reference to the "vicious circle" of HIV/AIDS and poverty, which can operate in mutually re-enforcing cycles, as HIV/AIDS depletes multiple forms of capital, whilst a lack of capital diminishes capacity to recover from other shocks and increases vulnerability to HIV/AIDS [20,67,70]. Therefore, HIV/AIDS in itself can create a trap for some households, possibly putting them in the position where the income they obtain from social grants merely sustains them. HIV/AIDS can thus impact the transfer of knowledge on land-based livelihood strategies to the youth, erode culture and traditional ways of life (such as farming) and impact labor availability for farming activities.

Table 6. HIV/AIDS experiences and presence of proxy indicators in households by site*.

Type of Impact	Lesseyton (n = 170)	Gatyana (n = 170)
Non affected %	42.4	34.7
Chronic illness and receiving free care %	45.9	47.6
Illness-related death in previous 10 years %	18.2	22.9
Presence of *de facto* orphans %	24.1	17.6

* Percentages do not total 100 as households may have multiple experiences. Source: Stadler 2012, unpublished thesis [30].

Table 7. Narratives from the life histories (n = 20) suggesting the long-term impacts of HIV/AIDS, crime, and the erosion of social capital, but also how family still tries to assist in times of stress.

Contextual Factor	Representative Sample Quotes from Life Histories
HIV/AIDS	"In the past, the people were healthy and fit and working. Funerals were not an everyday thing as it is the case now. Every weekend there is a funeral, it has become the norm. HIV/AIDS is rife. Some disclose their status and others do not." "The death rate due to HIV is very high." "Sickness is worse now. There was no HIV and no asthma." "Health is not good at all. People could reach 80 years and above; but now we die at age 40. We are a cursed generation, we think."
Crime	"We got a phone call on a Saturday at about 12 midnight to say Grandma was being attacked by thugs. They stabbed her 4 times. Grandma stayed with a 14 year old girl. They ran away with the girl. We searched for the girl and when we arrived we found the girl dead; they had raped her and broke her neck. We are disappointed in the police in the way the case was handled. In fact the case was dismissed. Justice had failed us." "Crime is rife. I cannot leave my chickens in the fowl run. When it is bed time, I fetch the chickens and they sleep with us in the house." "It was a quiet place when I first arrived here but now things have changed. There was no crime here and no murders."
Social capital and networks	"People in the community are unreliable. You cannot count on them." "It is not easy; very few people are willing to assist. If someone assists you with anything, they expect you to pay back." "I seek help from my sisters. The community does not help. They gossip about other peoples' problems so I do not bother asking them for help." "I borrow money from my relatives; they do give when in a position to."

Source: Clarke 2012, unpublished thesis [26].

5.2.3. Changing Perspectives and Values

Interviews with different actors in the two communities, including the youth, the elderly and mixed groups, suggested that the rural areas of the Eastern Cape are undergoing a process of social and cultural change as new values, goals and ambitions take root. As the above sections suggest, the youth do not appear to have a strong interest in systems of local production and prefer the idea of working in formal employment (Table 5). This has also been observed by other authors with respect to food gardening in other areas of the Eastern Cape [50,71]. Moreover, adult members of the communities felt that the younger generation has lost respect for traditional ways of life and the authority of their elders and is flaunting discipline, family and community values, rules and norms (Table 7) [26,30]. One of the biggest issues that emerged in our participatory social learning meetings in both sites was concern about the future of the youth (Figure 3). A lack of parental care and discipline were mentioned repeatedly in both sites as major factors contributing to community vulnerability. This can be partially linked to the HIV/AIDS epidemic as described above. At the same time, interviews with the youth suggested that they feel misunderstood and neglected. All groups, however, commented on the serious problems of substance abuse, truancy, crime and teenage pregnancy amongst the younger generation and saw this as a barrier to development, farming and lower dependence on grants (Table 5).

These trends, together with the general perception amongst the youth of a hopeless future, their own general lack of empowerment and few adult role models, are in line with findings of youth in marginalized communities elsewhere in the country [72], and emphasize the need to incorporate youth in community development and adaptation planning. Reflecting the modernization that is happening, discussions with youth groups in both sites indicated a desire for a strong emphasis on education, schooling and knowledge (Table 5), whereas these themes were largely absent from discussions with elderly groups. Knowledge and agency are important to counter-act the negative behaviors arising from not being able to imagine a bright, hopeful future [72].

In addition to changes in how the youth see their lives and futures, and the intergenerational tensions this causes, Bank [7] asserts that other tensions in household dynamics relate to the increasing influence and decision-making power of women, encouraged partly through state projects designed to support women's development through (off-farm) commodity groups. He contends that households have "crumbled under the pressure of change," and that this social fragmentation has serious consequences for any efforts to re-institute household-based activities such as family farming.

5.2.4. Urbanization and Shifts in Demographics

Migration has always been an important process shaping rural demographics in the Eastern Cape, but more recently much of this migration of both men and women has become permanent in that migrants rarely return to their rural homes on either a part-time or full-time basis [7]. Population statistics from the province show highly skewed, concave population pyramids for those municipalities that consist mainly of rural, communal areas. In these areas, the proportion of young and middle-aged adults who make-up the economically active sector of the population (the peak age for out-migration is between 25 and 39 years) is substantively less than in urban areas [7,64,73]. The Eastern Cape has the highest rate of permanent net out-migration of any province in South Africa, with most of these migrants going to the Western Cape, followed by Gauteng and Kwazulu-Natal. Some 73 percent of black Africans living in the Western Cape were born in the Eastern Cape [74]. As described above, the permanence of this migration means that there is little spare cash to send remittances back to the rural family [7]. The dominance of children, looked after by the elderly, in the rural Eastern Cape population and in our study sites is another factor affecting the ability of households to use their land productivity. Lack of labor as pensioners become less mobile was a common reason provided for withdrawal from arable farming in our study sites, particularly in the Gatyana site. Over the four years of our study, we observed several elderly households give up farming their fields due to their inability to continue with the physical work. The poor socio-economic conditions in the province described above is reflected in this pattern of outward migration.

5.2.5. Crime and Other Maladaptive Behavior

Crime was an issue that surfaced repeatedly in our interviews and workshops (Table 7, Figure 3). This is unsurprising as South Africa's incidences of criminal activity are one of the highest in the world [57]. Amongst men, women and youth in the two study sites, crime was seen as a major factor contributing to vulnerability, preventing productive activity including farming and small business establishment, and leading to mistrust and conflict in the community (Table 5, Figure 3, Table 7). The Lesseyton youth mentioned how the theft of communal taps was affecting water security, and women in Gatyana cited how it has become unsafe for them to walk far distances to collect natural resources or to cultivate fields. Women and girls, in particular, are vulnerable to rape and violence.

Crime was linked to a host of problems, including: unemployment, poverty, a failing education system and school drop-outs, lack of parental care and discipline amongst the youth, and alcohol abuse, amongst others. The absence of a parent/parents within a household can impact emotionally on a child, and may increase the likelihood of children (primarily boys) becoming involved in crime [75]. In both sites, crime was described as an inhibiting factor preventing people from investing in assets, notably cattle, fences and other farming equipment for fear that these would be stolen. Theft was also shown to be one of the main constraints to farming in a study by Perret [76]. In addition, the youth in both sites associated crime with the lack of recreational activities, peer pressure and social media. The youth believed that there were no real consequences for crimes committed and therefore people had no fear of repercussions. Crime was often described in terms of cyclic knock-on effects: crime leading to limited livelihood options, in turn driving people to substance abuse and other desperate crimes. The effects crime has on a community's social capital and livelihood options, not to mention the health and well-being of the individual members affected, quite clearly indicates how crime can be viewed as a persistent barrier to action and transformation and a contributor to the trap people are in. Bhorat and Kanbur [57] argue that there is a fair degree of policy inertia in terms of dealing with crime, and this is evident in some of the narratives obtained from the life histories (Table 7). Corruption (outside and within the community) and nepotism were also mentioned as important factors affecting access to services, jobs and assets, as well as impacting on community development. Crime and corruption thus form major barriers to agriculture, innovation, motivation and livelihood improvement (also see Table 5).

Alcohol abuse was also mentioned in several discussions as a key factor that kept people in poverty (Table 5, Figure 3). About a quarter of all households, regardless of income group, were purchasing alcohol every month (Table 8), although this was much lower in female-only households. It has been estimated that about one in four men and one in 10 women in South Africa experience symptoms of alcohol-related problems [77]. Passing time in the taverns was also mentioned as a serious problem amongst young men that prevented them from getting involved in productive and healthy activities. Alcohol is recognized as a driver of risk-taking sexual behavior in South Africa, with HIV intervention strategies that focus on responsible drinking showing positive results in reducing the risk of infection [78]. Alcohol abuse also increases a person's vulnerability if they experience social marginalization and discrimination as a result of their addiction [79].

5.2.6. Erosion of Social Capital

Seeking assistance from family and friends was one of the most common responses to shocks revealed in our sites [26]. However, there was also evidence that drawing on others for assistance is not as easy as it was in the past (Table 7). Adato *et al.* [62] explain that the erosion of social capital amongst black South Africans started with the social exclusion policies of the apartheid government, which barred people from opportunities to earn income, access to assets, decision-making, social services and community and family support. However, this loss of social capital seems to have accelerated in recent years. In both our study sites, community members in various discussions (Figure 3) mentioned how crime and corruption were undermining community trust and cohesion, and how HIV/AIDS was resulting in a strain on social networks. Generally, family members may assist in preventing destitution, but usually cannot do much more because they also lack resources (Table 8) [62]. A general decline in social capital was identified in our study sites [30], although female-headed households seemed to have more social capital than those headed by men [32].

Table 8. Differences in percentages of households purchasing alcohol.

		Lesseyton (n = 168)	Gatyana (n = 170)
Gender of adults in the household	Male only	39.1	32.4
	Adult male head with adult female/s	26.7	34.1
	Adult female head with adult male/s	19.2	25.6
	Female only	4.0	4.2
	P Value (Pearson Chi-Square)	0.008	0.002
Income category	Lowest income	19.5	20.9
	Low income	33.3	17.4
	Moderate income	25.0	28.2
	High income	20.5	26.8
	P Value (Pearson Chi-Square)	0.458	0.604

Source: Stadler 2012, unpublished thesis [30].

This situation is not unique to the Eastern Cape. Hebinck and Bourdillon [80] also observed a decline in kinship ties in rural areas of Zimbabwe, while Adato *et al.* [62] found that one third of the households they had surveyed in KwaZulu-Natal reported problems of conflict and distrust between families. Linked to the phenomenon of declining cultivation is the disappearance of communal labor-sharing work parties and after feasts, which were essential to plow and plant large areas [14,28,48]. Bank [7] and Dubbeld [49] also write about the fragmentation and loss of social capital within rural households as a consequence of several of the factors discussed above, which in turn affect collective family activities such as farming.

6. Conclusions: Traps, Transformations and Possible Future Trajectories

There are multiple factors impeding or making it difficult for people in the rural communal areas of the Eastern Cape, in particular female-headed households, HIV-impacted households, and those with low levels of education, to escape their current situation of poverty and vulnerability and to make choices regarding their livelihoods. The various stressors and changes identified, combined with the highly polarized legacy of the country, effectively hamper local people's ability to take action to better their situation and respond to new or future shocks and stresses. It thus could be argued that the majority of households are in a poverty trap, affected to some extent by a locked-in trap (with path dependency) at a higher level. In particular, a key consequence of the changes is that poor rural households in the Eastern Cape are becoming increasingly dependent on cash income, more specifically social grants (which in themselves are not secure), for their livelihoods and food security. Moreover, residents of these areas see these changes as placing a strain on their livelihoods, with consequences for savings and coping with future shocks. The decline in cultivation and the high dependence on purchased food can put poor people at the mercy of markets; something to be viewed with concern in the context of growing food demands and the projected effects of climate change on food prices [81,82]. Maintenance of small-scale farming systems, or at least some form of food self-reliance, is therefore seen by many to be a priority for food security, health, livelihood resilience, and adaptation to future climate change [52,83]. Given the multiple drivers of the decline in cultivation in the Eastern Cape, and the barriers to reviving it, as identified in this paper, new approaches that cater to different groups with different aspirations will be needed. These approaches could range from boosting household food security to small-scale farming, but not before the fundamental issue of land rights is addressed. As pointed out by Kepe and Tessaro [5], without transformation in communal area tenure systems, it is unlikely that much progress will be made with efforts to reinvigorate agricultural production. At the same time, it will be necessary to consider the potential impacts of changing weather patterns and future climate trends on production and the consequences of the increasing woody invasion of the landscape. Participatory scenario development could be used as a tool to explore different pathways to multiple futures [24].

The notion of people in the rural Eastern Cape being "stuck" is supported by work by Adato *et al.* [62], who argue that large numbers of South Africans are trapped in poverty without a pathway to upward mobility. Similarly, Bank and Minkely [59] talk about people in the rural Eastern Cape being caught-up in a process of "involution"; a situation in which, "Economic marginalization and disconnection inhibit social change because they lock people into a system of shared poverty, where the poor reinvent communitarian traditions of mutual dependence and risk-spreading in an effort to protect themselves from the effects of deepening poverty" ([59]; p. 26). They further go on to identify the rural Eastern Cape as a "case of the entrenchment of a livelihood system which has failed to either stabilize or transform itself into a new pattern under changing conditions." Such arguments are reiterated in the work of Cundill and Fabricius [84]. They show how the Eastern Cape's history, together with more recent experiences of poverty and mobility, has resulted in a situation characterized by high levels of vulnerability, low levels of institutional capacity and the reduced production of, particularly, provisioning ecosystem services through agriculture. They contend that these conditions represent a trap that will require a great deal of energy to escape, and that transformability must be at the focus of discussions about future trajectories in the region.

The complexity of livelihood change and the multiplicity of factors interacting at an immediate level in the study communities—some common and others uncommon, some emerging from within the community and others arising from wider spheres of influence—together with the mixture of economic and social drivers, highlight the importance of understanding local perspectives of change and stressors to plan for effective transformation. Like Levine *et al.* [85], we recognize that "There is no neat distinction between risk, shock, stressors, vulnerability, barriers and responses (whether coping, adaptive and transformative), because in the real world they can be one and the same" ([85]; p. 2). As our results have shown, a stressor or change can become a barrier, just as can a maladaptive response or an unintended transformation (e.g., decline in agriculture). Moreover, research from our study sites shows that escaping traps that result from multiple stressors and change are likely to require multiple transformations. Trying to address the complexity of these traps with simple, single transformations may perpetuate, or even exacerbate, these. For example, Ndlovu [27] indicates that the potentially transformative policy of initiating social grants for large segments of the poor population has led to reduced labor supply in productive activities including farming. However, results also suggest that this negative effect could be overcome with improved education. Conditions at our study site thus imply, in addition to clearer land rights, addressing problems related to health and HIV/AIDS, crime and other social welfare issues may also be pre-requisites for transformative change, as will addressing basic structural issues related to continued neglect. Though the focus of our work in understanding multiple stressors is a fundamental step in better comprehending, and reacting to, poverty and vulnerability in the Eastern Cape, more work is needed on assessing the impacts of multiple responses, such as social grants, education, land tenure reform, health programs, and crime prevention. Moreover, the dynamic context of vulnerability and transformation is an area that livelihood investigations, such as ours, rarely have high quality, longitudinal data to address. For example, in our work, most of our data were collected within three years, requiring our interpretations to be supplemented with secondary findings from other researchers that may not be directly comparable. Collecting panel data over decades would greatly help in developing a more nuanced understanding of the time dimension to multiple stressors and responses. This was highlighted in Chapter 13 of the 2014/15 IPPC report, where it is stated that the majority of studies "Do not focus on continuous struggles and trajectories but only offer snapshots. An explicit analysis of livelihood dynamics would reveal more clearly how people respond to series of climatic stressors and shocks over time" ([86]; p. 818).

Within this complexity of traps and changes in our study sites, we argue that, while some of the changes observed could be thought of as unplanned transformations, their future trajectories remain unclear and their desirability ambiguous given the uncertainty of the future. Within this context, introducing specific, as opposed to enabling, transformations could become risky. For example, despite the decline of cultivation, and the problems associated with this trend, we believe, as mentioned above, that to introduce policies that seek to re-establish agrarian livelihoods in these rural areas will need careful consideration and a suite of approaches so as to prevent these being counterproductive. Further, we do not know whether the resources that households would have to divert to agriculture would be worth the increased food security and reductions in cash expenditures on food that such changes may bring. Instead, if an enabling environment can be created, where people are educated, healthy, live in communities with minimal crime, and have tenure security and the rights to self-determination, then they will be in a better position to consider pros and cons of alternative livelihood options themselves, including reinvigoration of agricultural production as a livelihood diversification activity. Such an approach also helps to avoid the normative nature of specific pathways and aligns with Leach *et al.*'s [87] notion of multiple pathways to sustainability, where promoting numerous solutions through diversity helps to recognize the values, needs and interests of different people and places. Such an approach also keeps options open and allows people to change.

However, creating an amenable environment for making choices is not easy, and we agree with Bank and Minkley ([59]; p. 21) that to transform these areas requires:

> " ... more detailed analysis and understanding of the social, political and economic forces that have deepened and entrenched poverty in rural areas over the past decade. We need to know how existing inequalities are being maintained and reproduced. It is not enough to assume that the removal of 'obstacles' and the introduction of localized incentive packages will be enough to re-orientate an entire region with a long and deep history of poverty and under-development."

Indeed, deep political commitment to structural and discursive transformations that take into account aspects of distributional justice, that recognize the shifts in rural livelihoods and household dynamics, and that build on local visions, strengths and agency are needed. In the words of Matiwane and Chimera-Dan ([64]; p. 16), this entails "extraordinary measures by the government and all development partners ... ... in order to break the shackles of structural poverty and its consequences among the population of the province."

Acknowledgments: We are grateful to the International Development Research Centre (IDRC) Ecohealth Programme and the National Research Foundation (NRF), South Africa for funding. We acknowledge all members of the project team whose work contributed to this synthesis paper. In particular, we recognize the essential contributions of the following postgraduate students: Caryn Clarke, Leigh Stadler and Ross Shackleton from Rhodes University, and Patrick Ndlovu and Amy Trefry from University of Alberta (their theses are available online at their respective universities). Georgina Cundill, as the postdoctoral leader of the social learning process, made a substantive intellectual contribution to the project and the synthesis papers derived from it, while Monde Ntshudu expertly facilitated our close relationship with the communities of Lesseyton and Gatyana, whom we also thank most sincerely for their time and cooperation. Paul Hebinck was a collaborator for the follow-up work with field farmers, and we are working on a publication based on this research.

Author Contributions: Sheona Shackleton and Marty Luckert were co-leaders of the project and conceived and wrote this synthesis paper drawing on the work of other team members.

Conflicts of Interest: The authors declare no conflict of interest.

References

1. Faber, D.F.; McCarthy, D. Neo-liberalism, globalization and the struggle for ecological democracy: Linking sustainability and environmental justice. In *Just Sustainabilities: Development in an Unequal World*; Agyeman, J., Ed.; MIT Press: London, UK, 2003; pp. 38–63.
2. Schröter, D. Our vulnerability to changes in ecosystem services. In *Assessing Vulnerability to Global Environmental Change: Making Research Useful for Adaptation and Decision Making and Policy*; Patt, G., Schröter, D., Klein, R.J.T., De la Vega-Leinert, A.C., Eds.; Earthscan: London, UK, 2009; pp. 97–114.
3. Rockström, J.; Steffen, W.; Noone, K.; Persson, Å.; Chapin, F.; Lambin, E.; Lenton, T.; Scheffer, M.; Folke, C.; Schellnhuber, H.; *et al.* A safe operating space for humanity. *Nature* **2009**, *461*, 472–475. [CrossRef] [PubMed]

4. Fraser, E.D.G.; Dougill, A.J.; Hubacek, K.; Quinn, C.H.; Sendzimir, J.; Termansen, M. Assessing vulnerability to climate change in dryland livelihood systems: Conceptual challenges and interdisciplinary solutions. *Ecol. Soc.* **2011**, *16*, 3. [CrossRef]
5. Kepe, T.; Tessaro, D. Trading off: Rural food security and land rights in South Africa. *Land Use Policy* **2014**, *36*, 267–274. [CrossRef]
6. Andrew, M.; Fox, R. "Undercultivation" and intensification in the Transkei: A case study of historical changes in the use of arable land in Nompa, Shixini. *Dev. S. Afr.* **2004**, *21*, 687–706. [CrossRef]
7. Bank, L. On family farms and commodity groups: Rural livelihoods, households and development policy in the Eastern Cape. *Soc. Dyn.* **2005**, *31*, 157–181. [CrossRef]
8. Fay, D. The trust is over! We want to plough! Land, livelihoods and reverse resettlement in South Africa's Transkei. Ph.D. Thesis, Boston University, Boston, MA, USA, 2003.
9. Fay, D. Cultivators in action, Siyazondla inaction? Trends and potentials in homestead cultivation. In *In the Shadow of Policy: Everyday Practices in South African Land and Agrarian Reform*; Hebinck, P., Cousins, B., Eds.; Wits University Press: Johannesburg, South Africa, 2013; pp. 247–262.
10. MacAllister, P.A. *Building the Homestead: Agriculture, Labour and Beer in South Africa's Transkei*; Research Series/African Studies Centre No. 16; Ashgate: Aldershot, UK, 2001.
11. Hajdu, F. Relying on jobs instead of the environment? Patterns of local securities in the rural Eastern Cape, South Africa. *Soc. Dyn.* **2005**, *31*, 235–260. [CrossRef]
12. Hebinck, P.; Lent, P.C. *Livelihoods and Landscapes: The people of Guquka and Koloni and their Resources*; Brill Press: Leiden, The Netherlands; Boston, MA, USA, 2007.
13. Hebinck, P.; Van Averbeke, W. What constitutes "the agrarian" in rural Eastern Cape settlements? In *In the Shadow of Policy: Everyday Practices in South African Land and Agrarian Reform*; Hebinck, P., Cousins, B., Eds.; Wits University Press: Johannesburg, South Africa, 2013; pp. 189–204.
14. Shackleton, R.; Shackleton, C.; Shackleton, S.; Gambiza, J. Deagrarianization and forest succession in abandoned fields in a biodiversity hotspot on the Wild Coast, South Africa. *PLoS ONE* **2013**. [CrossRef] [PubMed]
15. Beinhart, W.; Brown, K. *African Local Knowledge and Livestock Health: Diseases and Treatments in South Africa*; Boydell & Brewer Ltd: Suffolk, UK, 2013.
16. Christensen, J.H.; Hewitson, B.; Busuioc, A.; Chen, A.; Gao, X.; Hold, I.; Jones, R.; Kolli, R.K.; Know, W-T.; Lpaise, R.; *et al. Climate Change 2007: The Physical Science Basis*; Contribution of Working Group 1 to the Fourth Assessment Report to the Intergovernmental Panel on Climate Change; Cambridge University Press: Cambridge, UK; New York, NY, USA, 2007.
17. Shackleton, C.M.; Shackleton, S.E.; Gambiza, J.; Nel, E.; Rowntree, K.; Urquhart, P.; Fabricius, C.; Ainslie, A. *Linking Ecosystem Services and Poverty Alleviation in the Arid and Semi-arid Lands of Southern Africa*; Nova Publishers: New York, NY, USA, 2010.
18. Reid, P.; Vogel, C. Living and responding to multiple stressors in South Africa—Glimpses from KwaZulu-Natal. *Glob. Environ. Change* **2006**, *16*, 195–206. [CrossRef]
19. Loevinsohn, M.; Gillespie, S. *HIV/AIDS, Food Security and Rural Livelihoods: Understanding and Responding*; Discussion Paper No. 157; IFPRI: Washington, DC, USA, 2003; Available online: http://ebrary.ifpri.org/cdm/compoundobject/collection/p15738coll2/id/71905/rec/2 (accessed on 28 June 2015).
20. Piot, P.; Greener, R.; Russell, S. Squaring the circle: AIDS, poverty, and human development. *PLoS Med.* **2007**, *4*, 1571–1575. [CrossRef] [PubMed]
21. Marks, S. An epidemic waiting to happen? The spread of HIV and AIDS in South Africa in a social and historical perspective. *Afr. Stud.* **2002**, *61*, 13–26. [CrossRef]
22. Misslehorn, A.A. What drives food insecurity in southern Africa? A meta-analysis of household economy studies. *Glob. Environ. Change* **2005**, *15*, 33–43. [CrossRef]
23. Shackleton, S.E.; Shackleton, C.M. Linking poverty, HIV/Aids and climate change to human and ecosystem vulnerability in southern Africa: Consequences for livelihoods and sustainable ecosystem management. *Int. J. Sust. Dev. Wld. Ecol.* **2012**, *19*, 275–286. [CrossRef]
24. Folke, C.; Chapin, S.F.; Olsson, P. Transformations in ecosystem stewardship. In *Principles of Ecosystem Stewardship: Resilience Based Natural Resource Management in a Changing World*; Chapin, F.S., Kofinas, G.P., Folke, C., Eds.; Springer: Amsterdam, The Netherlands, 2009; pp. 103–125.

25. Mucina, L.; Rutherford, M.C.; Powrie, L.W. *Vegetation Map of South Africa, Lesotho and Swaziland*, 2nd ed.; South African National Biodiversity Institute: Pretoria, South Africa, 2007.
26. Clarke, C. Responses to the Linked Stressors of Climate Change and HIV/AIDS amongst Vulnerable Rural Households in the Eastern Cape, South Africa. Master's Thesis, Rhodes University, Grahamstown, South Africa, 2012.
27. Ndlovu, P. Effects of Social Grants on Labour Supply and Food Security of South African Households: Is There a Disincentive Effect? Master's Thesis, University of Alberta, Edmonton, AB, Canada, 2012.
28. Shackleton, R.T. Deagrarianization and Forest Succession in Abandoned Fields over 65 Years: Gatyana, South Africa. Honors Dissertation, Rhodes University, Grahamstown, South Africa, 2012.
29. Shackleton, S.; Luckert, M.; Cundill, G.; Cobban, L.; Clarke, C.; Shackleton, R.; Ndlovu, P. Transformation and barriers in the context of multiple stressors: Understandings from two rural sites in the Eastern Cape, South Africa. In Proceedings of the Transformation in a Changing Climate, Oslo, Norway, 19–21 June 2013; pp. 136–148.
30. Stadler, L.T. Assessing Household Assets to Understand Vulnerability to HIV/Aids and Climate Change in the Eastern Cape, South Africa. Master's Thesis, Rhodes University, Grahamstown, South Africa, 2012.
31. Trefry, A.; Parkins, J.; Cundill, G. Culture and food security: A case study of homestead food production in South Africa. *Food Secur.* **2014**, *6*, 555–565. [CrossRef]
32. Shackleton, S.; Cobban, L.; Cundill, G. A gendered perspective of vulnerability to multiple stressors, including climate change, in the rural Eastern Cape, South Africa. *Agenda* **2014**, *28*, 73–89.
33. Bunce, M.; Rosendo, S.; Brown, K. Perceptions of climate change, multiple stressors and livelihoods on marginal African coasts. *Environ. Dev. Sustain.* **2010**, *12*, 407–440. [CrossRef]
34. Cundill, G.; Shackleton, S.E.; Sisitka, L.; Ntshudu, M.; Lotz-Sisitka, H.; Kulundu, I.; Hamer, N. *A Social Learning Approach to Building Capacity to Adapt to an Uncertain Future: A Descriptive Guide for Practitioners and Action Researchers*; Rhodes University/IDRC/Ruliv: Grahamstown, South Africa, 2014.
35. Walker, B.; Holling, C.S.; Carpenter, S.R.; Kinzig, A.P. Resilience, adaptability and transformability in social–ecological systems. *Ecol. Soc.* **2004**, *9*, 5. Available online: http://www.ecologyandsociety.org/vol9/iss2/art5/ (accessed on 10 April 2015).
36. Brown, K.; O'Neill, S.; Fabricius, C. Social science understandings of transformation. In *World Social Science Report 2013. Changing Global Environments*; ISSC and UNESCO: Paris, France, 2013; Chapter 9; pp. 100–106.
37. Chapin, F.S., III. Managing ecosystems sustainably: The key role of resilience. In *Principles of Ecosystem Stewardship. Resilience Based Natural Resource Management in a Changing World*; Chapin, F.S., Kofinas, G.P., Folke, C., Eds.; Springer: Amsterdam, The Netherlands, 2009; pp. 29–53.
38. Chapin, F.S., III; Folke, C.; Kofinas, G.P. A framework for understanding change. In *Principles of Ecosystem Stewardship. Resilience Based Natural Resource Management in a Changing World*; Chapin, F.S., Kofinas, G.P., Folke, C., Eds.; Springer: Amsterdam, The Netherlands, 2009; pp. 3–28.
39. Kates, R.W.; Travis, W.R.; Wilbanks, T.J. Transformational adaptation when incremental adaptations to climate change are insufficient. *Proc. Natl. Acad. Sci. USA* **2012**, *109*, 7156–7161. [CrossRef] [PubMed]
40. Marshall, N.A.; Park, S.E.; Adger, W.N.; Brown, K.; Howden, S.M. Transformational capacity and the influence of place and identity. *Environ. Res. Lett.* **2012**. [CrossRef]
41. Stafford Smith, M.; Horrocks, L.; Harvey, A.; Hamilton, C. Rethinking adaptation for a 4°C world. *Phil. Trans. R. Soc. A* **2011**, *369*, 196–216. [CrossRef] [PubMed]
42. Carpenter, S.R.; Brock, W.A. Adaptive capacity and traps. *Ecol. Soc.* **2008**, *13*, 40. Available online: http://www.ecologyandsociety.org/vol13/iss2/art40/ (accessed on 29 September 2015).
43. Walker, B.; Salt, D. *Resilience Practice. Building Capacity to Absorb Disturbance and Maintain Function*; Island Press: Washington, DC, USA, 2012.
44. Kofinas, G.P.; Chapin, S. Sustaining livelihoods and human well-being during social–ecological change. In *Principles of Ecosystem Stewardship. Resilience Based Natural Resource Management in a Changing World*; Chapin, F.S., Kofinas, G.P., Folke, C., Eds.; Springer: Amsterdam, The Netherlands, 2009; pp. 55–75.
45. Page, S.E. Path dependence. *Quart. J. Pol. Sci.* **2006**, *1*, 87–115. [CrossRef]
46. Greener, I. The potential of path dependence in political studies. *Politics* **2005**, *25*, 62–72. [CrossRef]
47. Hebinck, P.; van Averbeke, W. Rural transformation in the Eastern Cape. In *Livelihoods and Landscapes: The People of Guquka and Koloni and their Resources*; Hebinck, P., Lent, P.C., Eds.; Brill Press: Leiden, The Netherlands; Boston, MA, USA, 2007; pp. 33–66.

48. Schmidhuber, J.; Tubiello, F.N. Global food security under climate change. *Proc. Natl. Acad. Sci. USA* **2007**, *104*, 19703–19708. [CrossRef] [PubMed]
49. De Klerk, H. The Mutual Embodiment of Landscape and Livelihoods: An Environmental History of Nqabara. Master's Thesis, Rhodes University, Grahamstown, South Africa, 2007.
50. Dubbeld, B. How social security becomes social insecurity: Unsettled households, crisis talk and the value of grants in a KwaZulu-Natal village. *Acta Juridica: Marriage Land Custom* **2013**, *2013*, 197–217.
51. Aliber, M.; Hart, T. Should subsistence agriculture be supported as a strategy to address rural food insecurity? *Agrekon* **2009**, *48*, 434–458. [CrossRef]
52. Timmermans, H.G. Rural Livelihoods at Dwesa/Cwebe: Poverty, Development and Natural Resource Use on the Wild Coast, South Africa. Master's Thesis, Rhodes University, Grahamstown, South Africa, 2004.
53. Hendriks, S. The potential for nutritional benefits from increased agricultural production in rural KwaZulu-Natal. *South Afr. J. Agric. Ext.* **2003**, *32*, 28–44.
54. Hebinck, P. Investigating rural livelihoods and landscapes in Guquka and Koloni: An introduction. In *Livelihoods and Landscapes: the People of Guquka and Koloni and their Resources*; Hebinck, P., Lent, P.C., Eds.; Brill Press: Leiden, The Netherlands; Boston, MA, USA, 2007; pp. 1–32.
55. *Integrated Development Plan: 2012–2017*; Mbhashe Local Muncipality: Dutywa, South Africa, 2012.
56. Shackleton, S.E.; Hebinck, P.; Rhodes University, Grahamstown, South Africa. Unpublished data from a re-survey of households farming fields in Gatyana, Wild Coast. 2014.
57. Bhorat, H.; Kanbur, R. *Poverty and Well-Being in Post-Apartheid South Africa: An Overview of Data, Outcomes and Policy*; Working Paper 05/101; Development Policy Research Unit, University of Cape Town: Cape Town, South Africa, 2005.
58. Marais, H. *South Africa Pushed to the limit: The Political Economy of Change*; UCT Press: Cape Town, South Africa, 2011.
59. Bank, L.; Minkley, G. Going nowhere slowly? Land, livelihoods and rural development in the Eastern Cape. *Soc. Dyn.* **2005**, *31*, 1–38. [CrossRef]
60. Hagen-Zanker, J.; Morgan, J.; Meth, C. *South Africa's Cash Social Security Grants: Progress in Increasing Coverage*; ODI Working Paper; ODI: London, UK, 2011.
61. Eyal, K.; Woolard, I. Female labor force participation and South Africa's child support grant. In Proceedings of the CSAE 25th Anniversary Conference: Economic Development in Africa, Oxford, UK, 20–22 March 2011; Available online: http://www.csae.ox.ac.uk/conferences/2011-EdiA/papers/467-Eyal.pdf (accessed on 18 June 2015).
62. Adato, M.; Carter, M.R.; May, J. Exploring poverty traps and social exclusion in South Africa using qualitative and quantitative data. *J. Dev. Stud.* **2006**, *42*, 226–247. [CrossRef]
63. Özler, B. Not separate, not equal: Poverty and inequality in post-apartheid South Africa. *Econ. Dev. Cult. Change* **2007**, *5*, 487–529. [CrossRef]
64. Matiwane, M.; Chimere-Dan, D. *The People Matter: the State of the Population in the Eastern Cape. Research and Population Unit, Eastern Cape Department of Social Development*; South African Government: East London, South Africa, 2010.
65. Claassens, A. Law, land and custom 1913–2013: What is at stake? In Land Divided: Land and South African Society in 2013, in Comparative Perspective; Proceedings of the Land Divided Conference 2013, Cape Town, South Africa, 24–27 March 2013; p. 17.
66. May, J. Persistent poverty, asset accumulation and shocks in South Africa: Evidence from KwaZulu-Natal. In *Poverty and Policy in Post-Apartheid South Africa*; Bhorat, H., Kanbur, R., Eds.; HSRC Press: Pretoria, South Africa, 2006.
67. Wiegers, E.; Curry, J.; Alesandra, G.; Hourihan, J. Patterns of vulnerability to AIDS impacts in Zambian households'. *Dev. Change* **2006**, *37*, 1073–1092. [CrossRef]
68. Shackleton, S. *Forests as Safety Nets for Mitigating the Impacts of HIV/AIDS in Southern Africa*Forest Livelihood Briefs, 4th ed.; CIFOR: Bogor, Indonesia, 2006.
69. Barnett, T.; Whiteside, A. *AIDS in the Twenty-first Century: Disease and Globalization*, 2nd ed.; Palgrave MacMillan: New York, NY, USA, 2002.

70. Gillepsie, S.; Haddad, L.; Jackson, R. HIV/AIDS, food and nutrition security: Impacts and actions. In Proceedings of the 28th Session of the ACC/SCN Symposium on Nutrition and HIV/AIDS, Nairobi, Kenya, 3–4 April 2001; Available online: http://www.ifpri.org/themes/HIV/accscn2001en.pdf (accessed on 26 June 2015).
71. Møller, V. Attitudes to food gardening from a generational perspective: A South African case study. *J. Int. Relat.* **2005**, *3*, 63–80.
72. Campbell, C.; MacPhail, C. Peer education, gender and the development of critical consciousness: participatory HIV prevention by South African youth. *Soc. Sci. Med.* **2002**, *55*, 331–345. [CrossRef]
73. Hamann, M.; Tuinder, V. *Introducing the Eastern Cape: A Quick Guide to its History, Diversity and Future Challenges*; Stockholm Resilience Centre: Stockholm, Sweden, 2012.
74. Stats, S.A. *Census 2011 Municipal Report—Eastern Cape/Statistics South Africa*; Report No.: 03-01-50; Statistics South Africa: Pretoria, South Africa, 2012.
75. Coovadia, H.; Jewkes, R.; Barron, P.; Sanders, D.; McIntyre, D. The health and health system of South Africa: Historical roots of current public health challenges. *The Lancet* **2009**, *374*, 817–834. [CrossRef]
76. Perret, S. *Livelihood Strategies in Rural Transkei (Eastern Cape Province): How does Wool Production Fit in?* Working Paper 2002–20; University of Pretoria: Pretoria, South Africa, 2002.
77. Parry, C. South Africa: alcohol today. *Addiction* **2005**, *100*, 426–429. [PubMed]
78. Kalichman, S.; Simbayi, L.; Vermaak, R.; Cain, D.; Smith, G.; Mthebu, J.; Jooste, S. Randomized trial of a community-based alcohol-related HIV risk-reduction intervention for men and women in Cape Town South Africa. *Ann. Behav. Med.* **2008**, *36*, 270–279. [CrossRef] [PubMed]
79. Vermeulen, S.J. *Climate Change, Food Security and Small-Scale Producers*; CCAFS InfoBrief; CCAFS: Copenhagen, Denmark, 2014.
80. Khanal, N.R.; Watanabe, T. Abandonment of land and its consequences. *Mountain Res. Dev.* **2006**, *26*, 32–40.
81. Baiphethi, M.N.; Jacobs, P.T. The contribution of subsistence farming to food security in South Africa. *Agrekon* **2009**, *48*, 459–482. [CrossRef]
82. London, L. The "dop" system, alcohol abuse and social control amongst farm workers in South Africa: A public health challenge. *Soc. Sci. Med.* **2009**, *48*, 1407–1414. [CrossRef]
83. Hebinck, P.; Bourdillon, M.F.C. Analysis of livelihoods. In *Women, Men and Work: Rural Livelihoods in Central-eastern Zimbabwe*; Bourdillon, M.F.C., Hebinck, P., Eds.; Weaver Press: Harare, Zimbabwe, 2001; pp. 1–13.
84. Cundill, G.; Fabricius, C. Adaptive co-management under resource-poor conditions: Lessons from South Africa. In *Exploring Sustainability Science: A Southern African Perspective*; Burns, S., Weaver, A., Eds.; Sun Press: Stellenbosch, South Africa, 2008; pp. 537–568.
85. Levine, S.; Pain, A.; Bailey, S.; Fan, L. *The Relevance of Resilience*; HPG Policy Brief 48; ODI: London, UK, 2012.
86. Olsson, L.; Opondo, M.; Tschakert, P.; Agrawal, A.; Eriksen, S.H.; Ma, S.; Perch, L.N.; Zakieldeen, S.A. Livelihoods and poverty. In *Climate Change 2014: Impacts, Adaptation, and Vulnerability. Part A: Global and Sectoral Aspects. Contribution of Working Group II to the Fifth Assessment Report of the Intergovernmental Panel on Climate Change*; Field, C.B., Barros, V.R., Dokken, D.J., Mach, K.J., Mastrandrea, M.D., Bilir, T.E., Chatterjee, M., Ebi, K.L., Estrada, Y.O., Genova, R.C., *et al.*, Eds.; Cambridge University Press: Cambridge, UK; New York, NY, USA, 2014; pp. 793–832.
87. Leach, M.; Raworth, K.; Rockstrom, J. Between social and planetary boundaries: Navigating pathways in the safe and just space for humanity. In *World Social Science Report. 2013. Changing Global Environments*; ISSC, UNESCO, Eds.; OECD Publishing and UNESCO Publishing: Paris, France, 2013; pp. 84–89.

Section 4:

Beyond Smallholders

land

MDPI

Article

Spatial Distribution of Estimated Wind-Power Royalties in West Texas

Christian Brannstrom *, Mary Tilton, Andrew Klein and Wendy Jepson

Department of Geography, Texas A&M University, 810 O&M Building, College Station, TX 77843, USA; m.tilton83@gmail.com (M.T.); klein@geog.tamu.edu (A.K.); wjepson@geos.tamu.edu (W.J.)

* Correspondence: cbrannst@geos.tamu.edu; Tel.: +1-979-845-5923; Fax: +1-979-862-4487

Academic Editors: Claudia A. Radel and Jacqueline M. Vadjunec

Received: 15 July 2015; Accepted: 12 November 2015; Published: 2 December 2015

Abstract: Wind-power development in the U.S. occurs primarily on private land, producing royalties for landowners through private contracts with wind-farm operators. Texas, the U.S. leader in wind-power production with well-documented support for wind power, has virtually all of its ~12 GW of wind capacity sited on private lands. Determining the spatial distribution of royalty payments from wind energy is a crucial first step to understanding how renewable power may alter land-based livelihoods of some landowners, and, as a result, possibly encourage land-use changes. We located ~1700 wind turbines (~2.7 GW) on 241 landholdings in Nolan and Taylor counties, Texas, a major wind-development region. We estimated total royalties to be ~$11.5 million per year, with mean annual royalty received per landowner per year of $47,879 but with significant differences among quintiles and between two sub-regions. Unequal distribution of royalties results from land-tenure patterns established before wind-power development because of a "property advantage," defined as the pre-existing land-tenure patterns that benefit the fraction of rural landowners who receive wind turbines. A "royalty paradox" describes the observation that royalties flow to a small fraction of landowners even though support for wind power exceeds 70 percent.

Keywords: wind power; royalty; property; land use; income; Texas

1. Introduction

Explaining how biofuels affect land change is a key research theme in hybrid land change science that aims to incorporate insights from political ecology [1], but linkages between land change and utility-scale renewable energy are poorly understood. The rapid expansion of wind-power development in the U.S., primarily on private landholdings in a region from Texas to Minnesota, has produced unknown outcomes for livelihoods and land uses through the production of royalties for landowners who make private contracts with wind-farm operators. In dozens of rural communities, wind power has produced obvious impacts on land cover from construction of wind turbines and maintenance roads; however, these plainly visible land changes contrast with far less visible effects on landowner livelihoods and land cover, which have multiple potential behavioral, financial, and environmental determinants.

More than 97 percent of the ~60 GW of installed wind power capacity in the U.S. has been built on private land [2,3]. With 12,214 MW of installed capacity, Texas is the leading U.S. state, where 10 percent of its electricity comes from wind farms—virtually all sited on private lands—in the state's western region [4]. Wind turbines represent a land use that farmers and ranchers view as compatible with livestock grazing, hunting, or agriculture, and social acceptance is high [5–7]. Knowledge of the spatial distribution of royalty payments from wind energy is a crucial first step to understanding how income from renewable power may alter land-based livelihoods and may encourage land-use changes.

Here we use GIS-based methods and a royalty formula to estimate the amount and spatial distribution of wind-power royalties in Nolan and Taylor counties (Figure 1), a core area of windfarms in Texas. We show that royalty distribution among landowners is unequal, especially between two sub-regions, because of pre-existing "property advantage." We also identify a "royalty paradox:" a tiny fraction of the overall rural population receives royalties, but over 70 percent of the population support wind power. Spatial patterns of royalties should inform studies of land change in areas of energy development and studies of livelihoods possibly altered by energy royalties. Studies of wind power support or opposition should consider the flow and distribution of financial resources to landowners when attempting to understand the social acceptance of wind power.

Figure 1. Nolan and Taylor Counties, Texas.

1.1. The Property Advantage

One important spatial and land-based aspect of royalties is the "property advantage," which refers to unequal land-tenure patterns that may encourage unequal distribution of turbines and royalties. The distribution of royalties is important because it affects household wealth through increased land value [8]. More broadly, the spatial distribution of ownership of energy assets has attracted little critical attention, apart from recent work on shale gas in Denton, Texas [9]. Although Sowers included royalties into a "please-in-my-backyard" explanation for high apparent acceptance of wind power in rural Iowa [10], he did not estimate spatial distribution. Scholars of social, economic, and political aspects of wind-power expansion in North America have focused on other economic benefits, such as direct employment and indirect economic activity, arising from wind-power development, but livelihood implications are unknown.

More critical work reports possible tensions between landowners. In Ontario, Canada, wind turbines sited on private land generate income of at least $8000 per year, but neighbors without turbines receive no compensation. Walker *et al.* [11] report that wealthy farmers were getting wealthier because of lease payments and other farmers argued that lease payments helped sustain their farm operations. One-third of respondents to their study agreed strongly with the claim that "residents living near wind turbines, but without one on their land should also receive financial benefits" ([11]; p. 738). This finding resonates with a warning that energy payments could "undermine social capital" in U.S. rural communities because so few landowners received income from energy ([8]; p. 22). The energy revenue stream is highly concentrated because of the small number of rural households that own land.

Royalty payments are absent from studies regarding community support or opposition to wind power. Swofford and Slattery [5] measured public attitudes to wind power near a 100 MW wind farm in Texas, but their questionnaire did not prompt respondents to report whether they received royalties or lease payments. Abbott [12] found that public comments regarding wind farms in Washington state were positive with respect to impacts on agriculture, but it is not clear whether material benefits formed part of the view of wind power as "a new phase for extraction economies" ([12]; p. 981). Mulvaney *et al.* [13] reported that support for Indiana wind farms included the claim that turbines would help preserve farming or prevent farms from being turned into housing subdivisions, but they did not mention leases or royalties. Kahn [14] reported improvements in quality of life in places with wind farm development, specifically lower property taxes and increased school quality, but did not analyze royalty payments to landowners. Brown *et al.* [15] found an increase in county-level income and employment resulting from wind power, but did not indicate the specific impact of royalties and land leases on the estimated $11,000 increase in annual personal income per MW of wind capacity installed.

One possible reason for the focus on economic and employment aspects is reliance on the Jobs and Economic Development Impacts (JEDI) model, developed by the National Renewable Energy Laboratory, which includes a measure of landowner lease payments between $3000 and $7000 per MW per turbine per year (default = $6000). Several research teams have relied on this input-output model to estimate direct employment in construction and operational phases and indirect economic activity [16,17]. Greene and Geisken [18] applied JEDI to a case in Oklahoma, estimating nearly $400,000 in land lease payments for a 147 MW wind farm, but they did not report the number of landowners affected in a county nor the spatial distribution of the wind turbines on properties. Indeed, JEDI estimates are not spatially explicit, focusing at county or state levels rather than on individual landowners. Moreover, lease payments are considered synonymous with royalties, while in practice, leases for roads and other infrastructure represent a separate revenue stream from royalties, which are based on variable electricity output per wind turbine and electricity price.

1.2. The Royalty Paradox

Following from the "property advantage" relating to land distribution is the broader social and political significance of royalties. We describe this as the "royalty paradox," which other scholars have suggested. For example, Slattery *et al.* [6] reported high social support for wind power in the west Texas and Iowa, with more than 70 percent of respondents to a mail-out survey indicating that wind energy improved employment and the overall economy. But they noted that, ironically, only "relatively few residents" ([6]; p. 3699) received direct benefits in the form of royalty income from wind power. Wind turbines were "seen as the vehicle that will reverse economic decline" plaguing many rural communities ([6]; p. 3699). Economic benefits, in terms of royalty payments, may help explain why people in some parts of the U.S. are highly supportive of wind power; however, we do not know how these payments are distributed spatially.

Financial aspects of the "royalty paradox" have been suggested by many journalists and wind-power promoters. Just as owners of mineral rights receive royalties based on the extraction of hydrocarbon resources, owners of surface land benefit from wind power generated from turbines sited on their land. Landowners with turbines receive royalties estimated at between $4000 and $10,000 per wind turbine sited on their land per year, depending upon electricity output and contract terms [19,20]. Sowers [10] reported that landowners in Iowa received between $2000 and $2400 per turbine per year under 30-year contracts for turbines ranging from 0.75–0.90 MW capacity. However, scholars have not yet examined royalties from wind-power generation in spatial detail. Weber *et al.* [8] report that approximately 8 percent of farms in the U.S. Plains received an energy payment for wind or natural gas. They concluded that energy payments on a national basis reached 3.4 percent of farms and totaled $2.3 billion in 2011 compared to $11.8 billion in annual government payments to farmers, although important spatial and temporal uncertainties persist.

More generally, financial benefits are part of community support for wind power in cases where public input is highly limited and large landowners have influence on siting decisions. Sharing or distributing financial benefits helps predict support for wind farms [11,21,22]. Bohn and Lant [23] predicted that "local investiture" and "local control over siting" wind-energy projects will lead to strong local and community support for wind power, but it was not clear whether "local investiture" includes royalty payments made to individual landowners ([23]; p. 97). This view echoed Pasqualetti's [24] prediction that support for wind will increase as economic returns and property values increase, but the specific influence of royalties in economic returns was not elaborated. Walker *et al.* [11] indicate that determining how financial benefits from wind farms are distributed should demand attention of researchers and policy makers.

The property advantage and royalty paradox represent new ways to analyze how utility-scale renewable energy may alter land-based livelihoods of some landowners, and possibly encourage land-use changes. Determining the spatial distribution of royalty payments from wind energy and estimating their total value are important first steps toward this objective.

2. Research Design

2.1. Study Region

Our findings focus on Nolan and Taylor counties, a microcosm of a wider process that has made Texas the leading U.S. state in wind power. Nolan County (population = 14,700) had a median household income of $37,500 in 2009. Taylor County, where Abilene is the county seat, has 126,600 residents and a median household income of $40,400 [6]. Rural landowners in both counties are at the eastern margins of the Permian Basin oil fields that developed in the 1920s and 1930s [25,26] and have some familiarity with energy payments and leases, mainly through family history and legal expertise, but no published data or analysis on oil and gas royalties in the study region exist.

Writing in the 1960s, Abilene journalist A.C. Greene ([25]; p. 149) wrote that wind was the "greatest natural resource in West Texas" but it was also the cause, or the excuse, for most of what is "uncomfortable, abominable, and harsh about the country," creating difficult conditions for ranching and farming. This steady wind would later become characterized more optimistically as favorable wind climatology, which, coupled with buoyant demand for electricity, little political opposition to wind-farm construction, and the country's lowest permitting requirements, made west Texas the North American leader for wind power. Nolan and Taylor counties, Texas (Figure 1), received large wind-power investments dating from the 150 MW Trent Mesa site, built in 2001 by Enron's wind division. By 2009, the counties and adjacent areas had approximately 3000 MW of installed wind power. Installation of new turbines in Nolan and Taylor counties slowed after 2009, but completion of new transmission lines around 2011 helped increase efficiencies and provide incentives for future development. Sweetwater, Nolan County's largest city, aims to brand itself as the center of wind-energy development by promoting trade shows, encouraging manufacturing related to wind power, and hosting a wind technician training program at a local community college (Figure 2).

The large and rapid investments in Texas wind power rely upon legislation promoting renewable energy standards and electricity market deregulation, authorized in 1999, just a few days after then-governor George W. Bush declared his presidential candidacy. This statute resulted from Bush's desire to encourage wind—he told his top utility appointee to "go get smart on wind" ([7]; p. 121) in 1996—and Enron chief Kenneth Lay's desire for a deregulated electricity market in Texas, which helped create conditions for Enron's construction of the Trent Mesa wind farm, signaling the start of the wind boom in Nolan and Taylor counties [7]. Complementing Bush's stance on renewables, Governor Rick Perry in 2005 authorized Competitive Renewable Energy Zones (CREZ), which linked rural regions with the greatest wind energy potential to Texas's growing urban areas through a $7 billion transmission lines project capable of moving 18,500 MW [27]. These policies, combined with the federal

Production Tax Credit (PTC) for renewable power generation [28] and the fact that Texas offers simple permitting of wind farms, requiring no public consultation or deliberation [23,29], created powerful economic and political incentives for wind farms.

Figure 2. Wind turbine blade on Interstate 20 in Sweetwater suggesting the importance of wind energy to the city's identity and economy.

However, the rapid growth of wind power in Texas inspired work critical of subsidies that were "compensating the market for unfavorable production locations" for wind power and had resulted in negative electricity prices ([30]; p. 30). Zarnikau ([31]; p. 3912) noted that wind-generated power cost three times more than producing electricity from natural gas, adding that wind power in the Electricity Reliability Council of Texas (ERCOT) produced "adverse impacts on system reliability." Recent reports indicate that the negative prices reported in 2011 and 2012 disappeared in 2013 and 2014 because of ERCOT's switch to a nodal system and increased transmission capacity through the CREZ program [27,32].

At the county level, property tax abatements and a "wind welcomer" posture by public officials and landowners attracted wind farm construction to the region [33,34]. Many political elites view tax abatements as essential to attracting firms to establish wind farms, although some officials are skeptical of the long-term benefits on public finances and infrastructure, suggesting an impending "day of reckoning" ([33]; p. 847). County officials justify tax abatements to constituents by arguing that the positive economic benefits— increased employment and economic activity—are greater than property tax forgone through abatements. Economic and political elites in the study area confirm the influence of county, state, and federal policies for attracting wind-energy firms, but they also note the importance of the under-utilized 345 kV transmission line that carries power from gas-powered electricity generators in the Odessa and Midland through Nolan and Taylor counties to Dallas and Fort Worth [7].

Wind farms in the study region are sited exclusively on private properties. Firms establish land-use agreements with landowners, who usually pursue legal counsel as they seek to maximize their return on land leased for locating turbines, roads, substations, and maintenance facilities. During our previous research, key actors (large landowners, elected officials, business owners) told us that royalties were between $8000 and $10,000 per turbine per year [33]. Our earlier research led us to agree with Kahn [14] and Brown *et al.* [15], viewing the lease as a voluntary transaction for a land use that does not replace a more lucrative land use from the perspective of the land owner, and that the firm offering the contract prefers the site to other possible locations. The firm, of course, is influenced by incentives such as the federal PTC, which encourages rapid construction, and county-approved property tax abatements.

2.2. Data Sources

Location of wind turbines was determined by the point locations for individual turbines obtained in 2010 from the Federal Aviation Agency (FAA), which we confirmed by checking data from the American Wind Energy Association and the Texas Public Utilities Commission (PUC). Wind turbines must be registered with the FAA because they are considered to be a flight obstruction; they must also be registered with the PUC because they provide electricity to ERCOT, the grid operator. In addition, we visually compared the mapped turbine locations against aerial photography because wind turbines are obvious features of the landscape. Each turbine has a known "nameplate" or potential output measured in megawatts. We omitted turbines outside Nolan and Taylor counties and a small number of turbines within the counties owned by state agencies. Assigning turbines to land parcels in our two-county study region yielded 1701 turbines with ~2700 MW capacity.

Next, GIS was used to assign wind turbines to land parcels to identify which landowners had wind turbines. We obtained geo-located parcel data accurate to the 2010 tax year from Nolan and Taylor County Appraisal Districts. We eliminated urban properties using a separate GIS layer identifying the boundaries of Sweetwater, Abilene, and other cities, which resulted in 17,500 total non-urban parcels in the two-county region. Next, we consolidated parcels according to tax address where parcels of land having the same tax address were combined into a single landholding. This procedure yielded 8370 landholdings with unique tax addresses. A spatial join was used to connect wind turbines to tax parcels, which yielded 241 private landholdings with turbines in the two-county area.

2.3. Royalty Estimation

To estimate royalties generated by wind turbines we employed electricity output, wholesale electricity price, and royalty rate. The Texas grid operator, ERCOT, reports electricity production by wind farm in 15-minute intervals. However, use of these electricity production data would not materially improve our royalty estimates because electricity sales prices and the royalty rates are confidential between landowner and wind farm operator. Neither of the latter figures is available in precise terms. Moreover, our goal was to determine the spatial distribution of estimated wind-energy royalties rather than establish a precise figure for royalties, which is not feasible because of the private nature of contracts governing transactions between wind-farm operator and landowner.

Instead of actual electricity output we employ capacity factor (CF), which is the efficiency of the wind turbine expressed as a percentage. CF is a product of engineering characteristics (hub height, rotor characteristics, scheduled maintenance, and sweep area of rotor blades, among others), wind characteristics (air density, wind velocity), transmission capacity, and electricity price. Ironically, CF data are made available in real time to wind farm technicians and managers, but are not reported transparently to the public; rather, CF data are notoriously shrouded in secrecy and often over-estimated by wind farm operators [35].

We considered various CF estimates appropriate to our study region. A Nolan County based group, the West Texas Wind Energy Consortium, commissioned a study of economic impacts of wind power that relied on a 35 percent CF estimate as the "accepted norm for Rolling Plains wind energy projects" ([36]; p. 7). The Texas Comptroller General [37] reported that the mean CF for Texas wind farms installed in 2004 and 2005 was 39 percent, and cited at least one specific wind farm in the study region, Sweetwater-2, as having a CF between 35 to 40 percent. Bolinger and Wiser [28], based on data from 24 projects with installed capacity of 2275 MW, reported the U.S. average CF at between 33 and 35 percent for projects installed after 2003; however, Boccard [35] compared US installed capacity with output, arriving at 25.7 percent CF. A recent global estimate of wind power potential used a 20 percent CF [38]. The US average CF has remained at about 32 percent since 2000 [3].

In the Texas grid, transmission capacity lagged so far behind installed capacity that NextEra, a wind-farm developer, built a 230-mile transmission line that boosted capacity factor of a west Texas wind farm from 20 to 29 percent. Lack of transmission capacity resulted in curtailment, which was most severe in 2009, immediately prior to construction of transmission lines through the CREZ [39]. The trend is for capacity factor to remain at approximately 32 percent in west Texas because new wind farms are being built in areas with relatively low wind quality [3]. Even with the construction of new transmission lines, wind farm managers in the region reported informally to the first author the occasional use of "economic curtailment," a situation in which the wind farm operator orders turbines into neutral position because the price of electricity does not compensate the cost of operating the turbine.

The second and third variables (wholesale electricity price and royalty rates) used to estimate royalties may have less variability but they are still cloaked in secrecy because landowners negotiate confidential contracts with wind-power firms. The wholesale electricity rate is estimated at approximately $40 per MWh [36], although the figure of $35 per MWh for Texas was reported recently [3], which is likely to include a Renewable Energy Credit (REC) [28]. It is known that several wind farms have long-term contracts with utilities, while others sell on the spot market, so any single figure is unlikely to match actual prices received. However, we consider our figures to be representative of the most likely prices that are used to compute royalties. Moreover, imprecision on electricity price does not influence the spatial distribution of estimated royalties.

Nolan County landowners receive a royalty, normally 4 percent or a minimum guaranteed payment, on the basis of the wholesale price of electricity produced [36,37]. Key actors in wind power in Nolan County noted that the Trent Mesa wind farm was on a 1 percent royalty rate while contracts signed in 2009 had reached 5 percent [40,41]. Landowners with turbines report differences between "good" and "bad" contracts, in that "bad" contracts have a low royalty and do not carry a guaranteed minimum payment, and often were negotiated during the early 2000s, the "wild west" period, according to one landowner. Ranchers who were among the earliest to host wind farms obtained royalty rates and payment terms inferior to landowners who signed contracts later. One key actor in wind power refers to discussion about contracts as "Dairy Queen talk," the sort of discussion among rural landowners over lunch specials at Dairy Queen, a fast-food chain popular among farmers and ranchers in rural Texas. The reference to "Dairy Queen talk" suggests that variability exists among royalty rates, but that is not possible to determine actual rates without gaining trust among landowners. Therefore, we rely on the 4 percent royalty rate in our calculations.

To estimate royalty rate (per turbine per year) from wind power we created an equation relating MW capacity (CF), wholesale electricity price, and royalty rate:

$$\text{Estimated royalty (per turbine per year)} = \text{CF (MW)} * \text{Price (MW/h)} * \text{hours/year} * \text{royalty rate} \quad (1)$$

Using Equation (1) we computed a low royalty estimate (CF = 25%, wholesale electricity price = $35/MWh, and royalty rate = 4%), a high estimate (CF = 35%, wholesale electricity price = $45/MWh, and royalty rate = 4%). We calculated a mean royalty estimate per turbine from high and low estimates, then we applied this to the entire wind farm and aggregated to the landholding on which the turbines were located. Table 1 reports wind farms in the study region, their estimated electrical production and royalty rate per turbine per year. For example, the 67 turbines at Buffalo Gap would produce ~ 264,000 MWh per year at capacity factor of 25 percent, yielding gross revenues of $9.2 million at $35 per MWh. But at 35 percent capacity factor, and $45 per MWh, gross revenues would increase to $16.9 million per year. A fixed royalty of 4 percent would yield a mean, between high and low estimates, of $7726 per turbine per year (Table 1).

We used distance as a proxy for absentee ownership and land rental. Specifically, we determined driving distance from the *situs* (property) address, defined as the latitudinal and longitudinal location of the centroid of a single land parcel or aggregated parcels, to the tax appraisal (mailing) address. We assumed that the address listed for mailing the property tax statement is the same as the residential address, although it is possible that an attorney or financial advisor receives the tax correspondence for some landowners. The latitude and longitude coordinates of the parcel were entered into a Google Maps directions form, yielding turn-by-turn driving directions. To determine the driving distance for all 241 landowners with turbines, we wrote a C# script to query Google Maps for driving distance and time between two points excluding the turn-by-turn instructions. These distances were then added as landowner attributes.

Table 1. Summary of wind farms, capacity, and estimated royalty.

Wind Farm Name	No. Turbines	MW per Turbine	Total MW	Estimated Royalty per Turbine per Year ($)
Buffalo Gap	67	1.8	120.6	7726
Buffalo Gap 3	73	2.3	167.9	9873
Buffalo Gap II	139	1.5	208.5	6439
Callahan Divide	76	1.5	114	6439
Champion (Roscoe II)	54	2.3	124.2	9873
Horse Hollow II	125	2.3	287.5	9873
Horse Hollow	292	1.5	438	6439
Inandale (Roscoe IV)	143	1.0	143	4292
Roscoe	100	1.0	100	4292
South Trent Mesa	43	2.3	98.9	9873
Sweetwater	171	1.5	257	6439
Sweetwater 4a	124	1.0	124	4292
Sweetwater 4b	46	2.3	105.8	9873
Sweetwater 5	34	2.3	78.2	9873
Trent Mesa	101	1.5	151.5	6439
Turkey Track	113	1.5	169.5	6439
Total	1701		2688.1	

We interpret results based on our knowledge of the study region, obtained through key informant interviews conducted in 2010 and subsequent annual two-day field trips with undergraduate students enrolled in a "Geography of Energy" course taught by the first author. During these field trips, which included wind farm site visits and panel discussions, students inquired about royalties. Answers from key informant interviews and panelists, who included county commissioners and landowners, helped us determine the values for formula (1) and interpret results.

3. Results and Discussion

Total royalties for the two-county study region were estimated at ~$11.5 million per year, slightly less than the $12.264 million per year estimate of royalties accrued from 2500 MW installed capacity, which assumed a 35 percent CF, $40/MWh wholesale electricity price, and 4 percent royalty rate [36]. The mean annual royalty received per landowner per year was estimated at $47,879 (for 7 turbines on 274 hectares). The median annual royalty was $25,756 (for 4 turbines on 129 hectares), less than median household income in either county. Mean driving time from landholding to tax address was slightly more than 2 hours, although the median driving time was 38 minutes. The mean landholding for the sample was 274 hectares (Table 2).

Only 241 landowners out of ~8370 rural landowners in the study region have wind turbines on their land. This small fraction, less than 3 percent, of rural landowners shows how pre-existing land-tenure patterns, the "property advantage," enabled some landowners to obtain royalties from wind farms because of factors relating to size or relative location of their land before the development of wind farms. A similar low fraction of participation in energy assets has been reported for shale gas underneath Denton, Texas, where 2.4 percent of subsurface mineral (gas) value was owned by family trusts and individuals with Denton tax addresses [9].

Table 2. Summary statistics for estimated royalty, distance between landholding and tax address, landholding area, and number of turbines per landholding (n = 241).

	No. Turbines	Estimated Royalty Per Landholding ($)	Driving Time (minutes) to Tax Address	Landholding Area (hectares)
Total	1701	11,538,769	33,198	66,117
Mean	7	47,879	138	274
Median	4	25,756	38	129
Mode	2	29,619	21	65
SD	11	70,850	260	494

The total area of the 8370 landholdings was 452,800 hectares (mean = 54.1 hectares), and median size of 2.4 hectares; this compares to the ~66,000 hectares of the 241 landholdings that have turbines located on their land. Some landowners without turbines may have refused to sign contracts with wind-power developers, but many others may have had land parcels too small for siting turbines, or land with unsuitable topography, low wind potential, or poor road access.

Distribution of royalties by quintile is shown in Table 3. The top quintile concentrates 60 percent of royalties, with a mean of $146,189 (median = $100,444) per landholding per year. This group (n = 48) has a mean landholding area of 842 hectares. By contrast, the fifth quintile captures less than 4 percent of total royalties (mean = $7656, median = $8584), with 1.3 turbines per landholding on average.

Table 3. Landholdings with turbines divided into quintiles for number of turbines and estimated royalty per landholding per year.

Quintile and Metric	No. Turbines	Royalty per Landholding per Year ($)	Area per Landholding (ha)
Q1 Total (n=48)	1020 (60.8%)	7,017,091 (60.0%)	40,425 (61.1%)
Mean	21.3	146,189	842.2
Median	14	100,444	497.5
Standard Dev.	17.9	111,288	883.9
Q2 Total (n=48)	328 (18.5%)	2,129,078 (19.3%)	11,442 (17.3%)
Mean	6.8	44,356	238.4
Median	6	39,492	211.5
Standard Dev.	1.7	10,932	152.4
Q3 Total (n=48)	173 (10.8%)	1,249,539 (10.2%)	5334 (8.1%)
Mean	3.6	26,032	111.1
Median	4	27,473	86.5
Standard Dev.	0.5	4134	87.0
Q4 Total (n=48)	117 (6.7%)	767,913 (6.9%)	5307 (8.0%)
Mean	2.4	15,998	110.6
Median	2	17,168	70.0
Standard Dev.	0.5	3340	84.5
Q5 Total (n=49)	63 (3.3%)	375,148 (3.7%)	3609 (5.5%)
Mean	1.3	7656	73.7
Median	1	8584	65.0
Standard Dev.	0.5	2067	60.6

Regarding the spatial distribution of royalties per landholding, there is a clear divide between relatively large ranches (used for livestock grazing and hunting leases) south of Sweetwater and small farms, mainly growing cotton without irrigation, near Roscoe (Figure 3). We found significant differences ($p < 0.05$) for royalty per landholding per year by landholding area (Table 4). Roscoe landholdings had a mean annual royalty of $19,310, while Sweetwater landholdings had a mean royalty of $63,000 per year.

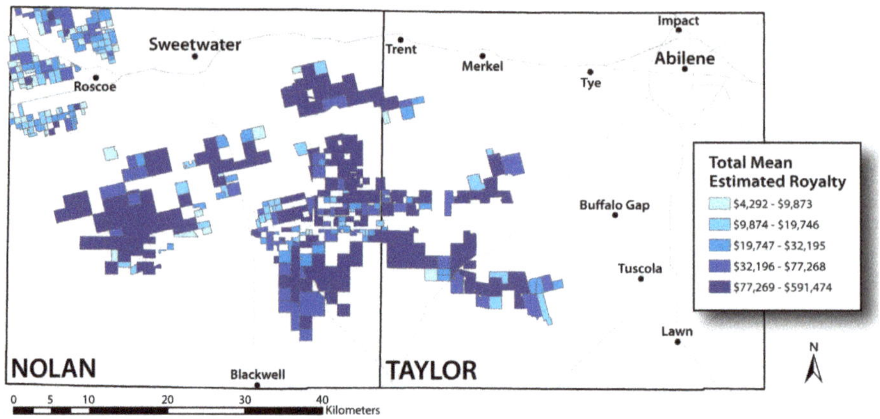

Figure 3. Spatial distribution of estimated royalties per landholder per year in Nolan and Taylor Counties.

The distribution of Roscoe royalties may be different than our estimates indicate. Around 2005, an entrepreneurial leader organized farmers as a "unitized" group of 120 landowners covering ~ 13,000 hectares [41]. Roscoe's many small landholdings, mainly devoted to rainfed cotton, were considered unattractive to wind developers. Unitization, a well-established practice in oil and gas production whereby small fields are grouped into one, and royalties are divided among members [42], helped make the region attractive to firms because they would negotiate one contract with a group, rather than with many small owners of rights. Royalties to all participants in the "unitized" group would include landowners without turbines, perhaps with land parcels too small or not well suited to the designs of the wind-farm developer, therefore slightly increasing the number of rural landowners receiving royalty payments.

Table 4. Descriptive statistics for landholdings with turbines in the Roscoe and Sweetwater areas. Royalty column refers to estimated royalty per landholding per year.

	Roscoe (n = 85)			Sweetwater (n = 156)		
	No. Turbines	Royalty	Hectares	No. Turbines	Royalty	Hectares
Mean	3.6	19,310	115.6	8.9	63,307	361.3
Median	3.0	12,876	65.0	4.0	32,195	182.5
Mode	2.0	8584	65.0	3.0	29,619	259.0
SD	3.1	15,857	97.6	12.9	83,351	593.2

We also determined the driving distance from landholding to tax address (Table 5) as a proxy for land rental and absentee ownership. We separated landowners with royalties into two groups, using 45 minutes driving time as a threshold, but found no significant difference in estimated royalty. Considering the tax address, we found that approximately 60 percent of total royalties accrued to landowners residing within Nolan and Taylor counties. It is possible that some of these landowners rent their land to ranch operators, but it is also possible that most landowners residing in Nolan and Taylor counties are the operators of their farm or ranch land. Around 30 percent of total royalties went to landowners residing elsewhere in Texas, and 7 percent of total royalties accrued to landowners residing outside Texas.

Table 5. Landholdings, turbines and estimated royalty per landholding per year according to location of tax address.

Tax Address	No. Landholdings	No. Turbines	Royalty ($)
Nolan County	105 (43.6%)	739 (43.4%)	4,867,684 (42.2%)
Taylor County	42 (17.4%)	312 (18.3%)	2,219,266 (19.2%)
Other Texas	79 (32.8%)	538 (31.6%)	3,652,962 (31.7%)
Other U.S.	15 (6.2%)	112 (6.6%)	798,857 (6.9%)
Total	241	1701	11,538,769

Estimating livelihood and land-use impacts is not possible with our findings, but theoretical claims and qualitative evidence offer several suggestions. First, Weber *et al.* [8] rely on a combination of life-cycle and behavioral frameworks [43] to argue that energy payments, such as wind royalties, may be consumed, saved, or invested. They assume that energy amenities or disamenities will affect land value, and, therefore, household wealth. Wind turbines are considered amenities and therefore add to land value. They argue that increased land value accounts for approximately 75 percent of total estimated private wealth from energy royalties. By comparison, consumption and farm investment are low, approximately 4.2 cents of marginal consumption for every dollar of energy payment.

Qualitative evidence from fieldwork offers other suggestions. Landholders offer many answers—except specific ones regarding actual investments—when approached on the topic of royalties. Previous work has suggested the existence of a particular moral economy in which wind power keeps farms and ranches in families, while landowners remain environmentally skeptical [33,34]. "Pay my notes (farm debt)" was the response of a cotton farmer in Roscoe when asked what he did with his wind-power royalties. For this farmer, landowners with turbines on cotton fields lost between 3 and 5 percent of farm area to pad site and access roads, making turbines attractive [41]. Similarly, a rancher outside Sweetwater joked that royalties allowed him to "maintain my hobby," by which he meant ranching in difficult climatic times. Farming or ranching is a "pretty tough gig" in the Nolan-Taylor region. Royalties help "pay bills and put kids through college." Landowners with turbines indicated that royalties helped families maintain their land, which had been in precarious state because of the declining cattle economy and lack of rainfall. But wind might also be just another type of energy income. Wind, according to one landowner, was "strictly money," echoing findings reported elsewhere [34].

It is not known whether life-cycle and behavioral frameworks [43] should be modified to particular conditions of west Texas. Perhaps the only published observation is that of Greene ([25]; pp. 24–25), who argued that west Texas landowners viewed income from cattle as more noble than income from oil, recounting the story of a rancher who "hated the idea that none of this new wealth [from oil] was his doing, as his cattle and his land deals had been." Greene also drew contrasts between the region's Christian morality ("God is the biggest industry in west Texas") and the "sinful humanity" and oilmen (a "villain, seen as a creature of lawyers and bankers") attracted by oil to west Texas ([25]; pp. 130, 163). It is unclear how broadly these assertions, if accurate, apply to current landowners, because as one landowner told us, "you do not gripe about the money that is coming in from oil wells, and you shouldn't be concerned about money coming in from the wind farms" ([34]; p. 859). It is also unknown whether a form of energy payment that is held in contempt, as Greene [25] suggested, is any different from wind royalties, which are viewed positively, in terms of consumption, savings, or investment. Whether households use royalties produced in this context any differently than models suggest is an open empirical question with implications for the vast areas of wind-power development.

Royalties from wind-power production accrue to hundreds of landowners in a vast area of central North America that hosts much of U.S. wind-power capacity. This area is spatially coincident with high reported social acceptance of wind power. The "property advantage" describes unequal distribution in royalties predicated upon land-tenure inequalities established before wind-power development. Because of their "property advantage," and without "unitization," some landowners will benefit significantly more than other landowners. The vast majority of people in rural areas who do not own land are excluded from royalties, but may benefit from wind power in other ways [14–16]. Future studies of social acceptance or rejection of wind farms should test for royalty effects, especially in rural areas where drought or poor commodity prices adversely affect farming and ranching, paying attention to particular moral economies that may not conform to models.

We also identify a "royalty paradox," the situation in which estimated royalties accrue to only a small fraction of rural landowners, even though support for wind power apparently extends into the general population, far broader than the small sub-population of landowners with turbines. We do not know how large payments to a small number of landowners translate into apparent high social acceptance. But it is likely that the landowners who benefit the most from royalties are prominent spokespersons for the wind economy, occupying key positions in organizations such as local school boards and other local civic groups that help influence opinion. They may be the "wind welcomers" previously described [33]. It is beyond the scope of this study to understand the behavior of local elites, but it is likely that landowners with royalties influence local discourse and opinion through varied means.

The influence of royalties on livelihoods and, by extension, land-change patterns, is not known. However, understanding the spatial distribution of royalties is an important starting point. A next step is to consider how landowners treat wind royalties, perhaps informed by a particular moral economy, and whether actual behaviors conform to the more well studied government payments that are independent of farm production decision making [43]. Even if a moral economy for wind royalties does not exist, it is possible that landowners at the lower quintiles treat royalties no differently than other sources of farm income—perhaps they just "pay notes," as one farmer told us. But at the upper quintiles, landowners may take different decisions with their royalty income, using royalties to fund capital expenses, invest in education of their children, or support their "hobby." Behavioral models may prove useful to offer estimates based on amount of royalty income and existing land cover (cotton or scrub vegetation), for example. An increase in land value among landholdings with energy payments, which Weber *et al.* [8] predict, may have numerous direct and indirect consequences that future research may determine. Certainly, the property value effect of wind power as an amenity may become apparent in property tax datasets and will figure highly in household wealth.

Our estimates of royalties are conservative because they do not include land leases for roads or maintenance facilities; nor does our estimated royalty include the signing bonus, which varies substantially among firms and landowner contracts. In addition, actual royalties could be higher because nameplate capacity may have been increased through software and hardware upgrades, either under manufacturer warranty or in the after-market period beyond warranty.

We cannot validate our royalty estimates because royalty rates are confidential between wind farm and landowner, and because wholesale electricity prices may vary considerably depending on contract terms between wind farms and utility. Future research aiming to model wholesale electricity price, coupled with analysis of actual electricity output per wind farm, could improve the accuracy of our estimates, but those estimates would still not have accurate estimates of the royalty rate.

The flow of wind-power royalties is contingent, of course, on the price of wholesale electricity, the output of turbines, and contract terms. Sales of land with wind turbines may slow, as farmers are reluctant to part with their royalty payments; however, some observers report the development of contract terms that allow land sellers to retain their royalties after selling the land. It is not yet known how landowners will respond when contracts approach the end of the projected lifetime of wind turbines, and whether re-negotiation opportunities may exist, especially as wind farms change ownership.

At a county level, future research could compare royalties that accrue privately with the cost of subsidies, in terms of property tax abatements, that are borne by the public in terms of property taxes assessed overall. County subsidies in the form of tax abatements may lower overall property tax rates, and help county officials politically, but they also encourage the flow of considerable benefits to a tiny minority of landowners.

Our findings indicate some challenges in conducting research aiming to link energy with land change and livelihoods in North America. The development of renewable energy in central North America, from Texas to Minnesota, is relatively recent, and while apparent impacts on land cover from the construction of wind turbines are plainly visible, possible effects on land use and landowner livelihoods are far more difficult to discern. The fact that wind power is located almost entirely on private landholdings means that landowners are "visible" in terms of tax addresses, but potentially high non-response bias will confront researchers aiming to determine the use of royalties through mail-out or online surveys. Qualitative and ethnographic methods may reveal important insights into the ways that renewable energy influences land uses through royalties, but require high levels of trust and long periods of field work.

4. Conclusions

Estimated royalties from wind power are unequally distributed among a small fraction of rural landowners who use the income in unknown ways to support their livelihoods. Landowners may try to overcome initial land inequality though cooperative agreements, known as "unitization," borrowed from oil and gas institutions. Royalties may help explain reported high support for wind power in the rural U.S., but livelihood and land-change outcomes of royalties are not yet known. We identified the "property advantage," the land-tenure patterns established before wind-power development that determine the unequal distribution of royalties. The "royalty paradox" refers to how royalty payments to a small number of landowners help produce high social acceptance, probably through the discursive and political actions of this small group of landowners. Future research should determine behavioral aspects of wind-royalty income and impacts on property value, which could influence land use and land cover.

Acknowledgments: We thank the Wind Initiative at Texas Christian University for supplying datasets and funding earlier research, and Nicole Persons for assisting in data collection in early phases of the research. Comments by Jackie Vadjunec, Claudia Radel, and anonymous reviewers improved earlier drafts.

Author Contributions: Christian Brannstrom designed the research, conducted research, analyzed the data, and wrote the paper; Mary Tilton conducted research; Andrew Klein conducted research and analyzed the data; Wendy Jepson designed the research and conducted research.

Conflicts of Interest: The authors declare no conflict of interest. The founding sponsors had no role in the design of the study; in the collection, analyses, or interpretation of data; in the writing of the manuscript, and in the decision to publish the results.

References and Notes

1. Brannstrom, C.; Vadjunec, J. Notes for avoiding a missed opportunity in sustainability science: Integrating land change science and political ecology. In *Land Change Science, Political Ecology, and Sustainability: Synergies and Divergences*; Brannstrom, C., Vadjunec, J., Eds.; Earthscan from Routledge: Abingdon, UK, 2013; pp. 1–23.
2. Public Land and Wind Energy. Available online: http://www.awea.org/Issues/Content.aspx?ItemNumber=858 (accessed on 25 July 2014).
3. Wiser, R.; Bolinger, M. *2012 Wind Technologies Market Report*; U.S. Department of Energy: Washington, DC, USA, 2013.
4. Wind Generates More than 10% of Texas Electricity in 2014. Available online: http://www.eia.gov/todayinenergy/detail.cfm?id=20051 (accessed on 7 June 2015).
5. Swofford, J.; Slattery, M. Public attitudes of wind energy in Texas: Local communities in close proximity to wind farms and their effect on decision-making. *Energy Policy* **2010**, *38*, 2508–2519. [CrossRef]
6. Slattery, M.C.; Johnson, B.L.; Swofford, J.A.; Pasqualetti, M.J. The predominance of economic development in the support for large-scale wind farms in the U.S. Great Plains. *Renew. Sustain. Energy Rev.* **2012**, *16*, 3690–3701. [CrossRef]

7. Galbraith, K.; Price, A. *The Great Texas Wind Rush: How George Bush, Ann Richards, and a Bunch of Tinkerers Helped the Oil State Win the Race to Wind Power*; University of Texas Press: Austin, TX, USA, 2013.
8. Weber, J.G.; Brown, J.P.; Pender, J. *Rural Wealth Creation and Emerging Energy Industries: Lease and Royalty Payments to Farm Households and Businesses*; Federal Reserve Bank of Kansas City: Kansas City, MO, USA, 2013.
9. Fry, M.; Briggle, A.; Kincaid, C. Fracking and environmental (in) justice in a Texas City. *Ecol. Econ.* **2015**, *117*, 97–107. [CrossRef]
10. Sowers, J. Fields of opportunity wind machines return to the plains. *Gt. Plains Q.* **2006**, *26*, 99–112.
11. Walker, C.; Baxter, J.; Ouellette, D. Beyond rhetoric to understand determinants of wind turbine support and conflict in two Ontario, Canada communities. *Environ. Plan. A* **2013**, *46*, 730–745. [CrossRef]
12. Abbott, J.A. The localized and scaled discourse of conservation for wind power in Kittitas County, Washington. *Soc. Nat. Resour.* **2010**, *23*, 969–985. [CrossRef]
13. Mulvaney, K.K.; Woodson, P.; Prokopy, L.S. A tale of three counties: Understanding wind development in the rural midwestern United States. *Energy Policy* **2013**, *56*, 322–330. [CrossRef]
14. Kahn, M.E. Local non-market quality of life dynamics in new wind farm communities. *Energy Policy* **2013**, *59*, 800–807. [CrossRef]
15. Brown, J.P.; Pender, J.; Wiser, R.; Lantz, E.; Hoen, B. Ex post analysis of economic impacts from wind power development in U.S. counties. *Energy Econ.* **2012**, *34*, 1743–1754. [CrossRef]
16. Slattery, M.; Lantz, E.; Johnson, B.L. State and local impacts from wind energy projects: Texas case study. *Energy Policy* **2011**, *39*, 7930–7940. [CrossRef]
17. Lantz, E.; Tegen, S. Variables affecting economic development of wind energy. In Presented at WINDPOWER 2008, Houston, TX, USA, 1–4 June 2008.
18. Greene, J.S.; Geisken, M. Socioeconomic impacts of wind farm development: A case study of Weatherford, Oklahoma. *Energy Sustain. Soc.* **2013**. [CrossRef]
19. BBC News. The world's largest wind farms. *BBC News*, 25 November 2010.
20. Shlachter, B. Rancher's windfall. *Ft. Worth Star-Telegram*, 29 January 2006.
21. Baxter, J.; Morzaria, R.; Hirsch, R. A case-control study of support/opposition to wind turbines: Perceptions of health risk, economic benefits, and community conflict. *Energy Policy* **2013**, *61*, 931–943. [CrossRef]
22. Warren, C.R.; McFayden, M. Does community ownership affect public attitudes to wind energy? A case study from south-west Scotland. *Land Use Policy* **2010**, *27*, 204–213. [CrossRef]
23. Bohn, C.; Lant, C. Welcoming the wind? Determinants of wind power development among U.S. States. *Prof. Geogr.* **2009**, *61*, 87–100. [CrossRef]
24. Pasqualetti, M.J. Opposing wind energy landscapes: A search for common cause. *Ann. Assoc. Am. Geogr.* **2011**, *101*, 907–917. [CrossRef]
25. Greene, A.C. *A Personal Country*; Texas A&M University Press: College Station, TX, USA, 1979 (1969).
26. Moore, R.R. *West Texas after the Discovery of Oil: A Modern Frontier*; Jenkins Publishing and Pemberton Press: Austin, TX, USA, 1971.
27. Fewer Wind Curtailments and Negative Power Prices Seen in Texas after Major Grid Expansion. Available online: http://www.eia.gov/todayinenergy/detail.cfm?id=16831 (accessed on 7 June 2015).
28. Bolinger, M.; Wiser, R. Wind power price trends in the United States: Struggling to remain competitive in the face of strong growth. *Energy Policy* **2009**, *37*, 1061–1071. [CrossRef]
29. Parker, B.D. *Capturing the Wind: The Challenges of a New Energy Source in Texas*; House Research Organization: Austin, TX, USA, 2008.
30. Baldick, R. Wind and energy markets: A case study of Texas. *IEEE Syst. J.* **2012**, *6*, 27–34. [CrossRef]
31. Zarnikau, J. Successful renewable energy development in a competitive electricity market: A Texas case study. *Energy Policy* **2011**, *39*, 3906–3913. [CrossRef]
32. Tao, H.Y.S.; Srivastava, A.K.; Pineda, R.L.; Mandal, P. Wind power generation impact on electricity price in ERCOT. In Proceedings of the IEEE Power and Energy Society General Meeting, San Diego, CA, USA, 22–26 July 2012; pp. 1–7.
33. Brannstrom, C.; Jepson, W.; Persons, N. Social perspectives on wind-power development in west Texas. *Ann. Assoc. Am. Geogr.* **2011**, *101*, 839–851. [CrossRef]
34. Jepson, W.; Brannstrom, C.; Persons, N. "We don't take the pledge": Environmentality and environmental skepticism at the epicenter of US wind energy development. *Geoforum* **2012**, *43*, 851–863. [CrossRef]

35. Boccard, N. Capacity factor of wind power realized values *vs.* estimates. *Energy Policy* **2009**, *37*, 2679–2688. [CrossRef]
36. New Amsterdam Wind Source. Nolan County: Case study of wind energy economic impacts in Texas. 2008. Available online: http://www.moakcasey.com/articles/viewarticledoc.aspx/Nolan%20County%20Case%20Study.pdf?AID=168&DID=288 (accessed on 2 March 2009).
37. Combs, S. *The Energy Report*; Texas Comptroller of Public Accounts: Austin, TX, USA, 2008.
38. Lu, X.; McElroy, M.B.; Kiviluoma, J. Global potential for wind-generated electricity. *Proc. Natl. Acad. Sci. USA* **2009**, *109*, 10933–10938. [CrossRef] [PubMed]
39. Wiser, R.; Bolinger, M. *2008 Wind Technologies Market Report*; Department of Energy: Washington, DC, USA, 2009.
40. Confidential interview with respondent 17, 20 May 2009.
41. Confidential interview with respondent 25, 2 October 2010.
42. Weaver, J.L. Unitization of Oil and Gas Fields in Texas: A Study of Legislative, Administrative, and Judicial Policies. Resources for the Future: Washington, DC, USA, 1986.
43. Whitaker, J.B. The varying impacts of agricultural support programs on U.S. farm household consumption. *Am. J. Agric. Econ.* **2009**, *91*, 569–580. [CrossRef]

land

MDPI

Article

The Quiet Rise of Medium-Scale Farms in Malawi

Ward Anseeuw [1,*], Thomas Jayne [2], Richard Kachule [3] and John Kotsopoulos [4]

1 CIRAD & University of Pretoria—The Center for the Study of Governance Innovation and the Post-Graduate School of Agriculture and Rural Development, University of Pretoria, Pretoria 0002, South Africa
2 Department of Agricultural, Food and Resource Economics, Michigan State University, Justin S. Morrill Hall of Agriculture, 446 West Circle Drive, Room 317c, East Lansing, MI 48824, USA; jayne@anr.msu.edu
3 Centre for Agricultural Research and Development (CARD), Lilongwe University of Agriculture and Natural Resources (LUANAR), P.O. Box 219, Lilongwe, Malawi; richardkachule@gmail.com
4 GovInn—The Center for the Study of Governance Innovation, University of Pretoria, Pretoria 0002, South Africa; john.kotsopoulos@up.ac.za
* Correspondence: ward.anseeuw@up.ac.za; Tel.: +27-12-420-50-22; Fax: +27-12-420-36-01

Academic Editors: Claudia A. Radel and Jacqueline M. Vadjunec
Received: 1 August 2015; Accepted: 8 June 2016; Published: 24 June 2016

Abstract: Medium-scale farms have become a major force in Malawi's agricultural sector. Malawi's most recent official agricultural survey indicates that these account for over a quarter of all land under cultivation in Malawi. This study explores the causes and multifaceted consequences of the rising importance of medium-scale farms in Malawi. We identify the characteristics and pathways of entry into farming based on surveys of 300 medium-scale farmers undertaken in 2014 in the districts of Mchinji, Kasungu and Lilongwe. The area of land acquired by medium-scale farmers in these three districts is found to have almost doubled between 2000 and 2015. Just over half of the medium-scale farmers represent cases of successful expansion out of small-scale farming status; the other significant proportion of medium-scale farmers are found to be urban-based professionals, entrepreneurs and/or civil servants who acquired land, some very recently, and started farming in mid-life. We also find that a significant portion of the land acquired by medium-scale farmers was utilized by others prior to acquisition, that most of the acquired land was under customary tenure, and that the current owners were often successful in transferring the ownership structure of the acquired land to a long-term leaseholding with a title deed. The study finds that, instead of just strong endogenous growth of small-scale famers as a route for the emergence of medium-scale farms, significant farm consolidation is occurring through land acquisitions, often by urban-based people. The effects of farmland acquisitions by domestic investors on the country's primary development goals, such as food security, poverty reduction and employment, are not yet clear, though some trends appear to be emerging. We consider future research questions that may more fully shed light on the implications of policies that would continue to promote land acquisitions by medium-scale farms.

Keywords: Malawi; medium-scale farmers; land; livelihoods; farmer trajectories; farm expansion; land acquisition; land consolidation; domestic investors

1. Introduction

After 40 years of relative inertia, African agriculture seems to have engaged a renewed structural transformation path. According to the most recent literature, the latter mainly concerns transformation in the large-scale farming and agribusiness sectors. Indeed, most of the attention on changing farmland ownership in sub-Saharan Africa over the past few years has been focused on large-scale foreign land investments [1,2], due in part to a rise in global food prices since the mid-2000s. These investments are significant, if not huge, in size, transforming the high-end of the sector and, as Anseeuw and Ducastel [3] note, initiating trends of corporatization and increased international financing of the agricultural sectors. Another stream of the literature focuses more on the transformation of the value-chains, in particular the dismantling of international commodity agreements and the liberalization of agricultural and agri-food markets, which are contributing to the restructuring and modernization of markets. Food markets are consequently becoming increasingly consumer-driven and vertically integrated, which leads to the tighter alignment of supply chains and promotes the emergence of fewer large farms [4,5].

Much less developed or analyzed in the literature are, however, the changes affecting other categories of the agricultural sector, such as the small- and medium-scale farmers. Recent national statistics—particularly in Africa—show nonetheless the rapid rise of medium-scale farmers in Africa [6,7]. Nationally representative smallholder surveys do not, however, shed much light on how rapidly medium- and large-scale farms are growing over time, how these farmers have acquired their farms, the prior status of the land they acquired, or the extent to which they are productively utilizing their farms. Nor do they inform whether these dynamics are related to the present renewed interest in agriculture and transformation of the value-chains (as described above), or whether they are more externally driven, or are endogenous processes of smallholder growth.

This paper endeavors to address this vacuum. The motivation is, therefore, to obtain a more detailed understanding of medium-scale farmers—how many there are, the size of their farms, their trajectories including background and socio-demographic characteristics, how they are acquiring their farms, as well as how rapidly they are changing the agricultural landscape and sector's structure overall. To do so, the paper presents the results of an in-depth empirical study implemented in Malawi. It is based on extensive qualitative and quantitative interviews and analyses of 300 farmers, the large majority of whom are classified as medium-scale farmers. Medium-scale farmers are defined here as farmers with holdings between 5 and 50 hectares (ha), occupying an intermediate position between small-scale, semi-subsistence production and larger-scale, more commercial farming.

The following section develops a conceptual framework to analyze medium-scale farmers as well as describes the data and methods underpinning the findings of the study. Section 3 reviews the case of Malawi, giving historical factors influencing the rise of medium-scale farms in Malawi, with particular reference to the country's land institutions. Sections 4 and 5 present the study's findings concerning the characteristics of medium-scale farm households and changes in farmland ownership patterns in Malawi. Section 6 discusses these results using the lens of our conceptual framework and presents agricultural trends and debates, before Section 7 highlights the main conclusions and policy implications of the paper as well as questions for future research.

2. A Framework for Analyzing Medium-Scale Farmers

2.1. Definitions

Medium-scale landholders/farmers are in this study defined as farmers with a total landholding of between 5 and 50 ha. These medium-scale entities are often characterized as occupying an intermediate position between small-scale, semi-subsistence production and larger-scale, more commercial farming. In accordance with this definition, small-scale farmers are farmers with a total landholding of less than 5 ha; large-scale farmers, with more than 50 ha. This definition often diverges from how government statistical organizations define farm-scale categories, which is typically on the basis of area under cultivation. Moreover, some African governments define medium-scale farms as those between 5 and 20 hectares. Our rationale for defining medium-scale holdings in the way we do is explained below in the methods section.

Three points related to this definition are relevant. First, although "medium-scale" and "emerging" are often used interchangeably, the "emerging farmer" term originally referred to previously disadvantaged farmers, particularly as a result of colonial government policies, who were attempting to make the transition to commercially based agriculture [8]. We deliberately retain the term "medium-scale farmers," both to avoid the debate over this emergent farmer narrative, and also to not automatically relate medium-scale to emerging. As made clear below, many medium-scale farms in Malawi may be neither emerging nor disadvantaged with respect to policy. Secondly, "medium-scale farmers" is also the term used by the Malawian authorities; although the size span retained for this category is generally 5 to 10 ha. The larger size retained for this study allowed for the inclusion of estate farms in the sample, to better account for potential new land and agrarian dynamics, and to use this work in comparison with other recent literature from the region [6,7]. Thirdly, the size used for this categorization is the total landholding of the farmer, whether under customary or public land occupation, lease or freehold. This allows us to directly examine the degree of land utilization by farms of different landholding size categories.

2.2. Literature Review

The international literature on medium-scale farmers remains rather limited. This is related to the ever prominent debate which situates smallholder vs. large-scale farming [9], neglecting the existence of a large diversity of farming and production systems in-between. The paucity of literature regarding medium-scale farmers is certainly also related to semantics and definitions, as "smallholder" and "small-scale farmer" are often used as catchall terms grouping all farmers that are not large-scale [10].

This being said, the aforementioned questions regarding Malawi's medium-scale farmers are situated at the crossroads of four interrelated bodies of literature, highlighting distinct conceptual frameworks. Each of them focuses on specific features related to the characterization of medium-scale farmers or of their trajectories.

The first body of literature examines concepts of technical efficiency and accumulation driving changes in farm structure over time [11]. On one hand, economic development is considered to be related to the technical improvement of the production processes and to increased labor productivity. Mellor [12] and many others showed that broad-based gains in productivity of small-scale farmers favored better development outcomes in terms of farm growth as well as poverty reduction, employment generation, and overall economic growth. As such, the improvement of farm efficiency and labor productivity is closely linked to capital accumulation trajectories of the production units. Indeed, according to these authors, the improvement of production techniques and of labor productivity allows for the generation of an increasing surplus of marketable production, enabling savings. These savings can then (partly) be reinvested in the means of production (e.g., farm equipment, modern inputs, and land), which further enhances labor productivity. More efficient farmers with favorable access to capital can engage in a virtuous growth and development cycle, established on technical improvement and capital accumulation [13], acquiring farms of less efficient neighbors after they migrate into non-farm jobs, leading to farm consolidation. Even though economists, in particular those who deal with agricultural development [14], have since enriched these founding theories, the supremacy of capital and technical efficiency as the main drivers of change in farmland ownership were rarely questioned. This accumulation pattern leads to agricultural mechanization, greater use of modern inputs, such as hybrid seed and fertilizer, and to increasing size of production units.

The second body of literature regards farm trajectories assessed through farm strategies, and more particularly diversification strategies (see [15–18], among others). Besides diversification undertaken to manage risk, cope with shock, or escape from agriculture in stagnation or in secular decline periods, farmers diversify for accumulation objectives. Reardon et al. [19] note that while diversification (especially driven by pull factors) is usually associated with an upward spiral of incomes and assets for the households, the diversification (by push factors) sometimes extracts households from poverty as well, though perhaps over a longer time period. Anseeuw and Laurent [20] show that, in many developing countries, a large number of farmers are constrained, at least in the early stages of their farming activity, to look for other sources of income in order to cover the costs of building up farm assets. Most of them engage in several independent activities or combine salaried and independent (including farming and non-farming) activities, in order to finance farm assets.

The third body of literature around medium-scale farmers is related to land dynamics and land acquisitions. The latter is particularly the case in the framework of the present growing interest in farmland [2]. This burgeoning interest in land has mainly led to an expanding literature focusing on foreign direct investment in land, foreign large-scale land acquisitions, etc. [1]. Lately, complementary works have shifted the focus onto what is called "indirect appropriation" [21] and the role of national institutions and domestic elites in the rush for land. The question of domestic elites, who are often urban based, directly relates to the issue of medium-scale farmers and to diversified accumulation patterns, through nonagricultural accumulation [7]. Related to this is, of course, the recurring debate in the development literature about the relative emphasis to place on the roles of endogenous growth of small-scale farmers vs. a more externally driven type of growth through non-farm accumulation, as depicted in the extreme case by the debate of small-scale vs. large-scale farms in fostering agricultural growth and economic development. Reviews [11] have re-affirmed the relevance of smallholder agriculture and its broader political economy effects. Together with the success of Asia's Green Revolution, when hundreds of millions of small-scale farmers rapidly adopted new technologies, this placed small-scale farm productivity at the center of the development agenda. At the same time, disillusionment with limited success of smallholder-based efforts to improve productivity in sub-Saharan Africa [22] and the apparent success of other countries, such as Brazil, in establishing a vibrant agricultural sector based on large farms have led many observers to view the development of large-scale mechanized farming as the appropriate path to modernization of the sector in Africa as well.

Lastly, the medium-scale farmer debate can be linked to the literature on the structural transformation in agriculture, which focuses on farm household activity and income structures in an evolving agricultural context marked by liberalization and trends of increasing economic integration [23,24]. This literature reviews the very different levels (and trajectories) of rural diversification among countries at various stages in the structural transformation process. According to this literature's framework, late developers, such as most countries of sub-Saharan Africa, are characterized by an incipient economic transition that is accompanied by an unfinished demographic transition (decrease of mortality rates with lingering high birth rates). This demographic structure leads to growing cohorts of new labor market entrants, which can result in opportunities for growth if the entrants can be accommodated, though if not, it represents a serious political and economic liability [25]. At farm level, Losch et al. [24] emphasize that large-scale integration of agricultural producers into the global economy has not happened. Most producers remain engaged in traditional crops—mainly staples—and in traditional marketing systems characterized by limited contractualization. They show that whereas poorer households diversify to mitigate risks, only households that are relatively well off can make larger investments and begin to specialize to take advantage of these new assets. In sub-Saharan Africa in particular, high levels of risk and limited economic opportunities constrain households' options and hence their economic returns. This situation can generate possible poverty traps for low-income farmers. If we believe that this story applies to much of Africa, then we might posit that the current farm structure in many countries on the continent will not generate broad-based income growth and hence we cannot rely on agriculture to generate the multipliers that lead to structural transformation [7]. There is a good amount of evidence suggesting this hypothesis stemming from the structural transformation literature could indeed be true, seeing how poverty rates have remained relatively high over the past several decades [26]. Considering these socio-economic challenges in light of the dramatic demographic shift underway on the continent, policy orientations should reintegrate these structural issues, stressing the importance of broad-based, inclusive agricultural growth. The latter generates demand for non-farm goods and services, which in turn can pull marginal farmers into better jobs in the non-farm economy, leading—once again—to structural transformation and demographic shifts.

2.3. Analytical Framework

In order to analyze the medium-scale farmers in Malawi, our analytical framework focuses on accumulation patterns and farm trajectories, examining how changes in farmland ownership patterns are associated with the socio-economic characteristics of households in different farm size categories. This leads to an in-depth analysis of farm and household livelihood strategies (where do the medium-scale farmers come from, are they full or part-time farmers, what are their sources of revenue, etc.). On the other hand, we assess the accumulation of farm and land assets by these farmers (how much land they own and/or cultivate, which land they acquired, how and when the land was acquired, etc.). Correlating these two (livelihood and land) will lead to a better understanding of how farm accumulation is happening in Malawi.

The contextualization of such analyses will then contribute to the broader literature discussed above, as related to large-scale land acquisitions and structural transformation in agriculture. Indeed, the livelihood approach will provide insight that contributes to the discussion on whether the farmers' trajectories are endogenous—growing internally out of smallholder agriculture—or, on the contrary, whether farm growth is externally driven through non-farm accumulation. These processes can then be related to the structural transformation processes in which these countries are engaged.

2.4. Data

Our analysis is based on a survey of 300 landholders, the large majority of which are "medium-scale" farmers. The survey is based on the implementation of a structured questionnaire, with closed and semi-open questions, leading to a largely descriptive methodological approach. Three hundred farm households were interviewed in the framework of this study. The sampling was done randomly through a sequential (every 5th or 10th farmer—depending on district) selection on an alphabetical list of the population of target farmers. Although the study focused on medium-scale farmers (5.01–50 ha), the results of the survey also show that 27 small-scale (0.01–5.00 ha) and 14 large-scale farmers have been included (Table 1). The latter mainly resulted from imprecise databases on which the farmers' selection and sampling were realized (see hereafter in Section 2.5 regarding the scope and limitations of the study). These small- and large-scale farmers have been retained, however, and used for comparative purposes allowing for a better understanding of specific characterizing elements related to these different farmer size groups. All farmers were selected from three districts—all situated in Central Malawi: Mchinji, Kasungu and Lilongwe. These districts were selected based on their progressive distances from the largest urban area in the region, the capital Lilongwe, allowing for the assessment of the urban impact on land-use dynamics. The district of Mchinji, straddling the eastern border of Zambia, represents the furthest point from the capital with the eponymous Lilongwe district being the closest. An equal sample per district was retained, with weightings made afterwards according to the proportion of medium-scale farmers per district. The full sample of medium-scale farmers retained is therefore 259, of which 90 were in Mchinji, 87 in Kasungu and 82 in Lilongwe districts (Table 1).

Table 1. Farming households interviewed in the framework of this research.

Landholding Size (ha)	Total (*N* = 300)	Mchinji (*N* = 104)	Kasungu (*N* = 99)	Lilongwe (*N* = 97)
0–5	27	6	7	14
5.001–10	168	46	57	65
10.001–20	73	32	24	15
20.001–30	9	4	4	1
30.001–40	6	5	1	0
40.001–50	3	1	1	1
+50	14	8	5	1

Source: Authors' medium-scale farm survey in Mchinji, Kasungu and Lilongwe Districts undertaken by Lilongwe University of Agriculture and Natural Resources (LUANAR), University of Pretoria (UP), and Michigan State University (MSU) (2014–2015).

The choice of these districts, and Malawi as a case-study in particular, is all the more relevant as the country has had a long history of estate farmers (lease and freehold farms, established since the Banda era onwards—see Section 3). A large proportion of the total landholdings and area under cultivation is in the hands of one percent of the nation's farmers (who control five or more hectares of land—see Table 2 [27,28]), and the vast majority of farmland in this category is controlled by people holding over 10 ha of land. It is also of interest as the country and these districts are characterized by huge land pressures, with a growing rural population and small and decreasing farm sizes over time.

Table 2. Distribution of landholdings and cultivated area by farm size in Malawi, according to the World Bank's 2013 Integrated Household Panel Survey in Malawi. Source: [27].

	Landholding Size				Total
	0–2 ha	**2.001–5 ha**	**5.001–10 ha**	**>10 ha**	
# of farms	2,355,461	217,757	18,446	10,230	2,601,893
% of total # of farms	90.53%	8.37%	0.71%	0.39%	100%
Area cultivated (ha)	1,496,465	567,556	113,241.6	776,407.6	2,953,670
Total landholdings (ha)	1,693,828	619,700.7	116,838.7	897,553.8	3,327,921
Mean farm size (ha)	0.72	2.85	6.33	87.74	1.28
Mean ratio of land cultivated to land owned	0.86	0.92	0.97	0.87	0.89
% of total landholdings	50.9%	18.6%	3.5%	27.0%	100%
% of total cultivated land	50.7%	19.2%	3.8%	26.3%	100%

Note 1: This table, i.e., the data and classification, is based on the nationally representative 2013 Integrated Household Panel Survey in Malawi, undertaken by the National Statistical Office. These findings may substantially underestimate the amount of land under medium-scale holdings in Malawi. It is well documented that population-based surveys such as the Malawi Integrated Household Panel Surveys and other Living Standards Monitoring Survey (LSMS)-type surveys, tend to under-report relatively large farms compared with census-based approaches for determining farmland ownership [27]. Moreover, the National Statistical Office indicates that the Integrated Household Panel Survey covers farms in customary lands only, and does not include the "Estate Sector," which purportedly contains roughly 30,000 farms encompassing over 1 million ha of land [29]. If this Estate Sector land were added to the ">10 ha" category in Table 1, it would show that roughly 44 percent of Malawi's farm landholdings are controlled by a small proportion of the population, with holdings greater than 10 ha. Note 2: The appendix shows the 2013 Integrated Household Panel Survey for the three selected districts.

From an agro-ecological point of view, the three districts are relatively similar. Falling within a semi-arid to sub-humid climate, production systems combine dryland crop production, often on individual plots (under traditional, leasehold or private ownership), with extensive cattle and small-stock grazing, generally on communal lands. Maize represents the major crop covering, on average, 76 percent of all cultivated land in the three districts. Other food crops include sorghum and millet. Cassava, potato, and sweet potato are also generally planted by small- and medium-scale farmers. The area planted with cassava, and often intercropped with maize, has increased sharply over the past 10 years, particularly in densely populated areas such as around Lilongwe. Kasungu district, on the other hand, is characterized by significant tobacco production as the dominant cash crop, providing 71 percent of export earnings, from the large-scale estate farms as well as from smaller-scale farming sectors [30]. Despite the input subsidy programs, the overall productivity remains low (e.g., average maize productivity is around 1.5 tons/ha). As it is with Malawi overall, the three districts are characterized by high population densities (Lilongwe rural has 216 inhabitants/km^2, Mchinji has 136 inhabitants/km^2 and Kasungu has 80 inhabitants/km^2). Farm sizes are thus small, with average sizes of 1.66 ha in Mchinji, 1.32 ha in Kasungu and 1.14 ha in the more populated Lilongwe district. According to government data, the urban and economic dynamics in the Lilongwe region result not only in overall smaller farm sizes on average, they also lead to relatively more significant consolidation as the region has proportionally more bigger farms and landholdings (1.10 percent of the total farms and 24.90 percent of the landholdings are bigger than 10 ha, compared with 0.39 percent and 2.63 percent, respectively, in Kasungu). Kasungu, on the other hand, is characterized by a larger number of estate farms [29].

2.5. Scope, Methods and Limitations

Our analysis is largely descriptive, in line with our objectives of documenting the changes in land controlled by medium-scale farmers and their defining characteristics. Our focus is on the medium-scale farmers, and we report data on small- and large-scale farmers for comparison purposes only. Emphasis is put, on one hand, on the categorization of medium-scale farmers and, on the other hand, on dynamics and trajectories of farmers and farm entities. We present descriptive statistics and, in some cases, the results of basic statistical tests for between-group comparisons. Some statistical analyses, however, are actually between-group (small-scale and large-scale) and sub-group (medium-scale), no doubt impacting the results of analysis of variance tests, but are included here for comparison purposes. Although this allows for some broader analyses and for empirical findings, we remain cautious about overall generalizations.

Such limitations regarding generalizations are also related to the lack of existing data. The study took into account several logistical challenges with respect to the identification of suitable farm households for the sample. First, none of the selected districts had up-to-date lists containing the sizes of landholdings within their respective jurisdictions. Lists of the population of farms with medium-scale landholdings therefore had to be prepared by Agriculture Extension Development Coordinators (AEDCs) from the Extension Planning Areas (EPAs). Malawi is divided into eight Agricultural Development Divisions (ADD) and 28 districts. The ADDs are further separated into 173 EPAs. Second, there was some confusion regarding the units of measurement, with some farmers and AEDCs mixing hectares with the smaller unit of acres. These challenges and the risks to the validity of the sample were recognized, precipitating the decision to return to all of the surveyed EPAs a few months after the initial survey to acquire the landholding size lists again and cross-check them with the lists used originally. This cross-check process was successful, affirming the integrity of the lists and accuracy of the farm data provided.

3. Land Policy and the Evolution of Medium and Large Farms in Malawi

Immediately prior to independence in the early 1960s, roughly 87 percent of Nyasaland's total land area was designated African Trust Land (Nyasaland was a British colony until 1964, when it became the independent Republic of Malawi). Less than 3 percent of the land area was freehold or government leasehold, with most of it being farmed by Europeans. The balance, roughly 10 percent, was public land used for State forest reserves and other public purposes.

Malawi was declared a republic on 6 July 1964, upon which it adopted a new constitution and Dr. Hastings Kamuzu Banda was elected the country's first president. His agricultural vision did not radically contrast with the colonial system, as it was mainly based on the establishment of, and support to, a large-scale farming sector. The main pillars of the rural economy promoted by the Banda regime were production of cash crops for export, and the export of migrant labor. The regime appreciated that the two could be facilitated by providing services of various kinds to commercial agriculture and neglecting traditional subsistence farming so as to reinforce its role as a reservoir of cheap labor [31]. The Banda government sought to transform legal tenure through a declaration in 1965 that economic farming systems were viable only when operated under a sound system of land tenure, based upon security of individual land ownership and negotiability of the title. As such, a portion of customary lands was converted into leaseholds on which *more entrepreneurial farmers* were provided with incentives and assistance to grow high value crops, especially tobacco [31]. The newly established estates had the further advantage of cheap finance collateralized by land acquired from smallholder farmers through acquisition of customary lands. Stambuli [32] observes that the estates also benefited from low interest rates and a favorable land tax regime.

Building on colonial development thinking, Dr. Banda fostered the creation of an elite class of Malawians that he thought would spearhead agricultural development and national growth, while excluding those individuals deemed to be less productive. Through this philosophy, the Agricultural Development and Marketing Corporation (ADMARC) "taxed" smallholder farmers through its pricing policies and used that income to promote estate farming for export [33]. Furthermore, because customary land was transferred to estates, credit to adopt improved technologies was less available to smallholders [32]. Soon, the term *mchikumbe*, meaning "master farmers", was being used to label the few farmers whose performance was approved of by the government extension workers, just as the case was in colonial days. Those classified as *mchikumbe* had easy access to loans for farm improvements, thereby reinforcing the advancement of these farmers.

During Dr. Banda's era, about a million hectares of land were converted to leasehold. This was land appropriated from customary land held by (or in reserve for) local smallholder farmers under the custodianship of local chiefs. Other advantages extended to estates included licenses to grow and sell burley tobacco (a privilege denied to smallholder farmers on grounds of technical incapacity, a need to control quality and prices, and its environmental impact), extension services, credit and marketing support. Such privileges were provided by the Ministry of Agriculture, the commercial banks, and ADMARC [31,32]. State bureaucrats, Members of Parliament (MPs) and party functionaries topped the list of beneficiaries. At the same time, senior civil servants were encouraged to go into commercial farming. Beneficiaries straddling the rural-urban divide moved profits from farming into urban real estate, retail trading and transportation.

Commack et al. [31] observe that the Structural Adjustment Programs implemented under pressure from the International Monetary Fund (IMF) and World Bank in the 1980s significantly disrupted the ways in which the system for channeling rents had worked under Dr. Banda. The IMF and World Bank policy of reforming the estate sector to reduce subsidies and make it more competitive, together with the shift against import-substitution industrialization, constrained the types of rent-creation and distribution that were formerly encouraged. As a result, inflation took hold. These and other economic factors had an impact on politics throughout the 1980s, resulting in dissent arising against Dr. Banda's regime from around 1991.

After the ousting of Dr. Banda during the multiparty elections of June 1994, the focus of economic growth and development shifted more towards small-scale businesses, the promotion of the informal sector of the economy and smallholder agriculture. With regard to the agricultural sector, firstly, the United Democratic Front (UDF)-led government introduced several agricultural programs, such as the Starter Park Program and drought mitigation programs, together with the encouragement of non-maize food crops. It was also under the UDF government that the Malawi National Land Policy was formulated in 2002. Among its key recommendations, the Land Policy sought to allow all customary land to be registered and protected by law against arbitrary conversion to public land. Furthermore, the traditional supervisory role of chiefs, clan leaders, headpersons and family heads in land matters was to be formalized and streamlined to allow uniform administrative procedures and transparency in all customary land transactions. Moreover, the policy stipulated numerous tenure system reforms, new rights for women, freezes on the amount of freehold land in Malawi owned by non-citizens, and prohibitions on the right of non-citizens to acquire title to new freehold estates [34].

Thereafter, the Democratic Progressive Party (DPP) regime of Bingu wa Mutharika and Bakili Muluzi (1994–2005) did not bring any revolutionary changes in land matters, but did implement several highly popular agricultural initiatives, most prominently the Farm Inputs Subsidy Program. During the Mutharika period, however, Malawi witnessed a spate of new land acquisitions from customary landholdings, primarily for the benefit of both foreigners and local elites. Chikaipa [35] observes that local communities, with support from civil society actors, have bemoaned and resisted most of the attempts on land grabs, but that government has remained "willfully" deaf, ignoring such concerns. Chikaipa [35] further notes that it is on record that no transaction has at any occasion been cancelled or put on hold due to the public outrage over the land acquisitions. Chikaipa [35] argues that as long as the power relationship between the state and local communities in the battle against land acquisitions is not equal, the resistance and opposition from the local communities and civil society actors are not enough to successfully thwart the land deal transactions.

This policy favoring land conversions from customary to titled land seems to be pursued by the current president, Peter Mutharika, the former president's brother. Although his track record with regard to land issues is not yet evident, he appears to be pursuing similar development strategies, with a few exceptions in terms of restructuring the public service operational framework. The Peter Mutharika administration does recognize the many challenges on land matters and has stated an intent to advocate for the enactment of all land-related Bills that are submitted to Parliament, and to conduct comprehensive public awareness campaigns on existing land laws.

The above agricultural political economy has had significant impact on the country's present agricultural and farm structures. As such, the most recent nationally representative survey of rural households shows that 30.5 percent and 30.1 percent of the total landholdings and area under cultivation are in the hands of one percent of the nation's farmers, who control five or more hectares of land (Table 2, [27,28]). The vast majority of farmland in this category is controlled by people holding over 10 ha of land.

4. Characteristics of Medium-Scale Farmers in Malawi

The medium-scale farmers of the three districts examined in this study predominantly occupy, lease or own 5 to 10 ha. They account for 65 percent of the total sample, with farms on landholdings of 10 to 20 ha accounting for 28 percent (Figure 1). This distribution varies significantly according to district however. Although the distribution among districts is similar for the 5 to 10 ha farm size category, Kasungu has the bigger farms, from 10 ha upwards. This is related to the estate development policy that the Government of Malawi implemented particularly in that region. On the other hand, Lilongwe is the district with, on average, the smallest landholdings. The latter is directly related to land pressures in higher population density areas [32,36] (Figure 2).

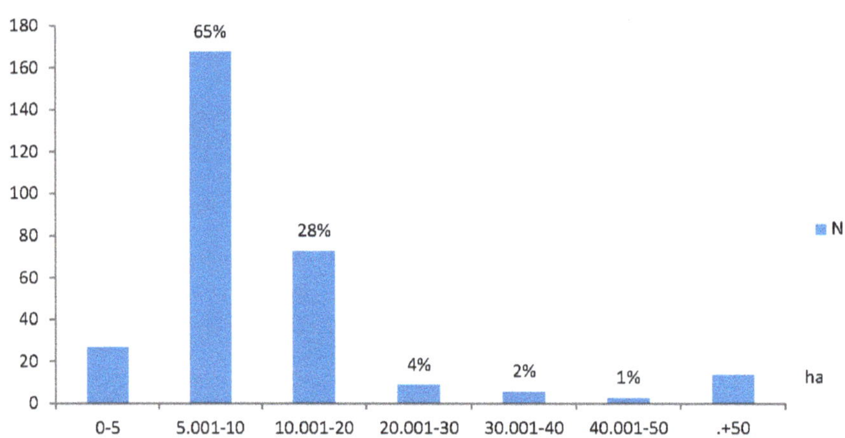

Figure 1. Numbers of medium-scale farmers sampled in Mchinji, Kasungu and Lilongwe, by farm size category (ha) (N = 300). Source: Authors' medium-scale farm survey in Mchinji, Kasungu and Lilongwe Districts undertaken by LUANAR, UP, and MSU (2014–2015). Notes: The percentages represent the proportion of farms in each of the categories of medium-scale farmers in the sample (N = 259). The number of farmers in the different land categories between 5 and 50 ha are significantly different (Anova p-value 2.84×10^{-88}).

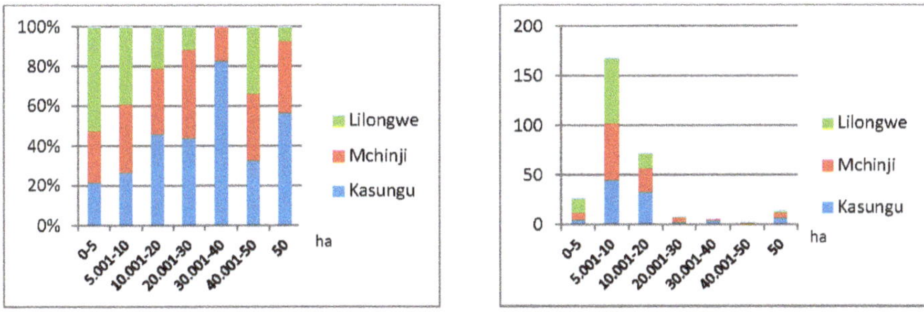

Figure 2. Numbers of medium-scale farmers sampled, per size category and per district, proportionally per size category and in absolute numbers per size category ($N = 300$). Source: Authors' medium-scale farm survey in the Mchinji, Kasungu and Lilongwe Districts undertaken by LUANAR, UP, and MSU (2014–2015). Note: There is a statistically significant difference in size categories between Lilongwe, Mchinji and Kasungu (Anova p-value 2.90×10^{-41}).

A large proportion of sampled farmers are part-time farmers, for whom farming is not a primary source of income and who, in general, do not reside on the farms. The latter amount to 40.2 percent of the total sampled farmers. This proportion rises with landholding size; it even reaches 69.2 percent of the +50 ha farms. Figure 3 shows a clear (positive) correlation between the sizes of the farms, owners primarily living elsewhere, and the origin of income used to acquire the farm. Land accumulation and medium-scale farm dynamics are thus certainly related to self-financing with non-farm income [7,20]. However, it would be inappropriate to conclude that these land dynamics are purely led by relatively wealthy, urban-based, segments of society: Most of the smaller sized, medium-scale holdings are indeed held by resident farmers, many of whom financed their land using income from farming. In fact, farming was the primary source of income for 59 percent (154) of the sampled farmers, but even among this group, 81.4 percent of them reported having had one or more jobs/incomes outside of agriculture at an earlier point in their lives (Table 3). This probably explains the relatively high age of the farmers and/or the relatively higher proportion of bigger farm owners who have a background, or whose father has a background, in the public sector (Tables 3 and 4).

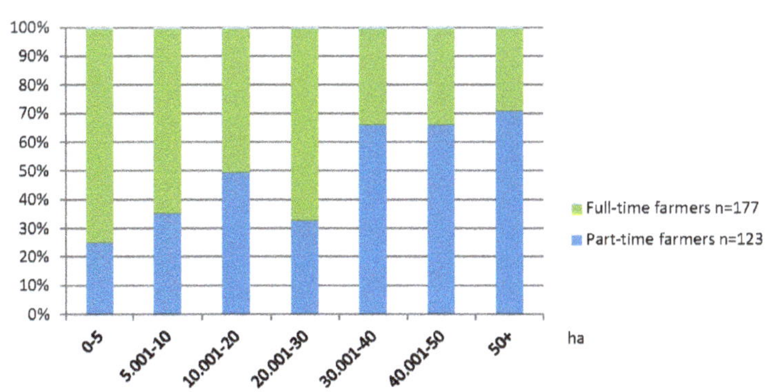

Figure 3. Percentage of full-time and part-time farmers, by farm size, in Mchinji, Lilongwe and Kasungu ($N = 300$). Source: Authors' medium-scale farm survey in the Mchinji, Kasungu and Lilongwe Districts undertaken by LUANAR, UP, and MSU (2014–2015). Note: There is a statistically significant difference in the distribution by land size between full-time and part-time farmers at a 5% level of significance (Anova p-value 0.021747).

Table 3. Multiple revenue sources of sampled medium-scale farmers * in Mchinji, Lilongwe and Kasungu. Source: Authors' medium-scale farm survey in the Mchinji, Kasungu and Lilongwe Districts undertaken by LUANAR, UP, and MSU (2014–2015).

		Small-Scale (0–5 ha) (N = 27)	Medium-Scale (5.001–50 ha) (N = 259)		Large-Scale (+50 ha) (N = 14)
			Full-Time Farmers: Farming as Longstanding Primary Source of Income	Part-Time Farmers: Farming Not as Longstanding Primary Source of Income	
Farmers N		27	154	105	14
% who have salaried jobs/regular income outside of farming		22.22	40.4	100.0	64.28
% who have never had a salaried job or regular income outside agriculture		74.1	18.7	0.0	30.8
% who are public sector employees		16.7	17.3	18.8	11.1
Father of household head's primary occupation	% in public sector	0	3.2	7.7	7.7
	% in salaried private sector jobs	3.7	5.2	13.5	15.4

Note 1: Part-time farmers (Figure 2 and Table 3) are defined here as not residing full-time on farm and as having multiple income sources, with farming not being a longstanding primary income source. Note 2: * Data for small-scale and large-scale farmers are presented here for comparison purposes.

Table 4. Farm characteristics by farm size (N = 300). Source: Authors' medium-scale farm survey in the Mchinji, Kasungu and Lilongwe Districts undertaken by LUANAR, UP, and MSU (2014–2015).

	Farm Sizes (ha)						
	0–5	5.001–10	10.001–20	20.001–30	30.001–40	40.001–50	+50
Farmer/manager is family member (%)	85.2	86.3	84.9	88.9	66.7	66.7	78.6
Average age of farmers (St. dev.)	54 (9.10)	53 (14.07)	55 (12.55)	58 (12.63)	55 (8.18)	50 (9.89)	52 (12.86)
Average ratio cultivated to total landholding * (St. dev.)	1.019 (0.99)	0.791 (0.31)	0.589 (0.33)	0.443 (0.33)	0.329 (0.24)	0.412 (0.41)	0.412 (0.31)
Average number of family laborers (St. dev.)	4 (1.73)	4 (2.27)	4 (2.69)	3 (2.78)	2 (1.60)	2 (1.53)	2 (1.83)
Average number of non-family full-time laborers (St. dev.)	0 (0.70)	2 (0.70)	4 (5.02)	6 (4.86)	6 (1.73)	4 (4.62)	25 (44.06)
Average number of non-family temporary seasonal laborers (St. dev.)	10 (10.82)	10 (8.34)	16 (19.83)	12 (9.81)	17 (8.37)	73 (115.63)	29 (77.68)
Farmers with tractors	0	0	0	0	0	0	5

Note: * Plots cultivated twice per year and rented land are included in the computation, which explains why the ratio exceeds 1 for the 0–5 ha category.

Overall, the large majority of medium-scale farmers in the sample, and especially the bigger ones, have or have had salaried jobs outside of agriculture or regular non-farm income. The situation thus has to be nuanced. The larger medium-scale farms have developed production systems effectively based on pluri-activity and multiple income sources, where the farmer is non-farm based, inducing an externally led development pattern through non-farm funds. The smaller medium-scale farmers (5–20 ha), on the other hand, although also significantly engaged in non-farm activities, seem to develop such strategies as a necessity to diversify in order to accumulate.

Income diversification also has significant consequences with regard to the production models developed across the full set of sampled farms, including the small- and large-scale comparison groups. Indeed, the lower farm size groups develop production models based on family farms, whereas the majority of the farms with more than 30 ha are of a more entrepreneurial nature. The former are mainly based on family labor, although they also do employ, on average, two to six full-time non-family laborers, as well as 10 to 16 seasonal laborers. The ratio of family labor to hired labor for the smaller farms varies between 1 per 2.5 and 1 per 5. The bigger farms, by contrast, are often run by non-family managers and the overall labor force is more dependent on non-family members. The family labor to hired labor ratio for the bigger farms decreases to 1 per 36. From a labor creation per ha point of view, in the survey, the number of full-time and temporary people employed per hectare of landholding is roughly 2 for farms under 20 ha and 1.2 for landholdings over 30 ha, suggesting a higher per land unit use of hired labor for smaller farms. Table 4 shows other distinctions between these two models.

Patterns of cultivation and productivity also vary by landholding size for the sampled farms. Based on the sampled farms in the three districts, the ratio of cultivated land to landholding size declines sharply with total landholding (Figure 4). For smallholder farmers (0–5 ha), this ratio is even higher than 1 (related to renting land and the use of multiple cycles on the same plot), indicating the degree of land pressure experienced by this category of farmers. This ratio falls to about 40 percent for the farms larger than 30 ha. These results show a decrease in land utilization as landholding size rises. The latter is also the case when the net value of crop output per hectare cultivated is measured. This measure of land productivity is the highest, on average, for sampled farms between 10 ha and 20 ha, decreasing for higher sizes, although there is great variability in net value of crop output per hectare cultivated at all levels of landholding size (Figure 5).

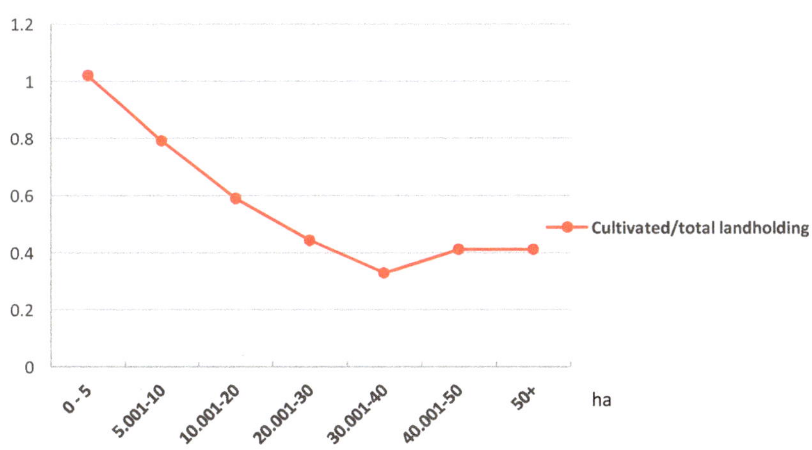

Figure 4. Cultivated land vs. total landholding ratio for the different farm size categories (N = 300). Source: Authors' medium-scale farm survey in the Mchinji, Kasungu and Lilongwe Districts undertaken by LUANAR, UP, and MSU (2014–2015). Note: The ratios of cultivated land to total landholding among the different land categories is statistically significant at a 5% level of significance (Anova p-value 2.14×10^{-46}).

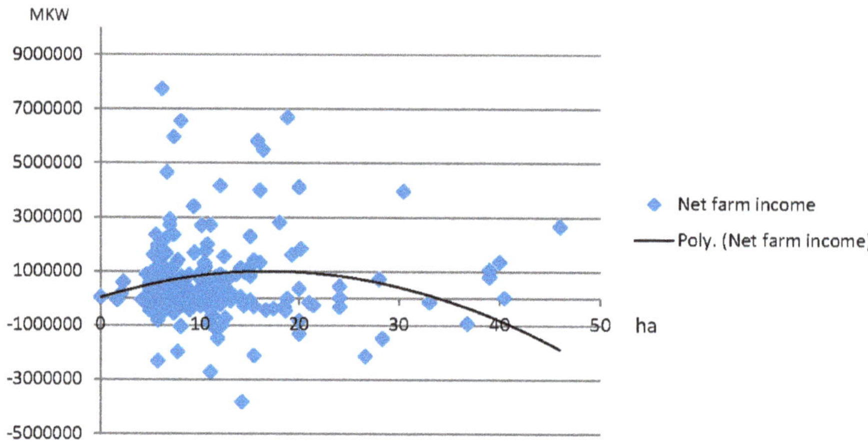

Figure 5. Productivity (net farm income) per hectare for the different sizes of farms (N = 300). Source: Authors' medium-scale farm survey in the Mchinji, Kasungu and Lilongwe Districts undertaken by LUANAR, UP, and MSU (2014–2015).

5. Farmland Ownership Dynamics

5.1. Rising Land Acquisitions of Medium-Scale Farmland

Figure 6 shows that the land under medium-scale holdings in the sample in the three districts has almost doubled between 2000 and 2015, from 2544 ha in 2000 to 4726 ha in 2015. As the reverse trend of medium-scale farmers becoming small-scale ones has not been identified during the survey and is rather unlikely (since it would entail a conversion of leasehold/private land back to traditional customary land) and as the origin of the present medium-scale farmers and their land is accounted for in this study (either conversion of traditional customary land or transfer from small- and medium-scale leasehold/private land to medium-scale farmers), it can be estimated that, just over the past decade since 2005, the land under medium-scale holdings has increased by 49.1 percent. If our sample in these three districts can be extrapolated to the whole of Malawi based on Table 2, we can conclude that roughly 300,000 ha have been newly acquired by medium/large-scale holders since 2005, slightly less than 10 percent of the total area of landholdings in Malawi. The rapid rise in medium-scale farms over the past decade is truly remarkable in a country where the majority of rural people face acute land scarcity and where household poverty is highly correlated with very small farm size [27].

Moreover, given our finding that 39 percent of the sampled medium-scale farms (and 48 percent of the land) are controlled by people primarily residing in urban areas, a significant portion of the farmland in the "over 10 hectare" category of Table 1 could be controlled by part-time farmers living in urban areas. If 48 percent of the 776,000 ha of customary land under cultivation on landholdings over 10 ha were controlled by urban-based households, this would imply that roughly 12.6 percent of the area under cultivation in customary lands is producing revenue that is generated in rural areas, but is spent by people living primarily in urban areas. Such estimates may rise further when considering the estate farm sector. This kind of farmland ownership structure implies that the multiplier effects of agricultural growth may therefore not be confined to rural areas; a significant portion of revenue from agriculture, and from subsequent expenditure and employment effects, may occur in urban areas. A better understanding of the relationship between surplus agricultural production and the location of expenditures from it may allow policy makers to anticipate how default patterns of land acquisition are affecting where the multiplier effects of agricultural growth are likely to occur.

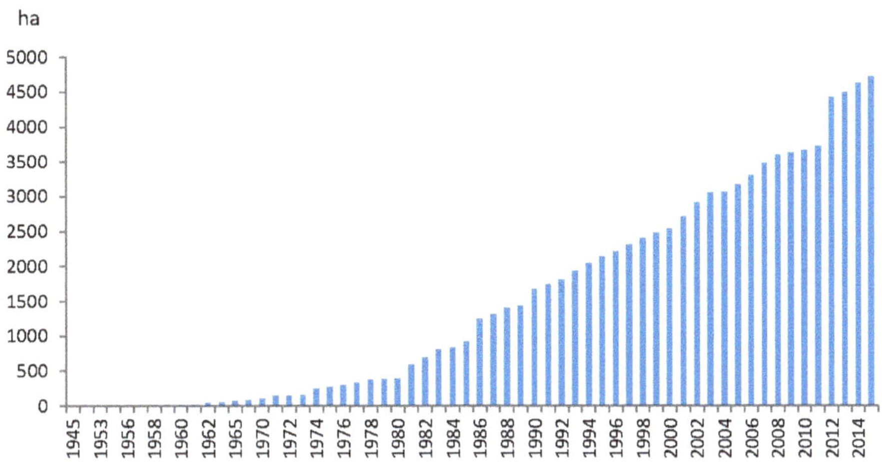

Figure 6. Cumulative land acquisitions by medium-scale holdings among sampled farms (N = 300) in the Mchinji, Kasungu and Lilongwe Districts (ha per year). Source: Authors' medium-scale farm survey in the Mchinji, Kasungu and Lilongwe Districts undertaken by LUANAR, UP, and MSU (2014–2015).

A very similar trend is occurring with field-level land acquisitions; they are also rapidly increasing. Among the 300 sampled farms, there were 89 field acquisitions that occurred before 1980; 272 fields were acquired between 1980 and 2000, and 211 fields were acquired since 2000. Interestingly, aside from the large-scale farms acquisitions (of which 50 percent represent acquisitions by a company, 31 percent by part-time farmers and 18 percent by full-time farmers), the large majority of the acquisitions are by private individuals. Their total landholding grew, on average, by 8 ha, representing an increase of 449 percent (Table 5). A significant increase in the size of acquisitions occurs after 2011. Although here, again, the large majority of acquirers were full-time farmers (66 percent of the acquisitions, with 31 percent being part-time farmers), with the major difference of one company acquiring 493 ha, illustrating the renewed interest in farmland since the rise in global food prices after 2008 (Figure 7).

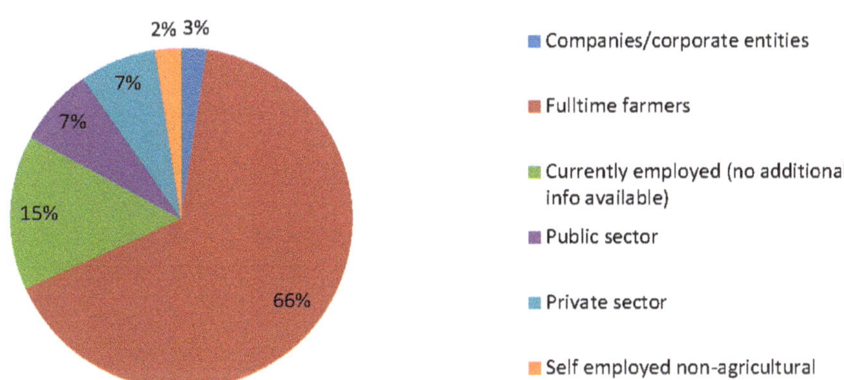

Figure 7. Land acquisitions, 2011–2014 (N = 300). Source: Authors' medium-scale farm survey in the Mchinji, Kasungu and Lilongwe Districts undertaken by LUANAR, UP, and MSU (2014–2015). Note: Land acquisitions include lease deals and customary land.

Table 5. Evolution of sampled medium-scale farmers' * landholding size, in Mchinji, Kasungu and Lilongwe.

Household Land	0–5 ha (N = 27)	5.001–50 ha (N = 259)		+50 ha (N = 14)
		Full-Time Farmers	Part-Time Farmers	
Average initial landholding size at time of formation (ha) (St. dev)	2.57 (4.25)	1.76 (9.67)	5.51 (9.14)	6.27 (14.71)
Average current landholding size (ha) (St. dev.)	4.25 (1.33)	9.67 (6.35)	11.21 (8.20)	135.33 (131.12)
Difference (%)	+65.37	+449.43	+103.45	+2058.37

Source: Authors' medium-scale farm survey in the Mchinji, Kasungu and Lilongwe Districts undertaken by LUANAR, UP, and MSU (2014–2015). Note: * Data for small-scale and large-scale farmers are presented here for comparison purposes.

These findings suggest that farmland ownership may be changing quite rapidly, at least on the basis of the three districts surveyed. Although the 5–10 hectare group is still the most significant category, the larger farms are rapidly increasing in number (Figure 8). This increase is particularly notable for the larger medium-scale farms.

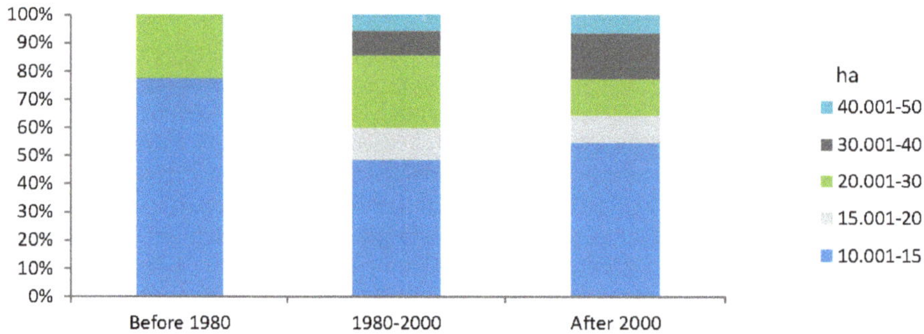

Figure 8. Increasing size of medium-scale landholdings in Mchinji, Kasungu, and Lilongwe (N = 91). Source: Authors' medium-scale farm survey in the Mchinji, Kasungu and Lilongwe Districts undertaken by LUANAR, UP, and MSU (2014–2015).

5.2. Former Use of the Land Acquired by Medium-Scale Farmers

The large majority of the land acquired by the sampled farmers over time is customary land formerly under the control of traditional authorities. This means that the rise of medium-scale farms is taking place on lands ostensibly reserved as a birthright for local subjects of the traditional authorities in the area. According to the survey respondents, just less than 40 percent of the customary land acquired by medium-scale farmers had been utilized by others prior to the acquisition. However, both the amount and proportion of land acquired by medium-scale farmers that was formerly customary land utilized by others has increased significantly in recent years (Figure 9). The high proportion of land acquired from formerly utilized customary land raises concerns about the possibility that some medium-scale farms are forming as part of a process of disenfranchisement of local rural people from land that was traditionally reserved for them (Figure 10).

After being acquired by the medium-scale farmer, much of the land is successfully converted by the owner to privately held land (Figure 11). The rise of medium-scale farms in the selected three districts of Malawi is associated with a transfer of land from customary tenure to state land. Such land transfers involve the tacit or explicit consent of those administering Malawi's customary and state lands. Most farmland acquisitions have followed a general pattern, whereby the potential investor approaches the local headman in his area, as well as the village chief/headman; after agreeing on terms, the traditional authority (usually the chief of the area) signs a consent form that essentially provides the traditional authority's blessings to transfer the designated land from customary to state tenure (and implicitly to privately held land). Thereafter, the farmer may succeed in acquiring a state title deed for a 99-year lease after the land has been surveyed and the owner pays the necessary fees. The leased land is officially state-owned land. These results raise questions as to whether the traditional and/or state institutions of land allocation in customary tenure areas may in some ways be distributionally regressive, transferring large quantities of land out of customary tenure (ostensibly for use by local smallholder residents), often to relatively wealthy landowners holding title deeds under long-term leaseholds. This is a question for further inquiry with potentially important implications for development in Malawi.

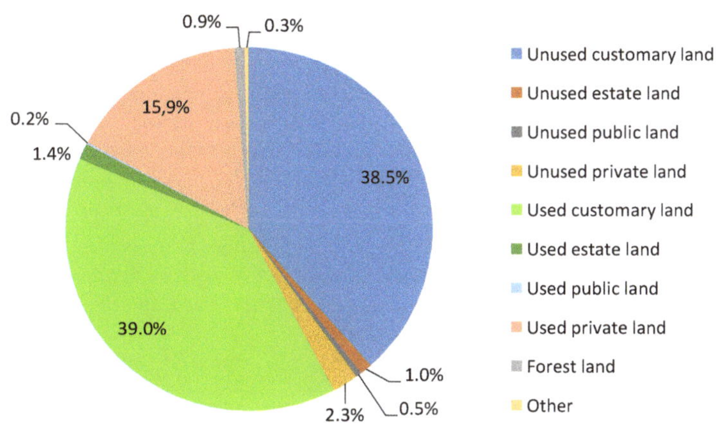

Figure 9. Prior land use and land tenure of lands acquired by medium-scale farms, Kasungu, Lilongwe, and Mchinji districts (in percentage) (N = 300). Source: Authors' medium-scale farm survey in the Mchinji, Kasungu and Lilongwe Districts undertaken by LUANAR, UP, and MSU (2014–2015).

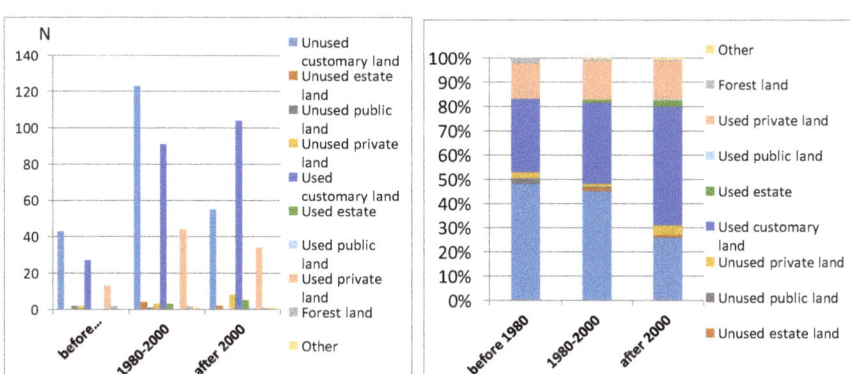

Figure 10. Types of land acquired and their evolution over time, in absolute terms and in percentages of cases (N = 300). Source: Authors' medium-scale farm survey in the Mchinji, Kasungu and Lilongwe Districts undertaken by LUANAR, UP, and MSU (2014–2015).

Although this process of tenure conversion has been promoted more or less actively by prior government regimes as described in Section 2 and might be associated with some form of economic transformation, the findings presented here highlight the point that a land commodification process is underway, whereby the stock of land under customary tenure systems is declining, the stock of land under freehold and leasehold tenure is rising, and land transactions are becoming more common.

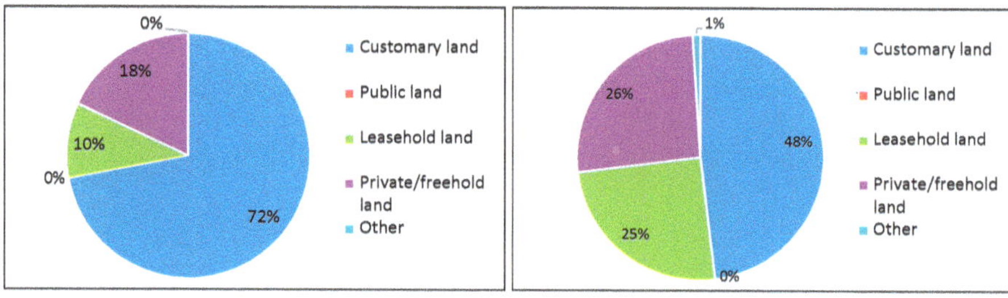

Figure 11. Tenure of the land when acquired (**left**) and after acquisition (**right**) (in percentages of number of cases) (N = 300). Source: Authors' medium-scale farm survey in the Mchinji, Kasungu and Lilongwe Districts undertaken by LUANAR, UP, and MSU (2014–2015).

6. Discussion

This study is motivated by the apparent revolutionary changes in the control of African farmland associated with the rise of medium-scale farms. Recent global policy attention to "land grabs" by international investors, while important in some countries, may have diverted attention away from even more rapid land acquisitions by national investors.

A recent study [7] shows a rapid rate of growth of medium-scale farms in three countries, but two of these, Zambia and northern Ghana, contain vast areas of unutilized or underutilized lands where new entrants could begin farming without necessarily dispossessing local people. Malawi is an interesting and important contrasting country where most of its farming population faces land scarcity and where small farm size is an obvious constraint on their livelihoods [27]. The Malawian government's existing strategies are officially intended to promote agricultural growth and food security for its 2.5 million or more small-scale farm households [36]. However, there are increasing concerns that both traditional and state land institutions have encouraged, and may be continuing to encourage, the transfer of land to non-local interests without due cognizance of how this is affecting the land access and food security conditions of current and future generations of rural people. A contrasting view is that medium-scale farms represent one of the few forms of economic dynamism and employment creation in Malawi's rural areas, are more productive users of land, are part of the farm consolidation process that normally accompanies economic development, and hence should be promoted. We reconsider these contrasting narratives after summarizing the main findings of the study.

The study highlights six findings. First, the recent pace of farmland acquisitions in the 5–50 hectare category in Malawi indicates that a relatively quiet but fundamental change in farmland ownership is underway. Though they constitute less than one percent of all farms, official government data indicate that farms over 5 ha now account for roughly 30 percent of the landholdings and area under cultivation in Malawi [27]. Almost 90 percent of this land is controlled by farms over 10 ha, with a mean farm size of 87 ha. The share of farmland under the control of medium/large farms would undoubtedly be significantly higher after accounting for Malawi's "estate" farm sector.

Second, in the three districts covered by our survey, the number of medium-scale farms appears to be growing rapidly. Land acquired by medium-scale holdings in the three districts of our study has almost doubled between 2000 and 2015. Over the past decade since 2005, land under medium-scale holdings has increased by 49.1 percent. If our sample in these three districts can be extrapolated to the whole of Malawi, we can conclude that roughly 300,000 ha has been newly acquired by medium/large-scale holders since 2005, slightly more than 10 percent of the total area under cultivation in Malawi's customary lands [27]. The rapid rise in medium-scale farms over the past decade is truly remarkable in a country where the majority of rural people face acute land scarcity and where household poverty is highly correlated with very small farm size.

Third, the vast majority of these medium-scale landholdings in the three case study districts are owned by Malawian African nationals. Over 39 percent of them were not primarily involved in farming prior to acquiring their farms. Even among the remaining 61 percent of medium-scale farmers, the prevalence of salaried employees in non-farm jobs is substantially higher than the average found within Malawi's small-scale farming sector. Approximately 40 percent of the sampled farms were owned by part-time farmers living in urban areas and hiring a manager to handle daily activities. Therefore, a significant portion of the revenue earned from agriculture, while generated in rural areas, may be spent in urban areas, contributing the multiplier and employment effects there and weakening such effects in rural areas.

Fourth, the rise of medium-scale farms in the selected three districts of Malawi is in most cases associated with a transfer of land from customary tenure to public land. Many of the medium-scale farmers (~25 percent) sampled in the three districts held title deeds with long-term leases. Such land transfers involve the tacit or explicit consent of those administering Malawi's customary and state land institutions. Given that land scarcity within many of Malawi's customary tenure areas appears to be a serious problem [31], the removal of land from the customary tenure system, where traditionally it would have been reserved for local community use, into state-owned land may have potentially important implications for Malawi's most important development objectives such as food security and poverty alleviation—implications which are as yet not fully clear.

Fifth, over the entire sample, 38 percent of the respondents indicated that the lands on which their farms are located were formerly utilized by others in the community prior to having been acquired by the current owner. However, among farms acquired prior to 2000, this same figure was only 23 percent, whereas since 2000 it has shot up to 69.6 percent, indicating increasing competition over, and dispossession of, land associated with the rise of medium-scale farms. In a dynamic sense, the distributional effects are unclear. It is possible that medium-scale farms, by producing surplus agricultural production and farm labor employment, may kick-start local growth multipliers in a rural economy that otherwise is somewhat stagnant. The medium-scale farms in the sampled areas of Malawi are found to have relatively high labor-to-land ratios compared to other countries in the sub-continent such as Zambia or Ghana [7], where use of labor-saving technologies, such as tractors for plowing and herbicides, has risen significantly in recent years.

Sixth, in terms of the net value of agricultural output per hectare, we find little evidence to suggest that medium-scale farms are more efficient than smaller farms. While our findings do not constitute a formal test of the inverse farm-size/efficiency relationship, we can conclude that there are no clear advantages in land productivity to medium-scale farms over small-scale holdings. Moreover, in the survey data, the ratio of area cultivated to total landholding size is found to decline progressively as farm size increases. Again, the distributional effects of transferring land from a system in which many owner-operators control the means and proceeds of their own production to a system in which that same land is controlled by a few, and the remaining people are converted to low-return farm laborers, may be a topic for the Government of Malawi to examine in much greater detail. In general, the findings of this study, while confined to three districts in Malawi, raise questions about the compatibility between the rapid pace of land transfers to medium-scale farms and the Government of Malawi's national policy goals related to poverty reduction, employment, food security, and agricultural productivity growth.

Lastly, and more broadly, as highlighted in the conceptual framework of this paper, these findings contribute to the development literature on agricultural and structural transformation. The high level of risks and limited economic opportunities may constrain households' options and hence their economic returns. This situation can generate possible poverty traps for the smaller-scale and low-income farmers. Malawi's current farm structure thus might not be able to generate broad-based income growth, and hence Malawi cannot rely on agriculture to generate the multipliers that lead to structural transformation [7]. It seems that there is at least circumstantial evidence to suggest that this hypothesis stemming from the structural transformation literature might indeed be true, seeing as how poverty rates have remained relatively high over the past several decades. Considering these socio-economic challenges in light of the land pressure in Malawi, policy orientations should reintegrate these structural issues, stressing the importance of broad-based, inclusive agricultural growth.

Warnings about the effects of non-broad-based agricultural growth and more particularly of rising land conversions from customary land to privately held land in Malawi are not new. Before his assassination in 2003, the Economic Advisor to the President of Malawi, Kalonga Stambuli [32] (p. 15), wrote that: *"I have seriously deplored the social injustice and economic marginalization associated with land conversion from communal tenure to leasehold tenure mostly enjoyed by the elite who also enjoyed a monopoly in the production of export crops. Most deplorable is the fact that the abundance of idle land among estates explains much of the low equilibrium trap to which our countries have been subjected. The economic hegemony of the agricultural elite was compounded by state enterprise expansion into the private sector, over-regulation, a stifling bureaucracy, and totalitarian politics. Inadequate amounts of land available to farmers remain a major constraint to supply response."*

7. Conclusions

This study shows a rapid rise of medium-scale farms in Malawi. Medium-scale farms are defined in this study as agricultural holdings between 5 and 50 hectares. The amount of land acquired by medium-scale holdings in the Mchinji, Kasungu and Lilongwe districts—the three districts examined in this study—is found to have almost doubled between 2000 and 2015. Only 54 percent of the medium-scale holdings resulted from small-scale farmers accumulating land and growing into medium-scale holdings; 46 percent of medium-scale farmers are (or were) pluri-active, non-full-time, and often urban-based. Our findings certainly question the premise that the rise of medium-scale farms in Malawi reflects dynamic agricultural growth and land accumulation among small-scale farmers. Even as medium-scale farmland is increasing rapidly, the majority of rural people in Malawi face acute land scarcity, and household poverty is highly correlated with very small farm size. Lastly, a significant portion of the land acquired by medium-scale farmers was under customary tenure and utilized by others prior to its acquisition, with current owners often successfully transferring the ownership structure of the newly acquired land to a long-term leasehold with a title deed. Farm consolidation, through land acquisitions often by urban-based people, is thus significantly altering the country's land structure and leading to a greater concentration of land assets. These results endorse similar findings in other countries in Africa [6,7].

These conclusions lead to two complementary, broader policy reflections and implications, both of which offer possible avenues for future research. Firstly, the results show that urban-based, local investors and their non-farm income sources may be contributing in important ways to agricultural sector investment and growth. They may also be altering the nature of rural-urban growth linkages [24]. Secondly, the growing focus on how best to exploit Africa's farmland has arguably diverted attention from the more central and enduring challenge of implementing agricultural development strategies that effectively address the continent's massive rural poverty and hunger problems, which require recognition of the growing land constraints faced by much of Africa's rural population. The prospects for inclusive agricultural growth [37]—and with it the ability to make serious progress in eradicating hunger and poverty—may likely depend on open discussions about how to ensure that sufficient land is made available for Africa's rural communities to enable them to contribute to Africa's economic growth processes, rather than be marginalized by more powerful actors—domestic or international—seeking land.

Acknowledgments: The authors gratefully acknowledge support for this study from the *Guiding Investment in Sustainable Agricultural Intensification in Africa* (GISAMA) program funded by the Bill and Melinda Gates Foundation, as well as from the *Food Security Policy Innovation Lab*, which is funded by United States Agency for International Development's Bureau for Food Security.

Author Contributions: The authors confirm that this paper is based on original work engaged in by the authors based on their medium-scale farm survey in the Mchinji, Kasungu and Lilongwe Districts, undertaken by the Lilongwe University of Agriculture and Natural Resources, the University of Pretoria, and Michigan State University (2014–2015).

Conflicts of Interest: The authors declare no conflict of interest.

Appendix

Table A1. Landholdings and area cultivated, by farm size in Kasungu district. Source: [27].

	Farm Size				
	0–2 ha	**2.001–5 ha**	**5.001–10 ha**	**>10 ha**	**Total**
# of farms	65,466	16,296	2360	333	84,455
% of total # of farms	77.52%	19.29%	2.79%	0.39%	100%
Area cultivated (ha)	54,972.42	53,659.74	15,104.71	3,615.17	127,352.1
Total landholdings (ha)	65,384.01	55,003.41	15,789.75	3,674.50	139,851.7
Mean farm size (ha)	1.00	3.38	6.69	11.03	1.66
Mean ratio of land cultivated to land owned	0.79	0.96	0.95	0.98	0.83
% of Kasungu landholdings that are cultivated	84.08%	97.56%	95.66%	98.39%	91.06%
% of total landholdings in Kasungu	46.75%	39.33%	11.29%	2.63%	100%
% of total cultivated land in Kasungu	43.17%	42.13%	11.86%	2.84%	100%

Table A2. Landholdings and area cultivated, by farm size in Lilongwe district. Source: [27].

	Farm Size				
	0–2 ha	**2.001–5 ha**	**5.001–10 ha**	**>10 ha**	**Total**
# of farms	239,992	21,242	1940	2938	266,111
% of total # of farms	90.18%	7.98%	0.73%	1.10%	100%
Area cultivated (ha)	130,288.2	47,649.83	15,852.23	20,072.63	213,862.8
Total landholdings (ha)	155,109.6	57,253.25	15,852.23	75,672.09	303,887.2
Mean farm size (ha)	0.65	2.70	8.17	25.76	1.14
Mean ratio of land cultivated to land owned	0.83	0.85	1 *	0.51	0.83
% of Lilongwe landholdings that are cultivated	84.00%	83.23%	100%	26.53%	70.38%
% of total landholdings in Lilongwe	51.04%	18.84%	5.22%	24.90%	100%
% of total cultivated land in Lilongwe	60.92%	22.28%	7.41%	9.39%	100%

* Only one household in the sample owns 5–10 ha. Note: excludes Lilongwe city.

Table A3. Landholdings and area cultivated, by farm size in Mchinji district. Source: [27].

	Farm Size				
	0–2 ha	**2.001–5 ha**	**5.001–10 ha**	**>10 ha**	**Total**
# of farms	81,676	8653	3668	0	93,998
% of total # of farms	86.89%	9.21%	3.90%	-	100%
Area cultivated (ha)	67,059.49	23,476.38	22,451.38	-	112,987.3
Total landholdings (ha)	77,190.08	24,666	22,619.95	-	124,476.4
Mean farm size (ha)	0.95	2.85	6.17	-	1.32
Mean ratio of land cultivated to land owned	0.83	0.94	0.99	-	0.85
% of Mchinji landholdings that are cultivated	86.88%	95.18%	99.25%	-	90.77%
% of total landholdings in Mchinji	62.01%	19.82%	18.17%	-	100%
% of total cultivated land in Mchinji	59.35%	20.78%	19.87%	-	100%

References

1. Deininger, K.; Byerlee, D. *The Rise of Large-Scale Farms in Land-Abundant Developing Countries: Does It Have a Future?* The World Bank: Washington, DC, USA, 2012.

2. Anseeuw, W.; Boche, M.; Breu, T.; Giger, M.; Lay, J.; Messerli, P.; Nolte, K. *Transnational Land Deals for Agriculture in the Global South*; Analytical Report Based on the Land Matrix Database; CDE: Bern, Switzerland; CIRAD: Montpellier, France; GIGA: Hamburg, Germany, 2012.
3. Anseeuw, W.; Ducastel, A. "Production grabbing"—New investors and investment models in agriculture. *QA Riv. Assoc. Rossi-Doria* **2013**, *2*, 37–55.
4. Reardon, T.; Timmer, C.P.; Barrett, C.B.; Berdegué, J. The rise of supermarkets in Africa, Asia, and Latin America. *Am. J. Agric. Econ.* **2003**, *85*, 1140–1146. [CrossRef]
5. Swinnen, J.F. *Global Supply Chains, Standards and the Poor*; CABI Publishing: Oxfordshire, UK, 2007.
6. Sitko, N.J.; Jayne, T.S. Structural transformation or elite land capture? The growth of "emergent" farmers in Zambia. *Food Policy* **2014**, *48*, 194–202.
7. Jayne, T.S.; Chapoto, A.; Sitko, N.; Nkonde, C.; Muyanga, M.; Chamberlin, J. Is the scramble for land in Africa foreclosing a smallholder agricultural expansion strategy? *J. Int. Aff.* **2014**, *67*, 35–53.
8. DAFF (Department of Agriculture, Forestry and Fisheries). *CASP—Comprehensive Agricultural Support Program*; Policy Document. DAFF: Pretoria, South Africa, 2014.
9. Collier, P. *The Bottom Billion: Why the Poorest Countries Are Failing and What Can Be Done About It*; Oxford University Press: New York, NY, USA, 2008.
10. Sourisseau, J.M.; Bosc, P.M.; Fréguin-Gresh, S.; Belières, J.F.; Bonnal, P.; le Coq, J.F.; Anseeuw, W.; Dury, S. *Diversity in Family Farming—Theoretical and Empirical Approaches to its Many Forms*; Rethinking Development Working Paper Series, No. 2014/2; University of Pretoria: Pretoria, South Africa, 2014.
11. Lipton, M. *Why Poor People Stay Poor*; Harvard University Press: Cambridge, MA, USA, 1977.
12. Mellor, J.W. *The New Economics of Growth: A Strategy for India and the Developing World*; Cornell University Press: Ithaca, NY, USA, 1976.
13. Boussard, J.M. *Economie de L'agriculture*; Economica: Paris, France, 1987.
14. Timmer, C.P. *Handbook of Development Economics*; Elsevier: London, UK, 1988.
15. Davies, S. *Adaptable Livelihoods*; Macmillan Press Ltd.: London, UK, 1996.
16. Ellis, F. Household strategies and rural livelihood diversification. *J. Dev. Stud.* **1998**, *35*, 1–38. [CrossRef]
17. Evans, H.E.; Ngau, P.M. Rural-urban relations, household income diversification and agricultural productivity. *Dev. Chang.* **1991**, *22*, 519–545. [CrossRef]
18. Reardon, T.; Delgado, C.; Matlon, P. Determinants and effects of income diversification among farm households in Burkina Faso. *J. Dev. Stud.* **1992**, *30*, 264–296. [CrossRef]
19. Reardon, T.; Berdegué, J.; Barrett, C.B.; Stamoulis, K. Household income diversification into rural nonfarm activities. In *Transforming the Rural Nonfarm Economy*; Haggblade, S., Hazell, P., Reardon, R., Eds.; Johns Hopkins University Press: Baltimore, MD, USA, 2006; pp. 115–141.
20. Anseeuw, W.; Laurent, C. Occupational paths towards commercial agriculture: The key roles of farm pluriactivity and the commons. *J. Arid Environ.* **2007**, *70*, 659–671. [CrossRef]
21. Fairbairn, M. Indirect expropriation: The role of national institutions and domestic elites in the Mozambican farmland grab. In Proceedings of the International Conference on Global Land Grabbing, Sussex, UK, 6–8 April 2011.
22. Collier, P.; Dercon, S. African agriculture in 50 years: Smallholders in a rapidly changing world? In Proceedings of the Expert Meeting on How to Feed the World in 2050, Rome, Italy, 12–13 October 2009.
23. Timmer, C.P. *A World without Agriculture: The Structural Transformation in Historical Perspective*; The American Enterprise Institute Press: Washington, DC, USA, 2009.
24. Losch, B.; Fréguin-Gresh, S.; White, E.T. *Structural Transformation and Rural Change Revisited: Challenges for Late Developing Countries in a Globalizing World*; World Bank Publications: Washington, DC, USA, 2012.
25. Eastwood, R.; Lipton, M. Demographic transition in sub-Saharan Africa: How big will the economic dividend be? *Popul. Stud.* **2011**, *65*, 9–35. [CrossRef] [PubMed]
26. Ellis, F. *Rural Livelihoods and Diversity in Developing Countries*; Oxford University Press: Oxford, UK, 2000.
27. FAO (Food and Agricultural Organization). *Country Pasture/Forage Resource Profiles—Malawi*; FAO: Rome, Italy, 2015.
28. Republic of Malawi. *2008 Population and Housing Census*; National Statistical Office: Lilongwe, Malawi, 2008.
29. Cammack, D.; Kelsall, T.; Booth, D. *Developmental Patrimonialism? The Case of Malawi*; Working Paper No. 12; Africa Power and Politics Program; Overseas Development Institute: London, UK, 2010.

30. Stambuli, K. Elitist land and agricultural policies and the food problem in Malawi. *J. Malawi Soc.—Hist. Sci.* **2002**, *55*, 34–83.
31. Mhone, G.C.Z. *Malawi at the Crossroads: The Post-Colonial Political Economy*; SAPES Books: Harare, Zimbabwe, 1992.
32. Mtewa, M. *Malawi Democratic Theory and Public Policy*; Schenkman Books: Cambridge, MA, USA, 1986.
33. Chikaipa, V. A critique of the government responses on oppositions and protests over land grabs in Malawi: A case study of the Chinese and Madonna's Raising Malawi project land leases in Salima and Lilongwe districts respectively. In Proceedings of the International Conference on Global Land Grabbing II, Cornell University, Ithaca, NY, USA, 17–19 October 2012.
34. World Bank. *2013 Integrated Household Panel Survey in Malawi*; The World Bank: Washington, DC, USA, 2013.
35. Lowder, S.; Skoet, J.; Raney, T. Making better use of agricultural census data to understand farmland distribution. In Proceedings of the World Bank Conference on Land and Poverty, Washington, DC, USA, 23–27 March 2015.
36. Ricker-Gilbert, J.; Jumbe, C.; Chamberlin, J. The impact of increasing population density on African agriculture and livelihoods: The case of Malawi. *Food Policy* **2014**, *48*, 114–128. [CrossRef]
37. OECD. *African Economic Outlook—Regional Development and Spatial Inclusion*; OECD: Paris, France, 2015.

MDPI AG

St. Alban-Anlage 66

4052 Basel, Switzerland

Tel. +41 61 683 77 34

Fax +41 61 302 89 18

http://www.mdpi.com

Land Editorial Office

E-mail: land@mdpi.com

http://www.mdpi.com/journal/land

www.ingramcontent.com/pod-product-compliance
Lightning Source LLC
Chambersburg PA
CBHW041754220926
51847CB00014B/194
* 9 7 8 3 0 3 8 4 2 3 4 2 3 *